MANAGING TO SUCCEED

MANAGING TO SUCCEED

STRATEGIES AND SYSTEMS FOR MANUFACTURING BUSINESSES

GRAHAM SMITH

Coventry University

Prentice Hall

London   New York   Toronto   Sydney   Tokyo   Singapore
Madrid   Mexico City   Munich

First published 1995 by
Prentice Hall International (UK) Limited
Campus 400, Maylands Avenue
Hemel Hempstead
Hertfordshire, HP2 7EZ
A division of
Simon & Schuster International Group

Typeset in 11/12pt Ehrhardt
by Hands Fotoset, Leicester

Printed and bound in Great Britain by
TJ Press (Padstow) Ltd

Library of Congress Cataloging-in-Publication Data

Smith, Graham (Graham S.)
    Managing to succeed / Graham Smith.
      p. cm.
    Includes bibliographical references and index.
    ISBN 0-13-230376-0
    1. Industrial management. 2. Industrial organization. 3. Success
in business. I. Title.
HD31.S585   1995
658—dc20
                                    95–22657
                                          CIP

British Library Cataloguing in Publication Data

A catalogue record for this book is available from
the British Library

ISBN 0-13-230376-0

1 2 3 4 5   99 98 97 96 95

CONTENTS

ACKNOWLEDGEMENTS

The author is grateful for and acknowledges the friendship and collaboration over many years of many colleagues in industry and at the universities of Birmingham and Coventry.

Dr Peter Hancell at Coventry University and Peter Townsend at North Warwickshire College contributed material which has been used in this book.

Special mention and acknowledgement is made of two co-authors who have brought to the book their expertise in areas beyond the direct experience of the main author.

Robert Mansfield has 34 years of industrial and teaching experience. He wrote the chapters on Innovation, Design, Logistics, and Purchasing and contributed some material on Accounting.

Trevor Pye has 29 years of industrial and teaching experience. He wrote the chapter on Marketing and contributed some material on Market Research and Personnel Management.

This book is based on the 15 years experience of the author in manufacturing industry and in teaching manufacturing management for 22 years.

Experience with the systems approach to manufacturing and management used in this book derives from part-time tutoring of Open University courses on systems for more than 20 years.

The bibliography at the end of the book is extensive and lists the source materials for readers wishing to read the original works.

The financial account from a thirteenth-dynasty Egyptian tomb, shown in Figure 19.1, is derived from Gardiner, A. H. (1927) *Egyptian Grammar*, Clarendon Press, Oxford, and is reproduced by permission of Oxford University Press.

Adding up of the experience of the author and co-authors reveals that between them they bring 100 years of industrial and teaching experience in manufacturing management to this book.

AUTHOR PROFILES

GRAHAM S. SMITH JP B.Sc.(Eng) DNCL M.Sc. C.Eng.
Fellow of the Institute of Management
Fellow of the Institution of Manufacturing Engineers
Fellow of the Institution of Electrical Engineers

The author has a London External degree in mechanical engineering, a postgraduate diploma in industrial administration and business, and a masters in manufacturing engineering and management.

The author has 15 years industrial experience starting with five years as an engineering apprentice in the steel industry.

This was followed by several years in Asia and Africa, working for a French company as an oil exploration engineer.

On return to Britain the author became a senior management consultant in manufacturing organization, job evaluation and industrial relations.

The author has been in higher education in manufacturing engineering and management for 22 years. For all of this time the author has been involved in designing and managing integrated M.Sc., B.Eng. and M.Eng. degree courses in manufacturing engineering and business management.

The author was for ten years the Director of Undergraduate School in the Department of Manufacturing Engineering at the University of Birmingham, England.

At the time of writing the author is Head of the Manufacturing Strategy Group in the School of Engineering at Coventry University, England, and a member of the university's Quality Assurance Committee.

The author has been involved for several years in the quality assurance of engineering degrees in Britain as an external examiner and through the process of professional accreditation of engineering degree courses by the Engineering Council. The author was on the Accreditation Board of the Institution of Manufacturing Engineers and later on the Accreditation Committee of the Institution of Electrical Engineers.

The author teaches manufacturing strategy, manufacturing system design, manufacturing management and manufacturing project management.

ROBERT MANSFIELD B.A. P.G.Dip.

Robert has a first-class honours degree and professional diploma in management from the Open University.

Robert is a freeman of the City of Coventry and has 30 years experience in the engineering industry, starting with an apprenticeship in precision engineering and going on to design and environmental testing in mechanical and electronic high-volume manufacturing industries.

At the time of writing Robert is at Coventry University and has four years experience of teaching manufacturing management, manufacturing system design and manufacturing project management.

TREVOR PYE B.A.

Trevor has a degree in economics and is obtaining further qualifications at master's level in engineering business management.

Trevor is a freeman of the City of Coventry and has 25 years experience in the engineering industry starting with an apprenticeship in the aerospace industry. Management experience has been in the fields of manufacturing system design and implementation, managing change and training.

At the time of writing Trevor is at Coventry University and has four years experience of teaching manufacturing management, project management, organization and strategy.

INTRODUCTION

OBJECTIVES OF THE BOOK

This book looks at organizations and shows what is needed for them to be successful. In this way readers will acquire an understanding of how organizations work and how they are managed to enable them to work effectively. Examples are given throughout the book and in the task sheets at the end of each chapter which give readers experience in applying some of the techniques introduced in the book.

This book aims to provide an insight into the roles and functions of management in industrial and other organizations. Its objective is to guide managers, and students who see their careers in management, to an understanding of organizations. The book examines the various roles of management. It shows the need for an understanding of organizations if management is to be effective and successful.

This book therefore examines all aspects of management and in particular shows the need for the different functions of management to understand how they need to interact together if the organization is to be successful. Given its breadth, the book does not take each topic to an advanced level. This is left to other books specializing in particular topics such as organization, marketing, design, manufacturing, purchasing, accounting and human resource management. In this book each topic is, however, covered in sufficient detail for the reader to be able to apply some techniques for the analysis and solving of problems in each functional area and to link these together for a better understanding of the organization as a whole.

The danger of using specialist books on their own is that the study of a single topic in depth, while necessary, is not sufficient for an understanding of the organization as a whole. This is likely to lead to organizational failure as management decisions are taken without considering the effect of such decisions on other parts of the organization.

Having a set of people who are experts in their own field but lacking understanding of the other functions with which they must relate reinforces their specialism and prevents an organization from operating effectively. They operate as 'islands' of understanding but with no bridges between them. Organizations which are full of experts are fragmented and decisions are myopic.

Experts know more and more about less and less; eventually they know everything about nothing.

This book, because of its broad base, will be of value for anyone who is beginning to manage a part of an organization. It is intended to provide a broad-based integrative view of management at the first level of study of the subject. Readers will find this book of value at work. Employees in a first management role are likely to be primarily concerned with a specific area of work for which presumably they are suitably experienced and trained. The value of this book is its broader view of management. It will show the roles and functions of different parts of the organization and how they need to interact. It shows how one function of management can contribute to the effectiveness of the organization as a whole but needs to interact with other functions for the success of the organization as a whole.

People taking on a role in a trade union will also find this book useful, not only as a guide to management of the organization with which they interact, but also for use in managing their trade union. This book is a reference manual to which readers will return as their management careers develop, it is not a book to be discarded.

For students, this book is intended to form a course text for a course unit or module in management. It is suitable for the management content of engineering courses, especially for those including manufacturing engineering. It is equally suitable for scientists, artists, doctors, teachers, priests and even politicians, since in all walks of life people work in and manage organizations. It is designed as the first study of management and has no specific prerequisites beyond those necessary for university entry or equivalent.

It is suitable for those who take only a single module of management within some other main subject. It is also suitable for students taking courses containing a substantial amount of management. For these students it is a first-level foundation on which to take other modules in management at a higher level. The further study of the separate functions of management will benefit from this initial understanding of their interaction.

The use of the word 'manufacturing' in the title may imply that this book is of value only for managing manufacturing organizations. However, it is of relevance for management in all types of organization. Although the detail and emphasis of management is different in different organizations, the objectives, principles and concepts are the same. All organizations have objectives and resources. They all need 'managing' if these resources are to be used to achieve the objectives. While the central part of the book is titled 'manufacturing', it is concerned with managing the people and processes involved in the 'primary task' of all organizations. This may be manufacturing but the same function is carried out in service industries such as banks, hotels, airlines, railways, hospitals, universities, government offices, courts, charities, religions and trade unions. The term 'manufacturing' itself applies not only to what are seen as the traditional 'metal bashing' manufacturing industries, for example making cars. Few people these days make their own bread, cakes, pizzas or other foods, they buy them.

> *If you make a pizza you are cooking – but if you make a million pizzas a week you are manufacturing!*

Clothing, cosmetics, petrol, motorways, houses, bricks, window frames, furniture and many foods are all manufactured products.

Not all organizations have profit as their objective but the pressure to make effective use of resources, referred to as cost-effectiveness, requires managers in all organizations to develop a similar understanding and similar approaches to its achievement. Where government provides funding it is important to be effective since taxpayers' money is being used. Charities too must ensure that most of the donations are used for the objective purpose and not spent on administration or ineffectiveness.

This book examines the reasons for the low productivity of British organizations in the past and shows how productivity can be improved. If Britain is to compete in the world economy it must design and manage manufacturing systems with world-class standards of productivity. This requires more graduates with an understanding of manufacturing and its interactions with other functions of management.

This book takes a positive and proactive approach to the problems of low productivity. It explains the causes and then uses a modern systems-based approach to develop better forms of organization, modern styles of management, an environment for changing attitudes and management systems which are needed to achieve high productivity and competitive advantage.

Readers will detect in this book the interest of the author in ancient history. Chapters are sprinkled with some background on the historical development of organizations and societies. This shows how many of the concepts of management have their origin in the ancient civilizations. Modern management can learn from this background.

THEMES OF THE BOOK

This book develops several themes as the basis of understanding organizations and designing management systems which will enable organizations to succeed. Few ideas in this book are new; what is new is the way the separate ideas have been brought together to provide a different way of looking at an organization. The main themes of the book are:

Organizations are systems

A diagram giving a visual picture of an organization is shown in Figure 1.1 in Chapter 1. It will be useful for the reader to refer to this diagram throughout the reading of this book. This theme continues throughout the book and provides an understanding of how organizations behave and how management can influence that behaviour. It also shows the need for integration of the many aspects of organization which are examined in the book.

Wealth creation

A second theme is that of wealth creation. Many organizations are wealth-creating systems. The processes of wealth creation are examined and the role of management in this process is identified.

Objectives of organizations

A third theme is that of objectives. All organizations have objectives, usually more than one. Profitability is an important objective of profit-orientated organizations, but there is a similar objective – cost effectiveness – for non-profit organizations. The book also explains other objectives such as added value, competitive advantage, productivity, customer satisfaction and employee motivation.

Win–win strategies

A fourth theme is the creation of 'win–win' situations as a way to achieve organizational success. This theme is referred to in most chapters of the book. 'Win–win' strategies occur in nature. They are needed too in organizations. 'Win–win' strategies look at all contributors in an organization and can be developed between customers, marketing, manufacturing, purchasing and suppliers and also in managing human resources.

Productivity as a common objective

The fifth theme is that productivity is the most important objective of all organizations. It is only by increasing productivity that organizational outputs can increase relative to inputs. It is the only way in which profits and wages can rise simultaneously. Productivity is therefore a common objective between the different participants in an organization. It is a means of achieving a win–win situation. Almost everything written in this book is about improving productivity.

Involving people in organizations

The final theme is that people are important and success depends on involving people. This requires the perception of the organization as a system and the creation of win–win strategies. Organization culture may need to change, jobs must be challenging and satisfying. This requires the delegation of decision-making responsibility and autonomy for small groups within the organization. Each group treats the previous group as a supplier and the next group as a customer.

This concept is developed in many parts of this book and is known by several different names. The concepts of strategic business units, internal market, plant-within-a-plant organization, win–win supplier relationships, profit centre and autonomous work groups are all examples of applications in different parts of an organization of the same concept – that of involving people and delegating responsibility – which will enable an organization to increase its productivity to world-class standards.

STRUCTURE OF THE BOOK

The book is structured as seven parts, each examining a particular function of management. At the end of the book is an integrating case study. In each chapter the objectives of each function of management are identified and related to corporate objectives. The necessary management systems are then developed to guide the organization to these objectives. Of necessity the presentation of material has to be in a linear sequence, as does any teaching based on the book. Despite this, attempts are made throughout the book to stress the links between the different functions of management. Covering each function separately is a limitation on the overall approach of this book which is to look at organizations as a whole. Each function is therefore looked at in the context of its relationships with the other management functions. A logical sequence is used which deals with issues in the order in which they need to be addressed in setting up a new organization. This sequence is used in the case study at the end of the book.

Managing organizations

The book starts with the legal formation of organizations and the development of their corporate objectives and strategy, and the structure of the organization needed to achieve these objectives. This part comprises Chapters 1–4.

Marketing management

This is followed in Chapters 5 and 6 by marketing. The concepts of marketing mix, market research and product demand forecasting are techniques included in this part of the book.

Product management

The product itself must be analyzed and designed to meet market needs and to be made efficiently. Innovation, quality and development of new products are essential for the continuing survival of organizations. These aspects are covered in Chapters 7–9.

Manufacturing management

Manufacturing forms the central part of this book in Chapters 10–14, but a similar function is needed in all organizations. This function of management identifies the resources – materials, machines, manpower and money – needed for the successful operation of the organization. It identifies the tasks required for the design and operation of effective operating systems. It identifies the information needed for effective management decision making and the measures of performance needed for the control of an organization.

Materials management

The management of the flow of materials and other inputs into the organization and the flow of products to customers has to integrate very closely with manufacturing. Logistics, purchasing and control of material stocks are covered in Chapters 15–17.

Managing financial resources

Accounting for the expenditure made by an organization is an essential part of the control of organizations and hence of management. Chapters 18–21 provide some basic procedures in analyzing cost data and presenting accounts and some basic techniques of analysis. Costs occur in marketing, design, manufacturing and materials. In this book the accounts are derived from manufacturing data. The main emphasis in accounting in this book is on management accounting, i.e. using accounting information for management decision making.

Managing human resources

The final part of this book, Chapters 22–25, addresses the management of the human resources of an organization. The feature that makes organizations different from other types of system is that people are a part of an organization. It is now increasingly being recognized that organizations will only be successful if the people within them want them to succeed and if they have the necessary skills and feel that they are valued and actively involved. This part provides some understanding of motivation and management style but also covers basic procedures and functions of personnel management and the management of conflict. The final chapter in this part, Chapter 25, develops a modern approach to human resource management which is needed if organizations are to succeed in a competitive world.

Case study

There is a case study at the end of the book which enables readers to apply most of the techniques introduced in the book. It takes readers through all of the decision stages involved in setting up an organization. While some of the calculations are orientated to manufacturing, e.g. determining the number of machines required to provide the capacity needed, similar calculations are needed in all organizations to determine the equipment, space and people requirements for the organization to achieve its objectives.

Law and information technology

Two further topics are covered in this book but are not dealt with in separate chapters. These are the law as it applies to organizations and management, and the use of information technology in management of organizations. These topics are important for effective management and are addressed within each part of the book. This is done so that they can be dealt with in context with the particular application to the particular function of management.

An understanding of the various aspects of law is essential to ensure that organizations operate within the law. In multinational organizations this understanding must extend to each country in which the organization operates. This aspect is, however, beyond the scope of this book.

Modern developments in information technology enable organizations to integrate their information and control systems and to operate more effectively. In manufacturing this includes the programming of machines and robots and ultimately the computer control of the factory. The factory as a system can be modelled on a computer using simulation techniques. This enables a manufacturing system to be 'run' on the computer to determine its capacity and performance. 'At a stroke' it can be redesigned until a desired performance is achieved. This is a new and powerful tool for the design of manufacturing systems. Success in management depends crucially on the availability and accuracy of the data on which models are built and analyses and decisions are made.

USE OF THE BOOK IN TEACHING

This book has been designed to be of use as a course material for a unit or module on management. It can be used flexibly with the time taken per chapter depending on the time available and the level of the students. There is a set of tasks at the end of each chapter. The amount of mathematics used in the book is deliberately minimal. This has not led to the avoidance of any necessary topics but rather their treatment in a non-mathematical way. Each chapter is designed to be reasonably independent of others so that if necessary some chapters can be missed.

An 'enterprise' style of teaching and learning is assumed in which students learn by doing, not by being 'spoon-fed'. The book cannot in itself develop all of a reader's enterprise skills but forms the basis for the development of these skills. The case study provides work for students to do in their own time and make presentations of their recommendations to fellow students.

Learning from this book is based on the learning principle expounded in about 500 BC by Kwan Fung Tzu (known in Europe as Confucius) that:

> What you hear – you forget
> What you see – you remember
> What you do – you understand

Each part of the book starts with a set of learning objectives, and each chapter contains tasks for students to do in their own time. Students are expected to spend some of their own time reading the book. The main learning by students comes from tutorials and directed reading in their own time.

The structure of the book is based on studying one chapter per week, thus covering a full academic year of about 25 weeks. The implied student workload for this mode of delivery is 5 hours per week, of which 2 hours per week are formal contact. A 1 hour per week lecture will normally present teaching material of a chapter. A seminar or tutorial slot of 1 hour per week can involve students in small group learning based on the task sheets.

It is not intended that a lecture will cover all of the material in a chapter but rather introduce its main concepts. Lecturing is not an effective way of learning.

> *Lectures transfer information from a lecturer's notes to the students' notes without it passing through the mind of either!*

In a lecture students are given a framework for thinking but must take responsibility for their own learning and develop their own learning strategy. An extensive booklist is provided to direct readers beyond this book for more detail and for development of the topics introduced here.

The case study at the end of the book can be used for assessment. The case study takes students through all aspects of setting up and running a small, fictional company. Most of the concepts and techniques introduced in the book are used in the case study. The case study gives students the opportunity to apply the concepts from each chapter and to integrate them by applying them all to the same case organization. There are 12 stages of analysis in the case study. The sequence and timing of tasks in the case study is matched to the sequence of topics in the book. The book includes all the guidance needed to do the case study.

PART I

MANAGING ORGANIZATIONS

LEARNING OBJECTIVES

1.1 Identify the purpose and typical objectives of an organization.

1.2 Understand the role of management in organizations.

1.3 Visualize an organization graphically.

1.4 Understand the reasons for the decline of manufacturing in Britain.

1.5 Appreciate the significance of productivity in achieving competitiveness.

1.6 Identify the paths to increasing productivity.

1.7 Identify the different legal forms of organization.

1.8 Recognize the 'stakeholders' in organizations, their inputs, outputs and objectives.

1.9 Understand the process of wealth creation and the definition of added value.

1.10 Explain the relationship between productivity, inflation and wage levels.

1.11 Define productivity, competitive advantage and profitability and their relationship.

1.12 Identify the need for and content of a corporate strategy.

1.13 Describe in outline the process of strategy formation.

1.14 Explain the concepts of boundary, component, subsystem and environment of a system.

1.15 Understand the process of systemic thinking.

1.16 Explain the concepts of synergy and symbiosis.

1.17 Visualize an organization as a system using diagramming techniques.

1.18 Identify the stages of control and apply the concept to managing an organization.

1.19 Identify the roles of management in planning, organising and controlling the resources of a organization.

1.20 Identify the major functional areas of management of a typical organization and the links needed between them.

1.21 Recognize the multiple and plural objectives within organizations and their compatibility and conflicts.

1.22 Draw an organization structure chart and be aware of the implications for responsibility, accountability, power and authority of position.

1.23 Relate communication and decision structures to the organization structure.

CHAPTER I

INTRODUCTION TO ORGANIZATIONS

OBJECTIVES OF ORGANIZATIONS

Organizations exist for a purpose. They are complex systems. They are created for a purpose. They comprise a group of people who invest something in the organization in expectation of getting something out of the organization. These inputs and outputs are not only the obvious monetary ones. People invest their time and their skill, their rewards may be the satisfaction they get from making their contribution. Different people in organizations have different objectives and personal motivation towards achieving these objectives. It is this feature of organizations which makes managing them complex. In order to manage organizations effectively these objectives and motivations must be understood and harnessed. Chapter 2 develops this theme. It is not easy to visualize how an organization works. To assist, a diagram can be drawn which attempts to show what an organization is and how its parts relate with each other. Figure 1.1 is a visual representation of an organization. An organization buying materials and making products, i.e. a manufacturing organization, is chosen to illustrate the idea of visualizing an organization, but the concept is the same for other types of organization.

Figure 1.1 does not pretend to show all aspects of an organization: that cannot be done in a single diagram. Each part of the diagram could be expanded in a set of further diagrams at a more detailed level. Such a diagram can show the main components of a system and the external systems with which it relates. It cannot easily show the nature or detail of the internal and external relationships, but does provide a visual representation of an organization which can provide a starting point for understanding these relationships. The concepts used in Figure 1.1 are developed in Chapter 3 but it is introduced here to illustrate the point being made, that an organization is a complex system.

It will be found useful to refer to this diagram at various stages of the book. In studying each function of management, the book continuously stresses the need for interaction between management functions if the organization is to be successful.

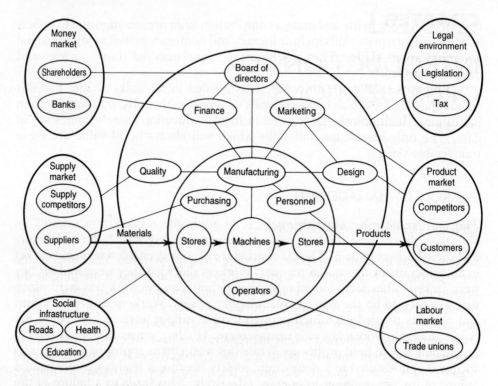

Figure 1.1 *A manufacturing organization as a system.*

Organizations and enterprise

This book is about enterprise. It looks at organizations that are known as enterprises, which people create in order to do something enterprising. A person managing an enterprise is called an entrepreneur, but all people involved in organizations need to be enterprising. They invest their time and skills and take risks. While only profit-making organizations may be called enterprises, other organizations need the same qualities or spirit of enterprise. A university, for example, is not a profit-seeking organization, but it is and needs to be enterprising in its management if it is to operate effectively, maximizing the quality and efficiency of its activities in order to deliver a quality service to its students and to society as a whole. The concept of enterprise invokes the idea of ingenuity, innovative thinking and people working together to achieve efficient operation and personal satisfaction and through this to ensure customer satisfaction. These attributes are required in all organizations.

Enterprise can also be seen as a set of skills, attitudes and personal capabilities needed in people who work with or in organizations. Since we all work with organizations of one form or another, we all need these enterprise skills. At their most basic, enterprise skills can be seen to be those skills needed to operate effectively in life and make a contribution to an organization and to society as a whole. Enterprise skills include literacy and numeracy as well as interpersonal communication skills. They include an awareness of how society and organizations work.

Enterprise skills are universal skills needed in all walks of life. For this reason they are known as common skills or transferable skills, transferable from job to job. Much knowledge learned in formal education soon becomes out of date. The only knowledge and skills which will always be of value are these transferable skills.

THE ROLE OF MANAGEMENT

The most basic role of management is to guide the organization towards its objectives. In the past this was easy. Managers decided what to do and simply told their workers to do it. The class structure of society enabled managers to act in an autocratic or dictatorial manner. Workers did what they were told or they were sacked. They were forced to work long hours and to work very hard, since that was seen to be the way to obtain higher profits. Workers who were infirm and not able to produce a high enough level of output were sacked. All of the power in organizations lay with management. Holding wages down was seen to be the only way to hold profits up. While this was seen as appropriate in a feudal society it is not suited to a democratic society because it leads to conflict and a failure of the organization to operate effectively. This leads to a failure of the organization to achieve its objectives. Although high profit is the objective, low profit is often the result.

In a democratic society the objectives of management are the same, but the role and style of management needs to be different. In a modern, democratic society the role of management is to create the conditions in which employees want to work. In this way the objectives of employees, managers and the organization can all be satisfied. All parts of the organization 'win', so this is described as a win–win situation. How to achieve win–win situations is developed in all parts of this book but particularly in Part 7.

Organizations undertake routine tasks in carrying out their day-to-day activities. Management must identify these tasks and determine the resources required to do them. This role of management manages the steady-state operation of the organization. In addition, however, an organization must develop and change with changes in technology and in the market. This is an additional role of management. The management of change requires a different approach which is developed in Chapters 14, 21 and 25.

MANUFACTURING IN A CHANGING WORLD

In the past Britain had an empire. This was a captive market for the products manufactured in Britain. The wealth and economic development of Britain was based on its empire and on its manufacturing industries. Manufacturing organizations were seen to be successful because they made profits and all that mattered was profit. Managers were managers because of who they were rather than what skills they had. Management was not taught at universities and most managers were not even graduates of any discipline. If they were graduates their degrees were mainly in Greek or Latin. Organizations in all sectors of the economy were inefficient, but this was not recognized at the time. Profits were made because there were no real competitors.

The latter half of the twentieth century has seen the economic development of many other nations, especially those of East Asia. These nations have created manufacturing industries which compete with Britain and Europe. In this economic climate it became apparent that British manufacturing industry was inefficient compared with these new industrialized nations. The productivity of the motorcycle industry in Britain and in Japan can be measured in terms of the output (the number of bikes produced per year), ratioed to the input (the number of employees). This data is shown in Table 1.1.

Table 1.1   Motorcycle productivity in Britain and Japan

Country	1956	1973
Britain	19	13
Japan	16	195

Productivity measured as bikes per employee per year.

Over this period of time productivity in Britain had fallen but in Japan had risen significantly. For this reason the British motorcycle industry collapsed. Productivity overseas was increasing significantly but in many industries in Britain it fell as organizations made and sold less with the same resources in a new economic climate involving overseas competition. Britain's share of the world market in manufactured goods fell from 11.2 per cent in 1969 to 7.7 per cent in 1983.

In 1984, for the first time since the industrial revolution, Britain imported more manufactured goods than it exported. Britain had lost its empire and was unable to compete in the modern world. Some industries had declined because of a reduced need for their products – shipbuilding and coal-mining for example. The new products of the electronic age – televisions, video recorders and microwave ovens for example – are in great demand but British industry is unable to make them competitively. All of these products were invented in Britain but Britain did not have the engineering or management skills to make them. British graduate engineers have the skills to design engineering products but not the skills to design factories to make them at a high enough level of productivity to compete in world markets. These products are now imported from East Asia, or at best assembled in Britain by organizations based in East Asia from components manufactured there and with profits exported there.

The importation of modern manufactured goods to satisfy the wants of the population has led to a reduction in the levels of employment in manufacturing in Britain, the decline of the economy and rising unemployment. The growth of the service sector – tourism and banking for example – has created some new jobs but not enough to replace those being lost in manufacturing. The economy of Britain, and other nations, is still and will continue to be dependent on its ability to manufacture many of the manufactured goods which its population wants, and to be able to compete overseas and so export to other nations. Only if a nation exports can its economy grow larger than that necessary to sustain its own population at poverty level.

Clearly, Britain cannot compete with the low-wage economies of East Asia in all areas of manufacturing. However, the most significant difference between British and East Asian industries is not the level of wages but the level of productivity. Wage levels in Japan are about twice those of Britain. Prices of British goods are higher than those from East Asia because the British price carries the wages of more people. For example, from the figures quoted above for the motorcycle industry in 1973, the price of a British bike contained 1/13 of a person's annual wage, but the price of a Japanese bike contained only 1/195 of a person's annual wage. This is the reason for products from East Asia being cheaper than British products. Only by increasing productivity can British and European organizations compete with those based in East Asia.

THE FUTURE OF MANUFACTURING

As indicated above, the lack of British competitiveness has been mainly due to lower productivity. The key to success therefore lies in the direction of increasing productivity. This is the foundation on which this book is based. So important is productivity that it should be seen as the most important objective of any organization. It is a concept equally applicable to profit and non-profit organizations. Terms already used such as efficiency, effectiveness and cost-effectiveness all mean the same as productivity and extend the concept to non-manufacturing organizations. For profit-seeking organizations the term profitability is used. This refers to the productivity of the financial resources of the organization.

Productivity can be measured in many ways. These are expanded in Part 4 of this book. The simplest definition of productivity, sufficient at this stage, is the ratio of output over input, a measure of what comes out of the organization compared with what goes into it. Productivity can be seen to be a measure of productive efficiency.

$$\text{productivity} = \text{output/input} \tag{1.1}$$

Poor productivity in Britain has led to the closure of many manufacturing organizations. Those that have survived are those that have recognized the problem and taken steps to reorganize themselves completely to achieve high productivity. With higher productivity, British organizations will be able to compete in at least some of the product markets in which they have not been competitive in the past.

There are many paths to productivity. Modern technology – electronics, automation and robots – has a part to play in increasing productivity, but this is not the only way. Productivity depends primarily on people. In the past the emphasis has been on increasing productivity through technology and people have been ignored. Those organizations that are succeeding in the modern competitive world are those that build their success on achieving productivity through people. Productivity depends significantly on the skills of management in designing and operating manufacturing and management systems. Productivity also depends crucially on the attitudes and motivation of all of employees and on how management systems enable them to be motivated and to be effective. Figure 1.2 shows the paths to productivity.

Higher productivity requires, for many organizations, a complete rethink of what they should be doing and how they should be doing it. The starting point is a recognition that an organization is a 'system' – a concept explained in Chapter 3. Success comes from identifying and focusing on the core tasks and core skills needed to provide the right products, or services, of the right quality, at the right price in the right market in order to attract, satisfy and retain customers. In many cases this requires changes in the form of organization, the style of management and the organization of the manufacturing and non-manufacturing operations.

In Britain and Europe the styles of management and attitudes of managers in the past have been a major constraint on productivity. These may have been appropriate to the feudal structure of society at the time but are not appropriate in a modern, educated democracy. The social and political structure of Britain, coupled with the attitudes of managers in the past, made Britain incapable of change. Management style was adversarial, managers believed they were fighting a battle and had to force their workers to work harder by issuing military-style commands. They saw holding wages down as the only way to hold profits up. The

Figure 1.2 *Paths to productivity.*

philosophy was one of 'we' win and 'they' lose. This is a win–lose strategy but it leads to a lose–lose result because of the conflict and alienation it causes. By the middle of the twentieth century Britain had become the museum of Europe with fossilized social structures and organizations clinging to an autocratic style of management. This archaic style of management can be described as neanderthal management, named after a early species of human which failed to adapt to change and became extinct about 30,000 years ago. Only recently have some British organizations changed from neanderthal management to a more modern style adapted to a modern society. The failure to do so earlier has been the cause of low productivity, bankruptcy, unemployment and takeover or replacement by Japanese organizations which use a modern systems approach to management.

Japan in the 1950s was rebuilding after the Second World War. Japanese managers read articles in British and American journals of new approaches in manufacturing and management in the fields of manufacturing planning and control, ergonomics, the development of mathematical analyses of manufacturing problems and the beginnings of group or cellular organization. As they attempted to build new industries operating at high productivity, they readily absorbed these Euro-American ideas, because they thought they needed to do so to compete with European and American industry, which, they believed, were using these new ideas. Japanese factories were built and organized on the basis of this readily available information. Only later did they realize that British industry was not using any of these ideas and was still operating with neanderthal management. These modern ideas were developed at universities and published in journals but were not read or implemented by British managers.

British managers in the past were not graduates and did not read journals. They were not aware of or were not interested in and did not absorb the new ideas which were known and available. Relationships between industry and universities were poor, and so industry did not benefit from the ideas developed at the universities. The best graduates went into teaching or the Civil Service. With the development of high productivity in Japanese industry in the 1950s and 1960s, British industry was unable to change and unable to compete. It therefore declined and many manufacturing organizations disappeared. Much of the cause of failure lay in the attitudes and neanderthal management style of managers.

Only in the 1990s has British industry begun to make the fundamental change in direction that it could have taken 25 years before. In the meantime so much of manufacturing industry has been lost. Much effort is needed to rebuild manufacturing in Britain to a level of productivity at which it can compete. Only by making, rather than importing, the modern manufactured products which consumers want can the European nations create the wealth, employment and standard of living which their citizens want.

The terms 'organization development' and 'business process re-engineering' have been given to this process of fundamental change. These are referred to in Chapters 10 and 25. The result has a name too – world-class manufacturing – meaning that the organization has the high level of productivity to compete and beat competitors. The different parts of this book address these and other modern approaches needed to rebuild manufacturing and other organizations in Britain to achieve high productivity and profit.

Only when a nation trains enough engineers and managers with an understanding of manufacturing organizations will manufacturing systems be designed and managed to operate at world-class standards of productivity. Britain has failed to produce enough engineers and grant enough status and pay for engineers to attract young people into the profession on which everyone depends for the products that make up a modern technological society. Few organizations in Britain are run by engineers and few politicians are engineers. Many organizations are managed by accountants. This contrasts with Europe and Japan in which most industrial organizations are managed by engineers.

In Europe the word for an engineer is 'ingenieur', spelt with an 'i'. This shows the meaning of the word engineer – a person of ingenuity. In Britain engineer is spelt with an 'e' and means an engine driver, mechanic or electrician. This generates a negative, low-status and low-salary image of the engineer, which contrasts strongly with the positive image, status and salary of an engineer in Europe and Japan. The poor performance of Britain and the success of Germany and Japan can be correlated with and attributed to the image of the engineer and the role of engineers in managing industrial organizations. People in Britain want the products of a modern society – cars, videos, mobile telephones, etc. – but do not want to be a part of creating and making these products. Only by creating wealth in Britain can its people afford to buy these products. Importing manufactured products has the effect of exporting jobs and wealth to other nations.

Only when a nation invests adequately in its educational system and produces graduates with the skills needed for creating modern enterprises will it create increasing wealth and employment. A nation that does not invest in its educational system is destroying its future.

Task sheet I   Introduction to organizations

1.1 Identify the country of origin of a range of products used in your household. Alternatively, examine a range of domestic products in a shop or catalogue store. You might consider televisions, video recorders, camcorders, radio, stereo and audio equipment, telephones, pottery, saucepans, furniture, glassware, cars, bicycles and clothing.

1.2 Identify the 'new' countries which manufacture domestic products and export them to Britain. What is the effect of this trade on the economies of Britain and exporting nations?

1.3 List the reasons for low productivity in Britain. If you are a student working in a group at a tutorial, discuss this topic in your group.

CHAPTER 2

STAKEHOLDERS IN ORGANIZATIONS

LEGAL FORMS OF ORGANIZATION

Types of organization

Organizations are created by people coming together for mutual benefit. There are different ways in which an organization can be created. The form selected will depend on the type of organization. It is important to ensure that the right type of organization is created for the intended activity. Organizations may have economic objectives but not all organizations have this objective; they may have educational, health or religious objectives.

There are many different types of organization. They differ in size, the types of activity they undertake and how they are legally established. Legal procedures differ from one country to another but they are not greatly different across the European Union. It is worthwhile identifying the different types of organization in Britain and classifying them in a way which enables an analysis of their structure and objectives.

Organizations may or may not be profit-seeking. They may or may not have shareholders who own shares. They may have paid employees or unpaid voluntary workers. The objectives of an organization depends on the type of organization it is.

Profit-making organizations with unlimited liability

- *Sole trader:* shopkeepers, landlords, window cleaners, painters and decorators, etc.
- *Partnerships:* groups of accountants, solicitors, doctors and other professionals.

In these organizations there is no separation of the wealth of the business from that of its owner or partners. The financial resource of the business is that of its owner or partners. Additional finance may be obtained as an overdraft or as bank loans in the name of an individual. In the event of bankruptcy the owner or partners may become bankrupt since they are liable for the debts of the business. The owner's personal assets such as a house will be used by a bank as security and may have to be sold in order to repay the bank.

In most countries the law does not impose legal procedures on the formation of such organizations. The law sees sole traders and partners as individuals. In any legal wrongdoing the individuals are charged or sued. A partnership agreement is not a legal requirement in Britain but is usually considered to be essential as a means of agreeing the distribution of revenues and costs and preventing disputes.

Formation of companies

Because of the risks associated with sole trading and partnerships, laws have been established to permit the formation of alternative forms of organization such as companies and to regulate the formation of companies. An individual, e.g. a consultant working alone, can form a company as a means of protecting personal assets. A company cannot, however, comprise a single person: it requires at least two directors and a company secretary. The latter is not a typist but the person responsible for the legal aspects of the company and known in America as a legal executive.

In the past in Britain companies could only be formed with the permission of the monarch. Worshipful companies, universities and professional institutions are formed this way. Such organizations are created by Royal Charter. It soon became a burden in a growing economy for all the documents for such company formations to have to wait to be signed, sealed and delivered by the monarch. This was not easy when, for example, King Richard I was out of the country for most of his reign. Eventually Parliament gave itself this role.

In Britain in 1503 the Ordinances of Corporations enabled Parliament to approve the creation of corporations. Companies formed this way were incorporated by an Act of Parliament. The East India Company, founded in 1600, and others such as the Hudson Bay Company, formed to exploit British colonies, were established in this way. A separate Act of Parliament was needed for each company. Such companies are Incorporated Companies, some of which still exist in modern times.

With the Industrial Revolution this too became a burden, so eventually Parliament passed legislation enabling companies to be established without needing a separate Act of Parliament, but instead under the framework of a Companies Act approved by Parliament. Many Companies Acts have been passed since this to modify the law in a changing society. The most significant early legislation was in 1855 which enabled the creation of the limited liability company. This legislation sought to encourage investors to invest in industrial and other enterprises by providing some legal protection for investors. It did so by separating the ownership of the company from the investors.

Profit-making organizations with limited liability

In Britain there are two types of limited liability company:

- Private limited companies (Ltd).
- Public limited companies (plc).

Most companies are private and recognized by the Ltd in their title.

In these organizations the main financial resource is from investment in shares by shareholders, although some additional finance may be from bank loans or other sources. These organizations are established as corporate bodies existing in law separately from their investors. The law sees the company as a corporate person, one that can be charged or sued as an entity. The shareholders do not own the business but own the shares and control the business by appointing the directors.

This separation of ownership limits the liability of shareholders such that in the event of bankruptcy the shareholders cannot be sued for the debts of the business. Shareholders may lose the money they have invested in the business but their personal assets are not at risk.

In public limited companies the public may buy and sell their shares in the business. Large companies and company groups are usually in this category. In private limited companies the ownership of shares is restricted and not open to the public. Smaller family businesses and wholly owned subsidiaries of public limited companies are usually in this category. Private sector businesses in manufacturing or services are normally limited liability companies but banks, insurance companies, hotels, airlines and private utilities are also in this category. Most limited companies are private limited companies.

Registration of companies

In Britain limited liability companies are established by registration with the Office of the Registrar of Companies. Within the European Union harmonization of company legislation is proceeding so that a company can be formed as a European rather than a British company. Registration requires the submission of four documents which identify the company. The information provided is available for public inspection.

1. *A memorandum of association.* This gives the name of the company, the address of its main office, the number and nominal value of shares and the main objective of the company. The name must include Ltd or plc to indicate a private or public company.
2. *Articles of association.* These give details of the company's financial year, rules for the appointment of directors and for conducting meetings.
3. *The names and signatures of the directors and company secretary.*
4. *The nominal value of total share capital.* This is the number of shares and the nominal value of each. The nominal value is often £1 and is the value of the

shares on the day they are issued. The actual value of shares will differ from this nominal value. The nominal value of shares appears in the balance sheet.

Directors of companies

Directors are appointed by shareholders at an annual general meeting. They form a board of directors which sets the objectives and policies for the company and reports to the annual general meeting of shareholders. Some of the directors have responsibility for carrying out the policy of the board. These are executive directors and are usually full time. They work at the company and may have a title such as Financial Director or Manufacturing Director. These directors are called executives because they execute, or put into action, the policy. They develop strategies and take decisions. In law directors are not employees. They receive a fee rather than a salary, but for all practical purposes can be seen as the senior managers of the organization.

Directors are required to make an annual report to the annual general meeting of shareholders on the state of the company. More detail of this requirement is covered in Chapter 19.

Other directors are appointed from outside the organization. They attend board meetings but do not otherwise work for the company. These are non-executive directors. They are in most cases part time. They are brought in because their skills or contacts are considered to be useful to the company. They usually receive a fee for their services. Politicians and retired executive directors may be a non-executive director of several companies.

Non-profit organizations

- Trusts, charities, clubs.
- Friendly societies, trade unions.
- Building societies.
- Co-operatives.
- Universities, schools.
- State-owned organizations.
- Professional institutions.

These organizations do not seek to make a profit, but they have a similar objective of cost-effectiveness. That is, they aim to make best use of the resources available to maximize the output (usually a service) for minimum resource or cost.

State-owned organizations

State-owned organizations are in the public sector, ultimately responsible to government and the whole of society. As they use investment from taxpayers they are under pressure to be cost-effective. This category includes government and local government departments, state schools and hospitals. The post office, police, fire services, courts, and road building and maintenance are examples of this type of organization. In some cases work is subcontracted to private sector organizations, but the management and funding is by the public sector.

Private sector non-profit organizations

The other non-profit organizations are in the private sector and are created and exist for the benefit of their members. They are established as corporate bodies by registration with the Charity Commissioner, the Registrar of Friendly Societies or the Trade Union Certification Offices. These offices are established to give some protection to members of these organizations and to ensure they are managed lawfully and in the interests of their members.

Notice that although building societies appear to be the same as banks because they do very similar things, in law they are differently established. Building societies exist to serve their members; banks exist to make profit for their shareholders. The assets of a building society belong to the account holders. In Britain the Abbey National Building Society recently converted itself into a bank and became a limited liability company. It had to buy itself from its account holders using shares from external shareholders. It cannot therefore now call itself a building society even though its activities are the same. The difference is that the objective is now to make a profit for shareholders, whereas previously the financial benefits went to account holders.

Co-operatives have some similarities with partnerships in that financial benefits are shared by their members. There are few in Britain but there are some in Spain. Israeli kibbutzim operate in a similar way. In Britain, Waitrose, John Lewis, Kalamazoo and Scott-Bader are co-operatives, along with the Co-operative manufacturing, banking, insurance and retail organizations. The largest operator in the funeral industry in Britain is the Co-operative Funeral Service. Co-operatives operate in the same way as companies, in that they make or sell something, but they do not pay a dividend to shareholders because there are no shareholders. Any financial surplus is retained in the business or distributed to employees, who in law are members rather than employees.

Private schools, private hospitals and private residential homes for the elderly are usually established as trusts or charities, but may alternatively be established as public or private companies.

In Britain the 'old' universities are established as universities by Royal Charter. A university is a group of people who are its members. Academic staff are members of the university. Universities are private sector organizations. They are not directly controlled by government even though most of their funding is from the government. The former polytechnics were public sector organizations which were a part of and controlled by local government. In the 1990s polytechnics were incorporated by an Act of Parliament and established in the private sector as higher education corporations funded directly by the government. Later they were permitted to use the title university but are still higher education corporations. They are now the 'new' universities. Academic staff are merely employees. All universities do the same things, but in Britain there are two different types of organization called universities. The culture and style of management is very different.

Most professional institutions are established by Royal Charter, enabling the members to use titles such as Chartered Engineer or Chartered Accountant. Some professional institutions are incorporated companies with trustees but no shareholders.

STAKEHOLDERS

An organization manufacturing a product or providing a service contains and interacts with several different groups of people, all of whom can be said to have a 'stake' in the success of the organization. Taking a manufacturing organization as an example, five major stakeholders can be identified. These are shown in Figure 2.1.

Some of these stakeholders can be seen to be within the organization but others are outside it. Suppliers and customers are clearly not a part of the organization but they do have a stake in it in the sense that they rely on it for their own wellbeing. For each stakeholder their inputs to and outputs from the organization can be examined and from this their objectives can be determined.

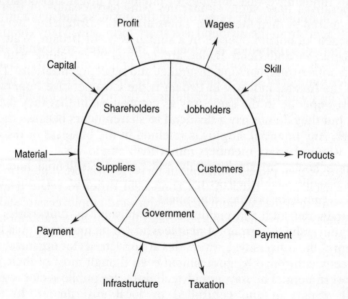

Figure 2.1 *The stakeholders of an organization.*

Suppliers

Suppliers supply materials or services to an organization. This is their product which they sell and from which they make their profit. Their stake is in the survival of the organization since it is their customer and from it they derive their revenue.

Customers

Customers buy the products or services of an organization. Their relationship with the organization is the same as that of the supplier but in the opposite direction. Their stake is in the survival of the organization since it provides them with their supplies, materials or services. Suppliers, manufacturers and customers can be seen to form a chain linking them together. There may be several links in the chain since the customer may carry out further processing and sell to a further customer along the chain. The concept of a supply chain is developed in Chapter 15.

Shareholders

Shareholders have invested money in the organization and seek from it a financial return on their investment. Their stake is in the survival and performance of the organization in order to protect their investment and the income they derive from it. In family businesses family members will be the main shareholders. They are also likely to be executive directors running the business. In larger, public companies most shareholders are financial institutions rather than people. Banks and pension funds have vast amounts of money to invest and do so by buying shares. Such shareholders are remote from the business and generally have little interest in it, apart from the dividend as a short-term gain. Many companies exist within a group of companies in which all the shares are owned by another company. This is part of a continuous process of concentration of shareholding into a smaller number of shareholders. This has been reversed, but perhaps only temporarily, by a wider public shareholding in the privatized former nationalized utilities.

Jobholders

Jobholders are the employees working in the organization. The term jobholder is not often used but is used in this context in order to relate it to the term shareholder and to show that both shareholders and employees are stakeholders. Employees invest their skills, experience and time in the organization in return for their wages or salary. Their stake is in the survival of the organization and in their job within it.

Government

Government have a stake in the survival of organizations as contributors to the wellbeing of the economy as a whole. Organizations provide the government with money through taxation and by creating employment avoid government having to pay so much in social security payment for the unemployed. Organizations benefit from government expenditure on education, health, transport and defence. Government or European Union money may be available as development grants.

Stakeholders' objectives

Each stakeholder naturally wishes to maximize what they receive from the organization in relation to their input to it. The objectives of the different stakeholders are therefore different. Many of the conflicts involved in managing organizations arise from these differing objectives. The inputs, outputs and objectives of the different stakeholders are summarized in Table 2.1. This table is a simplified view since it shows only the major aspects. On further analysis, other aspects will be found. Notice that managers are not specifically identified as stakeholders. They are employees of the organization with the role of guiding the organization in the direction of the objectives set by the shareholders.

Table 2.1 Stakeholders – inputs, outputs and objectives

Stakeholder	Input	Output	Objective
Shareholders	Investment	Profit	Return on capital
Jobholders	Skill and effort	Wage	Standard of living
Suppliers	Material	Payment	Maximum price
Customers	Payment	Products	Minimum price
Government	Infrastructure	Taxation	Economic growth

The process of wealth creation

Table 2.1 shows that the major interaction between the different stakeholders is the money flows between them. The stakeholders come together for their mutual benefit – the creation of wealth. Each stakeholder will seek to maximize their share of this wealth. For most organizations the only money inflow is from selling products or services; the other money flows are outflows, mainly the costs of providing the resources.

For profit-making organizations the traditional view of these money flows has looked at the organization from the perspective of the shareholder only as follows:

$$\text{sales revenue} - \text{total costs} = \text{profit} \tag{2.1}$$

Profit and wages are seen as opposites. To the accountant a wage is a 'loss' of profit. What is left for 'us', the profit for the shareholders, is what is left after the costs have been paid to 'them', the employees and suppliers. This creates 'us' and 'them' attitudes, which are socially polarized and politically divisive. This traditional view is based on conflict and reinforces the conflict between stakeholders. The style of management is based on the shareholders 'winning' and the employees and suppliers 'losing'. The conflict generated holds down productivity and profit.

For non-profit organizations sales revenue (or in the public sector, government grant from taxation) would be set to just cover costs.

For a manufacturing organization equation (2.1) can be expanded to:

$$\text{sales} - \text{materials} - \text{wages} - \text{taxes} = \text{profit} \qquad (2.2)$$

(Note that these five items reflect the five stakeholders.)

Sales is the revenue or money input from customers and is the only money inflow in the equation. All the other items are money outflows to the respective stakeholders.

A different perspective arises from reorganizing this equation to:

$$\text{sales} - \text{materials} = \text{wages} + \text{taxes} + \text{profit} \qquad (2.3)$$

This equation is numerically exactly the same as equation (2.2) but the terms are arranged differently. Each half of this equation is numerically equal to the other half:

left = net revenue = right = added value

left = wealth created = right = wealth distributed

Net revenue is the difference between the value of sales and the cost of materials and services provided by suppliers. It is the difference between what is received for what is sold, and what is paid for what is bought. Sales is the money received from customers, and materials is the money paid out to suppliers for both materials and services. This difference is the valued added by the organization and wealth created by the organization. It is also the wealth distributed by the organization.

This view sees customers as external to the organization and having an economic relationship with the organization. They do not share in the wealth of the organization. The government is also external, but is counted as a part of added value and shares in the wealth of the organization. This can be seen as the share of added value of an organization which goes to society as a whole. The concept of added value as being the wealth created by an organization is shown in Figure 2.2.

This view distinguishes wealth from profit. Wealth created by manufacturing arises from selling products for more than was paid for the materials. Manufacturing is the process of transforming materials into products, and in doing so adding value to them. The added value, or value added, is also the difference between the value of sales and the cost of materials and services. Manufacturing is thus the process of wealth creation. The wealth is created by the activities of people in organizations.

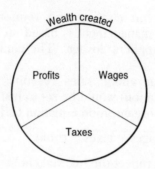

Figure 2.2 *Added value of an organization.*

The wealth created by adding value is distributed to three of the stakeholders. Profit is that element of wealth which goes to shareholders in return for their investment. Wages is that element of wealth which goes to jobholders or employees in return for their skill and effort, and taxation goes to government to provide the roads, health, education, police and defence systems that form the infrastructure of a society. The detailed calculation of added value is covered in Chapter 19 when costs have been analyzed in more detail.

Productivity

The process of wealth distribution, outlined above, shows the need for co-operation between the three stakeholders for their mutual benefit. Conflicts will, however, occur between the stakeholders because of their differing objectives.

Within the organization this is seen as the profit–wage conflict between managers who run the business on behalf of the shareholders and employees. This approach sees wealth as a fixed sum, and conflict is based on how this is to be distributed. This type of conflict is destructive in that output falls, employees lose money and shareholders lose profit – this is a 'lose–lose' situation in which both sides lose. Conflict arises over the share of a 'cake' of fixed size. The way to raise profit, in this perspective, is to hold wages down.

Productivity is the ratio of output to input. (Refer back to equation (1.1) in Chapter 1.) Output and input can be measured in many different ways and at different system levels. A simple measure at this stage would be the ratio of units produced (output) over resources utilized, i.e. capital and labour (inputs).

If productivity increases, then greater outputs can be achieved with the same resources. This greater output will create greater wealth and enable more wealth to be distributed to all the stakeholders. Larger profits and larger wages will be seen as beneficial by the respective stakeholders. Paying more taxation will never be seen as desirable but nevertheless most people benefit from government expenditure on social infrastructure such as education, health and roads. Increasing productivity increases the *size* of the cake, as shown in Figure 2.3.

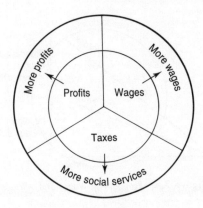

Figure 2.3 *Productivity increases wealth.*

Conflict over the *share* of wealth needs still to be resolved, but co-operation can lead to an increasing size or amount of wealth. Productivity thus enables profits and wages to rise simultaneously, and is the only way for this to happen.

Productivity is thus a common objective between shareholders and jobholders, and enables them to co-operate together in its achievement. This is a 'win–win' strategy since wealth is seen as a variable sum, not the fixed sum seen in the traditional style of management.

Productivity and wages

Naturally everyone would like their wages to be higher so that they can buy more goods and enjoy a higher standard of living. The standard of living is a measure of how much people can buy with their wages and combines the effects of wage and price increases. The standard of living increases when wages go up more than prices. The ability of organizations to pay higher wages depends on their productivity. Higher wages can be financed by higher productivity. In this situation the standard of living of employees increases.

Employees are concerned when they find that over a period of time they are not able to buy as much because prices have increased. The process of inflation, the rising of prices over time, therefore devalues wages. Employees want a pay increase merely in order to be able to retain their standard of living.

Wage increases that occur without an increase of productivity can only be financed by increasing prices, thus causing inflation. These relationships are shown in Figure 2.4.

The important point from Figure 2.4 is that the standard of living is closely related to the level of productivity. Wages can only go up more than prices if they are financed by increases in productivity.

Wages and productivity are determined at the organization level but inflation occurs at the national level. There is, however, a common pressure for wage

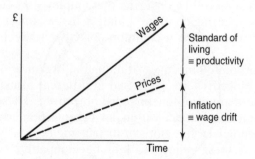

Figure 2.4 *Productivity, wages and inflation.*

increases in response to inflation, so there is a tendency for wage increases to be reasonably consistent across the economy in order to retain pay differentials which are perceived as fair.

Organizations that do not increase their productivity are likely to increase their prices and thus reinforce inflation. Wage increases not financed by productivity cause inflation. This type of wage increase is called wage drift. Only those organizations that increase their productivity are able to finance higher wages and avoid putting up prices and retain their competitive advantage. Ideally productivity is increased by increasing output but this requires market growth. Without this growth, productivity is increased by reducing inputs, e.g. by reducing the number of employees.

OBJECTIVES OF INDUSTRIAL ORGANIZATIONS

Some objectives of industrial organizations have been referred to above.

Profit, the difference between revenue and costs, can be misleading since it gives no indication of the magnitude of the profit compared with the investment. Profitability, the ratio of profit to the value of the investment, is the prime objective of the shareholders and often considered to be the prime objective of a business. Profitability can be measured as the return on capital employed (ROCE), return on investment (ROI), or return on total assets (ROTA). These are similar but slightly different measures of profitability. These and other financial ratios are covered in Chapter 20. Profit goes to only one of the stakeholders, so it is not a common objective.

Other objectives are necessary to identify the ways in which profitability is achieved. There is obviously no point in producing goods which cannot be sold. To sell products, customers must be satisfied. For goods to sell they must be in some way better than those of competitors. The business must establish competitive advantage in the product market. Competitive advantage comes from technological innovation, quality and productivity.

Productivity is achieved by ensuring that employee motivation is high and work is organized by focusing on value-adding activities and avoiding non-value-adding activities. Productivity is a common objective between the shareholders and jobholders.

The global corporate objectives of industrial organizations can be seen in Figure 2.5. These objectives are expressed in different forms and show that as well as financial there are also non-financial objectives. These objectives relate to the different functional areas within an organization, e.g. the marketing, manufacturing and financial functions of management.

While some of these objectives will be compatible with others, there will also be conflicts between objectives. This aspect is developed further in Chapter 3.

CORPORATE STRATEGY

A strategy is a set of long-term preferences. It is not an objective but a statement of an intended means of achieving objectives. For an industrial organization a corporate strategy will be established which defines the nature of the business which the organization is in. A statement of the sort of products or services which will be sold and in which markets; whether products are to be made or bought; how large the organization should be; where it should be located and what form of finance should be sought. These are strategic decisions arrived at by a process of corporate planning.

Corporate planning

Corporate planning (see Figure 2.6) starts with an analysis of needs. This involves an audit of current resources, current performance and problems, together with

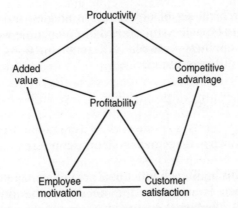

Figure 2.5 *A set of corporate objectives.*

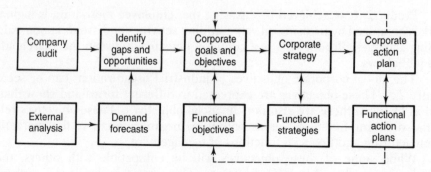

Figure 2.6 *The process of corporate planning.*

an analysis of the environment and markets within which the company operates, and some forecasts for the future of markets, technology and legislation.

SWOT analysis identifies the Strengths and Weaknesses of, Opportunities for, and Threats to the organization, and can be used to identify the 'gap' between where the company is now and where it wants to be in the future. SWOT must identify the STEP factors – Social, Technological, Economic and Political – which operate within the organization and its environment.

From this analysis, corporate objectives can be established and a corporate strategy formulated to achieve these objectives. Global corporate objectives are sometimes called a 'mission statement'. SWOT analysis is developed further in Chapter 5 where it is applied to an analysis of the organization's products and markets.

A corporate strategy is based on corporate objectives and will be used to set objectives for each functional section of an organization. From these functional objectives functional strategies will be established for each functional section to meet these functional objectives. This process of strategy formation is shown in Figure 2.6.

Thus a hierarchical set of objectives, strategies and action plans will be established, which, to be successful, should be compatible with each other. There will, however, be conflicts between different functions since the different functions will have different objectives.

Strategy as a framework for management

A corporate strategy forms a framework of functional strategies, as shown in Figure 2.7, within which management systems can be created to achieve corporate objectives.

In the rest of this book each functional area of management is examined and explained. Reference is made to the need for the different functions of management to work together to ensure involvement in each other's activities and from this the development of compatible objectives and strategies. Corporate

Figure 2.7 *A framework of corporate strategy.*

planning is the framework that enables this to happen. If an organization is to succeed, each function must contribute effectively to the achievement of corporate objectives. This does not just happen, it requires an effective process of corporate planning.

It is not possible to go further into the formation of corporate strategy in this book, but a foundation has been provided which enables an understanding of the way in which the different functions of management interact with each other. This can be developed in later studies at a more advanced level.

Task sheet 2   Stakeholders in organizations

Read the lecture notes, then undertake the following tasks:

2.1 Consider some additional 'stakeholders' in industry.

2.2 Identify their inputs, outputs and objectives.

2.3 Apply the idea of stakeholders to a university.

2.4 Apply the idea of money flows to a university.

2.5 Apart from manufacturing, which other activities create wealth?

2.6 Taking the wealth of an organization to be variable, identify ways of increasing it.

2.7 Consider alternative ways of measuring productivity.

2.8 Apply the concepts of corporate and functional objectives, strategies and plans to the case of a newly formed bicycle manufacturer.

CHAPTER 3

SYSTEMS

WHAT IS A SYSTEM?

Manufacturing and other organizations are complex. Understanding organizations and their behaviour is not easy using traditional forms of thinking. A modern approach to understanding organizations is a 'systems' approach. This chapter explains the basic concepts of a systems approach and applies these ideas to organizations – seeing them as systems. A systems approach is needed for effective management decision making. It is also valuable as a problem-solving approach. Figure 1.1 in Chapter 1 showed an organization diagrammatically. This view of an organization is developed in this chapter.

In the 1750s Adam Smith was Professor of Moral Philosophy at the University of Glasgow. He began to see patterns of economic behaviour of societies and developed the first 'laws' of economics which were based on a recognition that an economy was a system for which the behaviour could be analyzed. He recognized that a machine was a mechanical system and extended the idea to other areas. These systems he referred to as 'machines in the mind' – an early idea of a model of a system and of a perception of a system.

> A machine is a little system created to perform, as well as to connect together, in reality, those different movements and effects which the artist has occasion for. A system is an imaginary machine, invented to connect together in the mind those different movements and effects which are already performed.
>
> (Adam Smith, *Essay on Astronomy IV*, 1774, language updated)

SYSTEMS IN THE REAL WORLD

Ancient peoples began to recognize the cyclic changes of weather and the cyclic nature of the flowering of plants and birth of animals, and perhaps saw the relationship between them. They saw, even if they did not understand, systems of climate and of biological life. Things that they did not understand, such as thunder, human life and death, they attributed to gods.

34

The Sumerian and Egyptian astronomers were perhaps the first to recognize the different patterns of movement of stars and planets. The Celts too developed considerable knowledge of the solar system and built religious rituals around astronomically significant dates such as solstices and equinoxes. They recognized that the position and timing of sunrise depended not only on season but also on latitude. It must have taken thousands of years to accumulate this information, which was transmitted orally from generation to generation among an élite class of priests. They built a conceptual 'model' of the universe which explained these movements. Using this they were able to predict eclipses.

The Egyptian model of the universe was heliocentric (sun centred). By the time of the Greeks this knowledge had been lost, so the Greeks developed their own model which was geocentric (earth centred), and which also explained the behaviour of the universe. This model was accepted in Europe until a few hundred years ago. With the benefit of modern science the earlier heliocentric model of the solar system is now accepted. Although the Greek model worked, it was in fact based on a mistaken view. No model should be accepted as the truth but only as an attempt to understand a system. Modern astronomers have built system models of galaxies and have attempted to model the whole universe with components such as quarks, pulsars and black holes.

In the 1930s social psychologists studying human behaviour in organizations saw patterns of behaviour and developed the concept of 'socio-technical systems' to explain the behaviour of people in organizations. A socio-technical system is a way of describing the interaction between a group of people who interact with each other and with a set of machines. This is a way of seeing how a factory operates. The Human Relations School of management thought arose from these early ideas. This concept is developed in Chapter 24.

In the 1960s the complexity of modern engineering products in the aerospace and military industries rendered obsolete the traditional approaches to design and manufacture. A motor car contains around 3000 components, a spacecraft 2 million. Systems engineering developed to understand and cope with this level of complexity, and to use systems concepts for the design of products and manufacturing systems for their manufacture.

Governments try to manage a national economy. This too is a complex system. Its behaviour depends on the attitudes and actions of people in many countries. In order to understand how an economy works, a model of the economy is developed on a computer. This enables a government to test the effect of changes in tax and interest rates before they are implemented.

Gardeners want a weedkiller that will kill weeds in lawns without killing the grass. Ordinary weedkillers will kill grass as well as weeds, so something different was needed. Biochemists looked at the problem and developed what are now called 'systemic' weedkillers. They applied systemic thinking to the problem and created a new product wanted in the market. Systemic weedkiller is a plant growth hormone to which grass and weeds react differently because of their different leaf size. When this weedkiller is applied to a lawn, grass grows faster but weeds are forced to grow at an unsustainable rate and kill themselves from overgrowth. In this way weeds are killed but grass is not. A systems approach to the problem found a solution.

Environmentalists have recognized the interconnections between global energy consumption, pollution and climate change. An attempt has been made to construct a complete 'world model', linking population levels, food production, resource usage, pollution and climate. Such models can never be complete since the system is so complex that we do not fully understand it.

Even though the boundaries of knowledge have been pushed further back, it is most unlikely that we shall ever fully understand such complex systems.

The more we know, the more we know how little we know.

Linguists have noticed common features between different languages and have been able to show how language, as a system of communication, has developed and spread around the world. Languages have diverged and converged with the migrations and colonizations of the past. Religions have developed and changed similarly.

Archaeologists examine the remains of ancient civilizations. They can only examine the hardware left by these civilizations, but since 3000 BC this has included the remains of writing. By bringing together the knowledge of archaeology, language, religion and biology a better understanding is obtained than by studying each subject separately. This integrated approach benefits from the analysis of the interconnections between the different aspects and provides an understanding of the system as a whole and hence of each part of it. An understanding of the migrations of ancient peoples can be developed in this way. The analysis of the chromosomes of the cotton plant in America show that it is a hybridized form which is not natural but existed prior to Columbus. This infers migrations from Africa or West Asia to America which probably date back to the Phoenicians and/or the Egyptians. The cotton plant can be traced back to the Indus Valley civilization (in modern Pakistan) at about 3000 BC.

From these different beginnings a common theme evolved: that in all of these areas 'systems' exist, and that analysis and understanding required some understanding of how systems behaved.

In 1950 von Bertalanffy developed what he called a General System Theory as a way of understanding systems which could be applied in all areas of study. This integrated and holistic approach is now described as a 'systems approach' which is based on 'systemic thinking'.

SYSTEMS CONCEPTS

A system is a set of connected things. A single item is not a system, but when separate things interact together they form a system. Understanding each thing separately is necessary but not sufficient to understand the behaviour of a system. Traditional analyses used a 'reductionist' approach in that they reduced a complex system to its separate components as a means of understanding. A 'systemic' approach is 'holistic' in that it seeks to understand the system as a whole.

System components

Things can be materials, machines, money, people, or activities or ideas. They form the components of a system. They are shown on a system diagram as ellipses. Simple components are called elements. Components may also be subsystems (components of a given system but ones that are themselves systems comprising components). Thus there usually exists a nested set of systems at different levels. The choice and level of detail of components needed depends on the purpose of analysis.

System relationships

Connections are the interactions and relationships between the components of a system. They are shown on a system diagram as arrows. They may be physical, material, energy, money or information flows, but also include power, dependence, authority and influence. The total relationship between a pair of components is likely to include several of these of these different types of connection. Relationships are not just the flows between components. Some relationships exist wholly within the system, but others relate to other, external systems.

System boundary

A system boundary is a dividing line separating the things in a system from the things that are not in the system. It is shown on a system diagram as a large circle enclosing the components of the system. Defining the boundary of a system defines the system (its components and its objectives). A boundary should be selected to suit the purpose of a study and be wide enough to include all the factors relevant to the analysis.

The environment of a system is the set of other systems outside the given system. A thing is inside a system if it influences the behaviour of the system and is controlled by it. Something in the environment is outside the system and influences it, but is not controlled by it. A system influences things both inside and outside the system boundary.

Theoretically a system could have no relationship with anything outside itself. This would be a 'closed' system. Most systems, however, do interact with their environment and are called 'open' systems. Open systems operate within and interact with their environment. An analysis of system behaviour must take account of these interactions.

System inputs and outputs

Inputs and outputs of a system are the relationships between things inside and things outside a system. They are relationships which cross the boundary. Inputs and outputs may be any of the types of relationship.

Figure 3.1 shows diagrammatically the concept of a system. Figure 1.1 in Chapter 1 showed such a diagram applied to an industrial organization.

System behaviour

The behaviour of a system is the total set of outputs of a system in response to its inputs. Some of the outputs may be seen as undesirable and may lead to system failure. An analysis of system behaviour must include undesired as well as desired outputs.

System objectives

Objectives of a system are the set of desired outputs (or output/input ratios). Success occurs when objectives are achieved, but failure can occur if desired outputs are not achieved or undesired outputs occur. In analyzing outputs it is important to remember the undesired as well as the desired outputs. System design then focuses on maximizing the desired outputs and eliminating the potential undesired outputs.

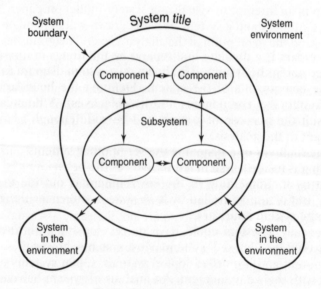

Figure 3.1   *A system map.*

Systems approach

A systems approach to a situation takes a holistic view and will use a system methodology (a set of stages of analysis) and systems diagrams to analyze the following:

- What the system is and what it does.
- What the objectives of the system are.
- What causes the behaviour of the system.
- Why and how a system failed.
- How to design systems to achieve objectives.

System modelling

A 'system model' is a diagram, a set of equations, a computer model or sometimes a physical model of a system. The diagram or equations show how the components interact together and how the behaviour of the system results from this. When designing a new system it is useful to construct a model in order to predict how the system will behave. With a model on a computer the configuration of the system can be quickly changed and the effect on the behaviour noted. In this way the behaviour of a planned system can be simulated prior to its design being finalized in order to obtain an optimal design. Spreadsheets can be used to build simple models.

System diagrams

A system map is an attempt to visualize a system model on paper. It is a useful way of trying to understand a system. What is drawn will depend on the purpose of the analysis and the perception of the analyst. Several diagrams may be needed, at different levels of the system with different boundaries, in order to show both detail and the overall interactions in a complex system. Some of the diagrams below are system maps. They represent a system graphically and show what the system is. A similar diagram, showing 'activities' instead of 'things', would show what a system does. In these diagrams a boundary is drawn as a large circle and named.

An additional way of diagramming system behaviour is to analyze the interactions between activities in a system. Activities are seen to cause other activities in other components of the system. A multiple cause diagram plots these causal relationships and helps the understanding of system behaviour. Causal relationships are cause–effect links and are not the same as casual relationships. System outputs are seen to be the end result of a chain or network of causes and effects. Of particular interest are circular relationships, since these form closed loops which reinforce their effect and affect and explain system behaviour. It is beneficial to label the relationships as positive or negative according to whether an increase in one activity leads to an increase or decrease in the other. A multiple

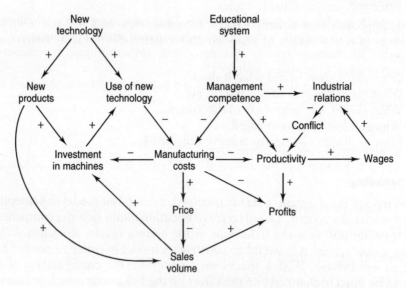

Figure 3.2 *A multiple cause diagram.*

cause diagram with signs on it is sometimes called a sign graph. Figure 3.2 shows a multiple cause diagram for a manufacturing organization. It shows quite different things from a system map for a manufacturing organization shown in Figure 1.1. The multiple cause diagram in Figure 3.2 shows three loops, all of which in this case are positive.

Systemic thinking

A systems approach is not a technique; it is a way of thinking, known as 'systemic thinking', and is a way of thinking based on a recognition and perception of the 'system' in a situation. This is not the same as 'systematic thinking', which is logical and orderly thinking that is necessary but not sufficient for a holistic view. Examples have been given above of the enhanced understanding which arises from an interdisciplinary and holistic approach to problems and situations. These are examples of systemic thinking. A major theme of this book is the application of this approach to the understanding of organizations.

SYSTEM TYPES

It is useful to classify systems in a way that is useful for understanding patterns of behaviour and control.

- Type I: natural systems – e.g. galaxies, solar system, life on earth, weather, water and carbon dioxide cycles.
- Type II: mechanical designed systems – e.g. cars, computers, telephones, buildings, spacecraft, a can of beans, all manufactured products.
- Type III: human activity systems – e.g. organizations, governments, universities, economies, cultures.
- Type IV: abstract designed systems – e.g. laws, languages, software and religions.

Type I systems:

- Behave but do not have objectives.
- Their behaviour can be analyzed and 'laws' derived which explain the behaviour.

Types II, III and IV systems:

- Have objectives.

Types II and IV are designed systems:

- Their objectives are set externally.
- They behave as they were designed.
- People design the system but are not part of it.

Types I and II are hard systems:

- They have physical components.
- They can be analyzed numerically.

Types III and IV are soft systems:

- They have human components.
- Their objectives are set internally.
- They are more complex, less well understood.

This book explores organizations which are type III human activity systems.

An example of a simple system

A car is a simple system: it comprises a few thousand components – all physical, inanimate things – mainly made of metal and plastic. A car is a type II mechanical designed system (the term 'mechanical' is generic and includes electrical and other technologies). All of its components are 'hardware'. A person designed it to achieve objectives (its specified performance) but no person is part of this system. Figure 3.3 is a system map of a motor car.

This diagram does not pretend to be a complete graphical representation of a car. There is a limit to how much can be shown on a diagram. It shows the major components of the system. The lines between components are an attempt

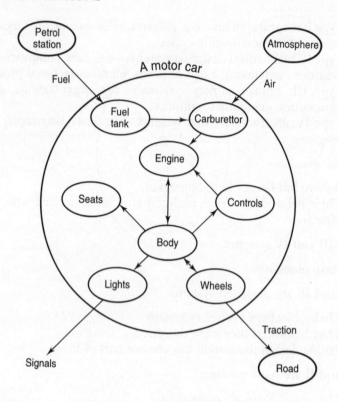

Figure 3.3 *A car as a system.*

to show a relationship, but clearly some labels and verbal description are necessary.

Clearly, to be of any use, cars are driven by people. This is a different system (and a different level of system). This system could be called 'A person driving a car'. This is a type III system, sometimes called a 'socio-technical system'. It has both technical and human components. Objectives are set internally by the human operator. The driver sets the objectives, uses skills and perception to control the car in order to achieve the desired outputs, adapting to a changing environment. The car is a component (and a subsystem) of this system. This different system is shown in Figure 3.4.

From these simple examples a General System Model can be drawn showing the conceptual components of a 'general' or typical socio-technical system. Socio-technical systems will generally comprise an objective-setting sub-system, a primary task subsystem and a control subsystem. The control subsystem will include performance monitoring and decision-making subsystems. Figure 3.5 is a system map of such a system in general terms. It is useful to look for these aspects when analyzing a system.

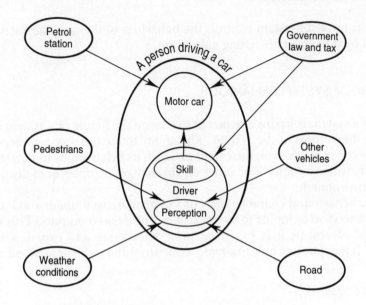

Figure 3.4 *A system map of a person driving a car.*

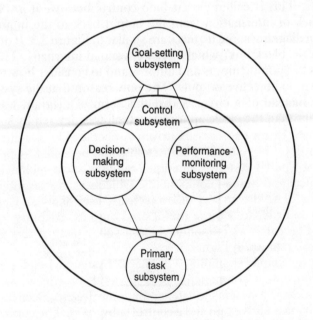

Figure 3.5 *A general system model.*

The control subsystem controls the behaviour of the task subsystem and is controlled by the objective-setting subsystem.

CONTROL OF SYSTEM BEHAVIOUR

Control of a system is having the power to change the inputs of a system in order to achieve the outputs of the system. The ability to control a system depends on the complexity of the system, the extent to which its behaviour is understood, and the extent to which its behaviour varies due to changes of inputs or of disturbances from the environment.

If the behavioural characteristics of a system are well understood, then the controller knows what inputs to use to achieve the desired outputs. This is called 'open-loop' control as it is based on prior knowledge and experience of the controller. It is applicable only to simple, structured and well-understood systems.

Feedback control

For more complex systems, the behaviour of the system is not fully understood. In such cases it is necessary to monitor outputs, compare them with objective outputs and then change (or actuate) the inputs to bring the outputs to the objective levels. This is called closed-loop control because it involves a closed loop of feedback of information from the output back to the input. Figure 3.6 shows a control diagram; some features are similar to Figure 3.5. It draws a system as a rectangle, or 'black box', which is not examined internally. The purpose of this diagram is to examine inputs and outputs and to consider how to ensure that outputs conform to objective outputs. The only reason that the system is shown as a rectangle instead of a circle is that this form of diagram was derived by engineers earlier than the circle approach, which was devised by sociologists.

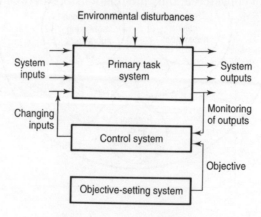

Figure 3.6 *A control diagram.*

The process of closed-loop control therefore involves three elements:

Monitoring of system outputs
+ Comparison with system objective
+ Actuation of system inputs to achieve outputs

Actuation means operating on the inputs, i.e. deciding what the inputs should be. This is the result of the decision-making subsystem. Inputs are changed if the output changes from its objective value. Feedback is negative if the input is changed in a direction opposite to that in which the output changes. For example, a decrease in output leads to an increase in input in order to restore the output to its objective value. Feedback is positive if the output is changed in the same direction as the output. Negative feedback causes a stabilizing, goal-seeking behaviour. Positive feedback causes an increase in the change in output. Since stability and the achievement of objectives is the purpose of many systems negative feedback is commonly used. Positive feedback in such a case will cause a system to fail – i.e. behave in a way that does not achieve its goals. However, when a change of system behaviour is the objective, then positive feedback is necessary. It should be noted that psychologists use the terms negative and positive feedback differently, and hence caution is needed in the use of these terms.

The behaviour of a system under negative and positive feedback is shown in Figures 3.7 and 3.8, respectively. In positive feedback the change in output can be downwards instead of upwards. If an upward change is desired, then a positive feedback loop which causes this would be called a virtuous circle, whereas a positive loop which caused output to fall would be called a vicious circle.

A control diagram is another type of system diagram. It identifies the main, or primary task, system but does not show any detail within it. It shows the inputs and outputs of the system. By convention, on this diagram a system is shown as a rectangle. This part of the diagram shows the basis on which an input–output analysis can be done. The system itself is seen as a 'black box' – its internal components are not considered, only the relationships between its inputs and its outputs.

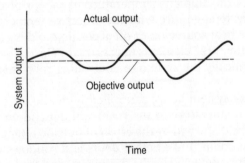

Figure 3.7   *The effect of negative feedback.*

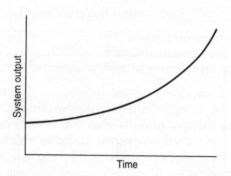

Figure 3.8 *The effect of positive feedback.*

In addition to the primary task system there is a control system, which is external to and has the function of controlling the primary system.

In closed-loop control the feedback path is a loop from the output of the primary task system through the control system to the input of the primary task system. A further input to the control system is an objective for the primary task system. This is usually set as the output of a higher-level system (the objective-setting system). In open-loop control the link from system output to control system input is missing. The objective-setting system controls the control system, so its role is sometimes described as 'double-loop' control. This has relevance in management since it describes the role of the board of directors.

An example of a simple control system

The use of a thermostat to control the heat in a room is an example of a simple control system. This is a mechanical (actually electro-mechanical) control system, but the concept applies to more complex socio-technical systems such as organizations. The thermostat has three functions. Firstly it is set manually to a desired temperature. This forms the objective for the primary task system. Secondly, it measures the ambient temperature of the air near to it. Thirdly, it compares the ambient temperature with the objective temperature and sends a signal which turns the heat source on or off according to this comparison. These three functions form the three elements necessary for closed-loop control. Figure 3.9 shows diagrammatically the system of controlling room temperature using a thermostat.

The primary task system being controlled is the boiler and radiators used to heat a room. The temperature in the room will depend on the setting of the thermostat but it can be disturbed by the effect of outside temperature, how many people there are in the room, and whether doors and windows are open. A fall in room temperature will be detected by the thermostat and the heat turned on. This will cause the temperature to rise, and this rise will be detected. Because of the

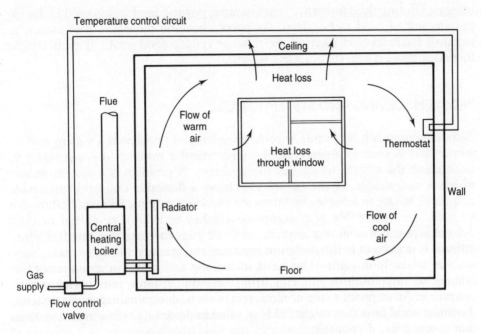

Figure 3.9 *The control of room temperature.*

distance between the radiators and the thermostat, there is a delay between the heat coming on and the rise of temperature being detected. This means that the temperature can only be controlled between limits of temperature rather than at precisely the objective temperature. The temperature in the room therefore oscillates between these limits as the thermostat turns the heat on and off. A thermostat uses negative feedback because it changes the switch on the heat source in a direction opposite to that in which the output varies from the objective. (If the room temperature goes up too much the heat source is turned down; if it goes down too much the heat is turned up.)

Controlling room temperature with open-loop control would involve deciding in advance for how many hours the heat source should be turned on each day. This would take no account of any changes in outside temperature and would only be satisfactory if all the factors affecting temperature were fixed and known in advance. Clearly, in practice this would not be very satisfactory.

An example of positive feedback can be seen in the instructions given with a jar of baby food. The amount of food to be given to a baby depends on the weight of the baby. The greater the weight of the baby, the greater the amount of food given; the greater the amount of food given, the greater becomes the weight of the baby. As the weight (output) increases, so the amount of food given (input) is increased in the same direction as the output. This is a positive loop of mutual causation (A causes B and B causes A). Thus the weight of babies increases during their formative years because positive feedback is used to decide the amount of food to be given. As people become teenagers they switch to negative feedback and increase exercise or reduce food intake if their weight increases above a self-set objective weight.

PROBLEM SOLVING AND SYSTEM FAILURE

A systems approach is of value in problem solving. A problem is a system with an output that is seen as a problem. To understand a problem it is necessary to understand the system which has the problem. A problem is a system output which is undesirable, or the failure to achieve a desirable output. Undesirable outputs or failure to achieve objectives are therefore systems failures. Failure can be total and irreversible (e.g. aeroplane crashes or death), or partial or even debatable (e.g. not achieving as much profit or wage increase as wanted). Failure analysis is important in the design of products and manufacturing systems. Such analysis is rarely thought of or used in looking at failures in management or failures of organizations but can bring valuable insights into organizational behaviour. An important role of management is to design management systems. To understand how they might fail is of value in designing management systems that succeed, i.e. do not fail.

ORGANIZATIONS AS SYSTEMS

Organizations are type III, human activity systems. Such systems will have objectives. People are a part of the system and will have their own objectives. The objectives of the system will be set by some of the people in the system. Organizations are goal-seeking and adaptive systems, adapting themselves to changes in their environment. The boundary of an organization, when seen as a system, can be drawn in different positions which represent different levels of system. For example, a manufacturing organization can be seen as a factory with a boundary coinciding with the physical boundary of the site.

Alternatively an organization can be seen as a financial system, an employment system or a manufacturing system. These are all valid yet different views of the organization, each having a different boundary and comprising different components. Whether the shareholders are part of the system or in the environment of the system depends on which system is being considered. They are a component of the financial system but not of the manufacturing system. Figure 1.1, which applies the concepts of system diagramming to a manufacturing organization, should be referred to at this point. It should now be possible to see that a manufacturing organization is a system. Figure 1.1 does not pretend to be a complete representation of a manufacturing organization as it is limited to the amount of information that can be put on a sheet of paper, but it is a start in the direction of seeing organizations as systems. This approach is fundamental to a modern understanding of manufacturing, a theme developed in Part 4 of this book, and of organizations as a whole.

Objectives in organizations

Organizations do not usually have a single or unitary objective but a complex set of objectives. Identifying and understanding this set of objectives is essential for understanding and managing the behaviour of the system. A set of objectives for an organization was introduced in Figure 2.5 in Chapter 2. The concept is expanded here, so that an organization can be viewed as a system. Managing an organization involves managing a complex set of objectives, some of which are compatible and others of which conflict.

System (corporate) objectives

- Are set by its 'owners', or a higher-level system.
- Exist on different timescales (tasks, objectives, goals).
- Exist in different domains or functional areas (financial, manufacturing, marketing and human resources).
- Exist at different levels (individual, group, company).

Examples of corporate objectives in different departments are profitability, productivity, market share and employee motivation. Thus an organization has multiple objectives. Since each of the objectives above can be combined with others, a complex set of objectives exists as shown in Figure 3.10 for a manufacturing organization. Objectives would be similar in a non-profit organization.

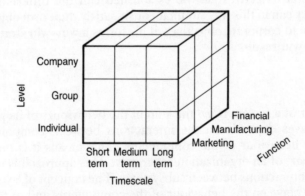

Figure 3.10 *The multiple objectives of an organization.*

Individual objectives

- Investors – security, power, financial return.
- Employees – job security, job satisfaction, wages.
- Managers – managers run the organization to achieve the objectives of the investors.

The role of managers is to run the organization to achieve the objectives of the investors, but they are also employees and so have objectives as employees.

In this way it can be seen that an organization has plural objectives, meaning that different groups in the organization have different objectives. The pluralism of objectives is an important concept for understanding the behaviour of people in organizations.

Common and conflicting objectives

Some of the objectives will be compatible with each other, some will be common between different groups, and others will conflict between groups. Conflict may be horizontal (between different managers, functions or departments) or vertical (between managers and employees). Much of this conflict will be destructive and may cause the system to fail to achieve its objectives. Some conflict can be constructive. Competition is an example of conflict that is usually considered to be constructive.

Behaviour of organizations

The behaviour of an organization arises from the activities of the people in it, each attempting to use the organization to achieve their own objectives. The behaviour that results depends on the extent to which the set of objectives is common or in conflict, and on the relative power of the different groups. If, and only if, common objectives can be established can the different groups work together as they can in this way simultaneously satisfy their own objectives in ways that contribute to corporate objectives. This forms a win–win strategy since both sides obtain a win result.

Synergy

The behaviour of a system is *not* the sum of the behaviours of its parts. The way a system behaves depends on the interactions between components as these determine the behaviour of the components. This means it is not sufficient to analyze each part of an organization separately. This approach is reductionist as it ignores the interactions between subsystems. The concept of 'synergy' explains the difference between the behaviour of the components and of the system as a whole. Synergy can be expressed as the following equation:

$$2 + 2 = 5 \tag{3.1}$$

This illustrates the meaning of synergy: the behaviour of the system is greater than the sum of the behaviours of its separate parts. In organizations the different functional areas of management have different roles and objectives, but by interacting can create a larger output than if they worked in isolation.

Symbiosis

Symbiosis is a relationship that is not merely mutually beneficial but essential for survival. The word originates in nature but the concept is applicable to relationships within and between organizations.

Lichens grow on rocks. They appear to be a species of plant similar to moss. In fact they are not a species at all, but are two separate species living together in a symbiotic relationship. In lichens the species are a fungus and an algae. Similar relationships exist between fungi and tree roots. Some animals, including humans, depend upon bacteria for digestion, the bacteria depending on their hosts for their survival.

Flowering plants need insects for pollination; the insects need nectar for food. Plants also package their seeds in fruit in order to attract birds and other animals. The fruit is eaten but the seeds are protected and indigestible, and are deposited elsewhere with a package of organic fertilizer. In this way plants are able to colonize different sites. These are examples in nature of evolutionary strategies leading to win–win results.

In organizations all of the functional areas of management are needed, but they also need the other functions if they are to be effective. Symbiotic relationships are based on win–win strategies and need to be developed between management functions if the organization as a whole is to function effectively. Taking the stakeholder perspective, the relationship between shareholders and jobholders is symbiotic: they both need each other to achieve their own objectives. Symbiosis at the component level leads to synergy at the system level.

The role of directors and managers

In a small organization the investor, the director, the manager and the operator may be the same person. In larger organizations these roles, and the people doing them, are different and separate.

The role of directors is that of goal setting. They set the objectives for the organization and the strategy for achieving them. The role of managers of an organization is to control the behaviour of the system and thus guide it towards the objectives set for it. An alternative word for a manager – executive – derives from their role, which is to execute, or carry out, the strategies set by the directors. Managers form the control loop, and directors, by setting objectives, exercise double-loop control.

In managing an existing system on a steady state, day-to-day basis, managers design management systems and take decisions on inputs and on the activities of people. This uses negative feedback to maintain behaviour at an objective level.

A further role of management is to develop and change an organization. This necessitates an understanding of human attitudes to change and the involvement of employees in the change if it is to be successful. This uses positive feedback to enable the change to occur and be accepted.

The style of management

This refers to the style of social and interpersonal relationship between an organization and other organizations, and also between a manager and those reporting to the manager (the latter is also referred to as 'leadership'). This is an aspect of system design that is crucial to the success of the organization. Through a 'systems' approach a manager gains a better understanding of the complexity of an organization and particularly its human aspects. A style of management should not just happen; it needs to be designed and to be appropriate for the type of organization. This aspect of management is developed in Chapters 24 and 25.

Task sheet 3   Systems

Read the chapter and apply the idea of a system to a university.

3.1   List the components of a university looked at as a system.

3.2   List the major systems in the environment of the university system.

3.3   Classify the system using the classification system in the lecture notes.

3.4   Draw a system map of a school or department in a university.

3.5   List the objectives of the following people:
   (a)   Students
   (b)   Lecturers
   (c)   Clerical staff
   (d)   The university director
   (e)   Industrial/business organizations
   (f)   Government

3.6   Consider the main money flows through the system.

Apply the idea of a system to an industrial organization. Use the system map of a typical manufacturing organization shown in Figure 1.1 which identifies the main components of the system and of its environment.

3.7   Identify and list the objectives of the system and of its subsystems.

3.8   What is the role of management in such an organization and how does it carry out this role? (Apply the control diagram, Figure 3.6, to a manufacturing organization.)

CHAPTER 4

ORGANIZATION

THE NEED FOR ORGANIZATION

Organization means here the process or arranging how things are done, not the organization itself, which is the existence of a group of people brought together to do something. Organization is concerned with making things happen effectively, i.e. making the organization behave in a specific way. It is therefore concerned with what tasks need to be done and how the total set of tasks can be split into smaller units.

Organizations need objectives and they need managing. This chapter considers the role of management in managing organizations and the management systems (organization, decision and communication structures) needed for organizations to be effective. Chapter 12 goes into more detail in manufacturing and factory organization. Chapter 25 considers styles of management and leadership.

THE ROLE OF MANAGEMENT

Many people have written about the role and functions of management. The Egyptians, Assyrians and Greeks wrote about management and some of their writings survive. The growth of industry in the nineteenth century led to writings on how an organization could be managed effectively.

Frederick Taylor, writing in the 1880s, took a scientific approach to the design of work, which he sought to optimize for maximum output. He saw the need for training and the benefit it brought in greater output. He suggested that pay be related to performance and developed a measure of performance based on a standard time for a task. He also recognized the mutual dependence between managers and workers, pointing out that working together was the only way to increase prosperity. The term 'scientific management' is used to describe this style of management. The following quotation from Taylor is an early statement on the need to establish a win–win strategy:

The principal object of management should be to secure the maximum prosperity for the employer coupled with the maximum prosperity for each employee.

Taylor also recognized that productivity was a path enabling prosperity or wealth to increase. Taylor's work in work measurement, management style and payment systems is referred to in further detail in Chapters 11, 24 and 25.

Over the years managers misused some of Taylor's concepts, particularly 'piecework' payment systems, which were used to force workers to work harder as a means of raising profit. Considerable conflict arose and the growth of trade unions can be attributed to this style of management.

Max Weber noted that organizations operate on the basis of rules and procedures. He identified different types of authority: the rational based on position, the traditional based on the person, and the charismatic based on the personality of the manager. He introduced the word 'bureaucracy' to describe a hierarchical organization with centralized, rational authority. He saw this as an efficient form of organization. This view has to be seen in the context of its time, the beginning of the twentieth century. Over the years the word bureaucracy has come to mean the opposite, an organization so restricted by rules and procedures that it is inefficient. The word is used to describe government offices and similar organizations that operate in this way and are considered to be very inefficient because of this form of organization.

Henri Fayol identified the functions of management as planning, organizing, commanding, co-ordinating and controlling. He wrote in France in 1916 but his work was not translated into English until 1949. Fayol also expounded fourteen principles of management. Other writers have referred to management functions such as designing, operating and controlling systems.

Mary Parker Follett, writing in 1949, recognized the similarities between different types of organizations and developed common principles of management based on a role of co-ordination. She suggested that in a democratic society the role of management is to create a situation in which people want to co-operate.

Joan Woodward in the 1950s recognized that the form of organization needed to be related to the type of work done. Organization structure would depend on the range of the processes and the variety and volume of products. Chapter 12 explains the different types of manufacturing system to which the organization structure should relate. Technology was seen as affecting both organization structure and the form of control that management was able to exert.

The people discussed above are only a few of the many who have written in the past about management and organization. A more advanced study of the topic would find many more. Chapter 24 introduces some more modern writers on organization, with the emphasis on the human aspects of organizational behaviour, and Chapter 25 develops a modern approach to the effective management of organizations.

Apart from the writers mentioned above, two more are worthy of mention.

They come into a different category since their writings are satirical and critical, but nevertheless they have a serious message.

C. Northcote Parkinson formulated some views on organizational effectiveness which are known as 'Parkinson's Laws'. These laws of organizational behaviour can be summarized as follows:

1. The time taken to do a task rises to meet the time available.
2. The number of office staff in any organization will rise by 5 per cent per year irrespective of the amount of work done.
3. The time taken to reach a decision is inversely proportional to its value or importance.
4. Incompetent mangers surround themselves with subordinates who are even more incompetent.
5. The best way to obtain promotion is to marry the director's son/daughter.
6. The opening of a new and expensive building is a sign that an organization is about to collapse.

Laurence Peter formulated the 'Peter Principle', which was based on a study of why many organizations seem to have so many incompetent managers. He suggested that people who are competent in their current job are promoted to a higher job, but those who are not competent remain where they are. This view can be expressed as:

1. In every organization every employee will rise to a level at which they are incompetent.
2. Over a period of time every post will become occupied by a person incompetent to do the job.

Promotion was seen by Peter as being due to either push or pull. Some employees push themselves for promotion by doing things that are of little value to the organization but aggrandize themselves. Others are pulled into higher positions by nepotism.

THE FUNCTIONS OF MANAGEMENT

In a sole trader organization one person possesses all of the knowledge, takes all of the decisions and does all of the work. In larger organizations the total task will be split between several or many people who are the employees of the organization. This causes a separation of the functions of different people in an organization into groups such as shareholders, directors, managers and operators. This process is known as the division of labour. It complicates organizations as it creates different groups of people with different objectives and some power to influence the behaviour of the organization. Given that the division of labour is necessary in a organization, there are many ways in which the tasks of management and labour can be split.

Typically, managers will be given responsibility for particular resources or for liaison with particular external systems. For an industrial organization these would be, for example, design, marketing, manufacturing, quality, purchasing, personnel and finance – the subsystems identified when perceiving the organization as a system. Within each functional area, resources would be established to carry out the tasks required to achieve the functional objectives. Each function would need to interact with other functions to achieve the objectives of the organization as a whole.

According to the size and nature of the organization, teams of managers and operators exist within a functional area, which may itself be split into subfunctional areas or subsystems. The detailed structure varies from one organization to another, depending on their major activity and the range of activities they undertake. Some manufacturing organizations design their own products; others do not. Some transport their products themselves; others subcontract this function to other organizations. Some make many of their own components; others buy most of them and concentrate on the assembly of their products. Some products are made in the expectation of future orders, but others are only made after an order has been received. These features will influence the detailed structure of the organization and the way the organization is managed.

There was in the past a tendency for organizations to merge or takeover other organizations. This was often done in order to exert control over suppliers by absorbing them. This is known as vertical integration, i.e. integration along the chain of suppliers and manufacturers. Alternatively it was done to exert more control over product markets by absorbing competitors. This is known as horizonal integration in that it is a merge of organizations at the same stage in the supplier – manufacturer–customer chain. This attempt to obtain better control was aimed at achieving better efficiency and higher profits. Notice that the terms vertical and horizontal integration between organizations have different meanings than when the same terms are applied within an organization (see Chapter 13).

These days it is recognized that the increased complexity of organizations created by mergers may have some negative effects. The organization may be operating on many sites, perhaps in different countries with different cultures. The range of products and technologies is likely to be much wider and may be beyond the skills and experience of the managers. For this reason the modern approach is more likely to be one of demerger. Each organization is then focused on a narrow range of products and technologies in which it has expertise. Such organizations are more likely to be effective. For this to work, however, organizations must establish good relationships and trust with other, separate organizations on which they depend, such as their suppliers and customers.

ORGANIZATION STRUCTURE

The 'division of labour' outlined above creates a structure of relationships between management functions, and also, since people undertake the tasks of an organization, a structure of relationships between the people doing these tasks. The organization structure is therefore a social structure. This is usually a hierarchical structure and should be related to the hierarchy of objectives stated in a corporate plan. The structure can be viewed as a pyramid with one person at the top and an increasing number of people at each level below the top.

Every organization will have an organization structure of some kind. Less often will it be recognised that the organization structure is a management system, and one that has to be properly designed if it is to be effective. Clearly the organization structure must match the organization and its activities. In many organizations the structure is static and was probably imposed a decade or so earlier to suit the whim of the then managing director. It probably shows people rather than job titles and usually adorns the board-room or the foyer. Organizations that fail to adapt their organization structures to changing needs will find themselves becoming increasingly ineffective.

An appropriate organization structure will enable effective and speedy communication between different parts of the organization. It is both the cement which holds an organization together and the lubricant which enables it to operate flexibly and effectively. A poor organization structure will prevent communication, interaction and involvement. It will prevent an organization from being effective and from achieving its objectives. An organization structure for a typical manufacturing organization is shown in Figure 4.1 but would be similar for non-manufacturing organizations. This structure is a process structure because the different departments are responsible for the different processes or functions of management. Alternative structures are considered below. A more detailed structure for the manufacturing function is given in Chapter 10.

A large organization may operate on many different sites, perhaps in different countries and in different markets. While such an organization could be organised as a single company with operating divisions, it is more likely to comprise a group of separate subsidiary operating companies in which all the shares are held by a single 'holding' company. The structure of such a holding company is shown in Figure 4.2. The holding company is likely to be a public company (plc) but the subsidiaries would be private companies (Ltd). This structure can arise as organizations take over other organizations but retain the original names. Complex structures are created when organizations buy shares in other organizations that buy shares in other organizations. Companies may be internally split into divisions which may be based on different sites or produce different products. Divisions are convenient business units for effective management but are not legally registered as separate organizations and do not have to publish their accounts. They may have considerable autonomy within the organization and will produce accounts internally for management decisions. Such units may be referred to as *strategic business units*.

A different structure is created when two or more organizations set up a joint-venture company in which they own the shares between them. Joint-venture

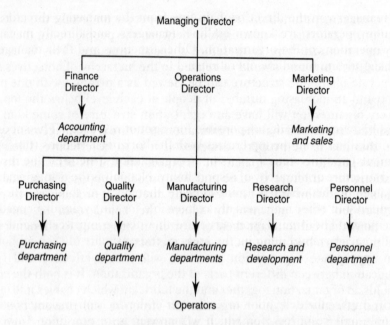

Figure 4.1 *Organization structure of a manufacturing organization.*

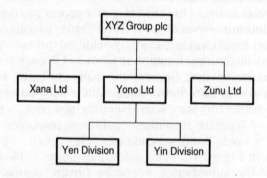

Figure 4.2 *Structure of a group of companies.*

companies are set up by hitherto separate organizations that wish to co-operate, for example on a new product or in a new market. Each relies on the other and usually complements the skills of the other. Working together may be necessary for success and shares the risk when developing new products or markets.

RELATIONSHIPS WITHIN ORGANIZATIONS

The vertical lines in an organization structure diagram are the lines of command.

Those managers on the direct or main line from the managing director to the production operators are known as 'line' managers, and directly manage the primary operations of the organization. Other functions and their managers are seen as supporting these line operations.

Responsibility

Responsibility means being 'responsible for' a set of resources. The subdivision of responsibilities from the top creates both the horizontal sections (functions or departments) and the vertical layers of an organization structure. The result is a pyramid structure or hierarchy of responsibility. Not only is the managing director responsible for the whole pyramid beneath that post, in this case the whole organization, but other managers are responsible for the pyramids beneath each of their posts. The organization structure therefore comprises nested sets of pyramids, each pyramid defining the areas of responsibility of each jobholder.

Authority

Authority implies the delegation to a manager of the decision-making power over the people and resources for which a manager is responsible. Authority is necessary in order to carry out the tasks necessitated by the exercise of responsibility. Areas of authority and responsibility must therefore be the same. Organizational failure is likely to occur if a manager is given responsibility but not the requisite authority to do the job. Authority is delegated to a manager by a manager one level higher in the hierarchy. This type of authority derives from a person's position in the organization structure. There is a different type of authority based on knowledge, for example a person being an authority on a subject. Weber distinguishes between power, which can be imposed, and authority, which is voluntarily accepted. Managers have both power and authority.

Accountability

Accountability is the requirement to report to a higher manager. A manager is accountable to the higher manager for the effective use of resources provided by the organization and for which the manager is responsible. Accountability and responsibility are therefore oppositely directed powers in a vertical 'line of command'. The term 'responsible to' someone is an older form used to mean accountability, but should be avoided since it confuses the concepts of responsibility and accountability.

Groups in organizations

Since people do tasks in an organization, the structure of tasks or functions relates to the people doing the tasks. Except for the managing director at the top and operators at the bottom, all managers are members of two groups. Each is the manager of people within the pyramid of which this manager is the head, but each manager, along with a group of other managers will be accountable to a single manager at the next higher level in the hierarchy.

Viewed this way the organization can be seen, in Figure 4.3, as a set of overlapping pyramids. This group – a manager and a group of people who report directly – is the most basic, yet most important, group in an organization. Effective management is based on managing the group and managing the relationships that the group has with other groups. It is worth re-emphasizing that an organization, at its most basic, is a group of people. Understanding the human dimensions of organization is therefore crucial to the successful management of an organization, which is a prerequisite for the success of the organization.

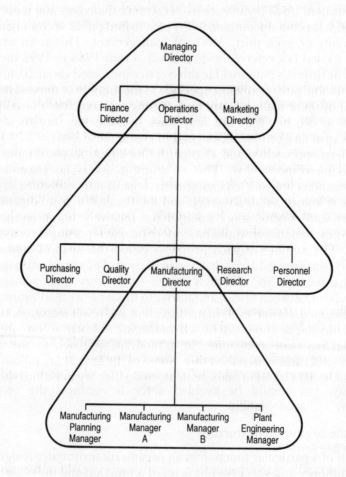

Figure 4.3 *Pyramidal groups within organizations.*

TYPES OF ORGANIZATION STRUCTURE

The shape of organizations

There is no one correct type of structure for an organization. The managers in an organization must design a structure that is appropriate. This is likely to have several layers or levels of management – but how many are needed? The larger the organization, the more managers and levels are likely to be needed. More managers means more costs, so a balance is needed. A 'tall' organization would have many levels relative to the number of operators. In this type of organization, planning and decisions are centralized and the jobs of operators reduced to the minimal manual or clerical tasks. In some organizations the number of people reporting to a manager is as low or two or even one. This results in a tall and top-heavy organization. Such organizations need more managers and more levels of management. It was not uncommon in the past to find public sector organizations with ten, twenty or even thirty levels of management. This is an example of Parkinson's second law referred to previously. From 1988 to 1992 the number of managers in Britain's National Health Service increased about 20-fold in four years, a remarkable rate of growth, especially as the number of nurses and hospital beds fell. All of these managers are busy, sitting on committees or telling other people what to do, but they are only busy because they are there and the organization structure has been designed to create these jobs.

A modern approach would change the jobs that people do and delegate decision-making responsibility. This structure requires far fewer managers. Some organizations are now changing their organization structure in this way. The process is known as 'de-layering' and the result is a 'lean' organization. A 'shallow' or 'lean' organization would have relatively fewer levels. In lean organizations each manager would have more people reporting directly from one level below. The number of such people is called the 'span of control' of the manager. It is usually considered that a span of control of a senior manager should be between three and six, although at the supervisory level managers often manage groups of between fifteen and thirty. In this type of organization, decision making is delegated. Fewer levels of management and fewer managers are needed as the operators are to some extent self-managing. Clearly, a lean organization is more effective than a tall one, so a modern approach to the design of organization structures is based on this. Some of the largest organizations in the world have only eight levels of management, so smaller organizations should need less than this.

Position in the organization structure

The location of a particular function in an organization structure is significant to its power, authority and success. The level of a function and the other functions with which it relates, horizontally, under a common manager determine the importance of a function, and the level at which decisions about that function are taken and grouped with other functions. In the past, personnel and quality management were seen as simple tasks that required only modest skills. These functions were often placed towards the bottom of the structure, and the people in charge did not usually have the title of manager. In such a position these functions were seen as not important, and in any conflict with, for example manufacturing, the views of personnel and quality were overridden.

With the increasing importance now being seen as necessary for these functions, their position has become much higher within the organization. This gives a voice and authority for these functions since they report at a higher level. It also ensures that conflicts are not swept under the carpet but are discussed and resolved to the benefit of the organization.

Functional or process organization

In organizations there are two basic ways in which the total set of tasks can be divided to form the organization structure. Managers can have responsibility for a process or function. The structure shown in Figure 4.1 is functionally split. In functional organization each department is responsible for a single function, a set of similar tasks. People employed in a department are skilled in that function, but often understand little about the rest of the organization. Often they do not see how their function relates to other functions. They are responsible only for one function or process; they do not see how work is progressing through the organization as a whole. This leads to organizational fragmentation, prevents interaction and synergy, and holds down productivity. This is made worse by educational specialization since it narrows the skill base of employees to their own specialism. They know more and more about less and less!

In functional organization, problems of lack of communication and understanding arise. The organization is usually inefficient and everyone blames everyone else for the problems. They do their own bit then push the work 'over the wall' to the next department. The walls between departments are not only physical; they are psychological. Additional people are employed to improve communication, but if the basic structure is inefficient the appointment of more people will not make it more effective.

Product organization

The alternative is for the split to be based on products. In this case departments are responsible for all functions of management associated with a given product. The support functions are placed in product departments. Each department sees progress at all stages of work from input to output, from the materials coming in, to the goods going out of the organization. Employees in a department interact with other employees in the same department doing different functions, they learn from each other and see how each contributes to the overall productivity of the department. A symbiotic relationship develops, leading to individual and organizational learning. Figure 4.4 shows a product organization structure.

A disadvantage of this form of organization is that it requires every support function to be in every operating department. For example, each department would have its own accounting and personnel staff. This would require more staff, so in practice some functions remain centrally and organized functionally. A suitable balance must be determined.

The concept of processes and products is valid in non-manufacturing organizations: in a university, for example, a lecture in a subject is a process and a degree programme is a product; a student is the material brought in and converted into a graduate.

If organization is based on processes, a manager would have responsibility for a process and all products which required that process would pass through the department. If the structure is based on products, a manager would have responsibility for several if not all processes involved in the manufacture of the product.

There are advantages and disadvantages to both of these approaches. These forms of organization must match the physical layout of equipment. These aspects are expanded in Chapter 12. A structure must be chosen that best enables the success of the organization.

Matrix organization

A third variant is a 'matrix' structure which combines both process and product management and has both process and product managers. This requires more managers but is justified on the grounds that it results in improved communication and better control. Each employee below the level of process and product

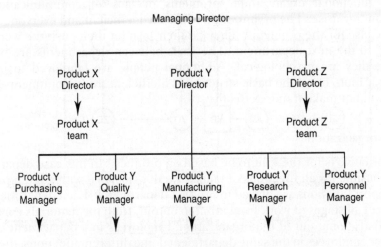

Figure 4.4 *Product organization structure.*

managers has two managers: a process manager and a product manager. Each manager can issue work instructions to an employee and dictate when the work is to be done. This can lead to conflicts when an employee is required to do two tasks at the same time. Instead of introducing more managers, a modern approach would improve communication in other ways and create self-managing work groups.

Figure 4.5 shows how managerial responsibilities are grouped in a matrix organization structure. It is a three-dimensional structure that is not easy to show on two-dimensional paper. This figure is a plan view, from above, as opposed to the 'side' views of Figures 4.1–4.4.

Project organization

Another type of structure is used for project management. A project is a set of activities done once only rather than repetitively. The introduction of change,

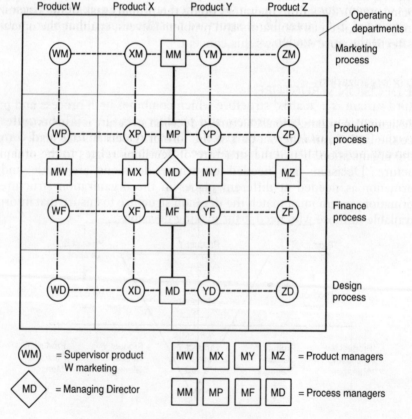

Figure 4.5 *Matrix organization structure.*

such as the setting up of a new factory or manufacturing system, is a one-off activity organized as a project. The organization of projects is explained in Chapter 14. A project organization structure is similar to matrix management in that an employee may have two managers – one functional and one for the project. Employees are seconded from their normal job to a project team for the duration of a project. A project manager leads a multidisciplinary project team with the range of skills needed to implement a project.

The functions of innovation and design are such that their normal activities are sets of one-off activities. Some marketing activities are also of this type. They are usually organized as projects.

The upside-down organization

This is more a concept of the role of management than a formal organization structure. The traditional organization is a pyramid; this concept turns the pyramid upside down. Instead of managers being at the top, issuing orders to the operators, they are at the bottom. This perspective sees the role of management as being supportive of the operators doing the primary task of the organization. Managers are seen as facilitators and problem solvers, enabling the operators to be effective. Figure 4.6 shows this concept.

THE DECISION STRUCTURE

In designing an organization structure, a decision structure is also being designed, since the latter is an inherent part of the former. Decisions are made by people in the organization, hence the structure of decisions relates to the organization structure. Decisions are needed on where decisions are made and what information is needed at different places in the organization structure. The information system must match the decision structure to ensure that information is available to those who need it for making decisions.

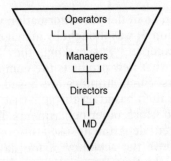

Figure 4.6 *Upside-down organization.*

Decisions on important, long-term and strategic matters are taken centrally, at the top of the organization. Decisions on day-to-day, short-term operations are delegated down to managers at lower levels. Deciding the best level for decision making is important for the success of the organization. Bureaucratic and 'autocratic' organizations tend towards centralized decision making, whereas more 'democratic' organizations seek to involve people and do so by delegating decisions.

In modern, 'lean' organizations decisions are delegated; this tends to create more interesting and challenging jobs for people in the organization. Fewer managers are required because most employees are taking decisions themselves.

The delegation of decisions can create autonomous groups. This concept is referred to in several chapters in this book. See strategic business unit earlier in this chapter; the internal market, Chapter 5; plant within a plant, Chapter 12; profit centre, Chapter 18; and job design, Chapter 25. These all express a common idea of active participation of employees, challenging jobs and delegated responsibility. With such delegation there will still be a central core for policy and strategic decisions. There may also be some central services such as personnel. Being central does not imply higher authority, just efficiency of organization. Centralized services are the servants not the masters of the wealth-creating functions.

THE COMMUNICATION STRUCTURE

Formal communication within an organization arises from the lines of responsibility and accountability. Person A can only tell person B what to do if B is on the same vertical line below A. In theory, and in a bureaucracy in practice, information and communication can flow only along these lines of command. This is the way in which government offices operate. Most organizations cannot operate effectively with such limited communication, so additional informal communication links need to be established. In an organization of any size it is an important and major task to design the information system and communication structure.

Information is power, so in the past information was kept at the centre, i.e. at the top of the organization. It was thought by managers that if employees knew how much money the company was making, they would ask for more for themselves. If they knew what new products the company was developing, they would tell the competitors. Such attitudes are based on, and create, mistrust. Information systems were then based on this mistrust. Each department had its own information system to which other departments did not have access.

In some cases, different departments used different data for the same thing because they did not believe the accuracy of the data being used by another department. For example, the time for a task is needed by at least two different departments; manufacturing need it in order to schedule work; accounting need it in order to cost the task. Personnel may also need this data for the payment system. Each department would use its own, but different, data.

When decisions are delegated, more information is needed at lower levels in the organization. For employees in one functional area to be effective they need to know what is going on in other parts of the organization. New types of information system are needed if such organizations are to have the information they need to be effective. Modern information technology using computer-based databases can provide a common database and common data for use by the different functions within the organization. Individual employees can access the database and obtain a more informed base for decisions. In this way the information system enables more integrated and better decisions to be made within the organization, and more people are involved in decision making.

Task sheet 4 Organization

4.1 Draw an organization structure diagram for the university (or for the school or department in which you are studying). You may not be aware of the complete structure, particularly at the top of the organization, so concentrate on the structure for the school or department.

4.2 Identify the channels of communication you have as a student with the university and school or department (or as an employee with your employing organization).

4.3 Given that the total task of an organization has to be split into smaller parts, consider the different ways in which a university could be split.

PART 2

MARKETING MANAGEMENT

LEARNING OBJECTIVES

2.1   Offer a definition of marketing.

2.2   Describe the evolution of the marketplace and the role of marketing in organizations.

2.3   Identify the core concepts of marketing.

2.4   Understand the meaning of needs and wants.

2.5   Describe the relevance of Maslow's hierarchy of needs to marketing.

2.6   Describe the four elements of marketing mix.

2.7   Identify the factors involved in determining price.

2.8   Explain the product lifecycle.

2.9   Discuss what is meant by market segmentation.

2.10 Explain the importance of the customer in marketing.

2.11 Construct a hierarchical market segmentation model.

2.12 Define the objectives of the market research process.

2.13 Describe the six classifications of market research.

2.14 Explain why desk research is of importance in industrial marketing.

2.15 Illustrate the limitations of statistical methods of forecasting product demand.

2.16 Distinguish between extrapolative and causal models used in forecasting.

2.17 Use graphical methods to derive forecasts for level, trend and seasonal demand data.

2.18 Explain the advantages, disadvantages and areas of application of different forecasting techniques.

CHAPTER 5

MARKETING MANAGEMENT

INTRODUCTION TO MARKETING

What is marketing?

This is a question often asked by many managers in commerce and industry. To many, not directly involved in the process, it is often misunderstood and is mistakenly used to describe the process of selling. Selling is a part of marketing but marketing embraces far more than selling. At the most basic level, marketing can be considered as providing the information to consumers which they require to make a purchasing decision. The objective being to encourage consumers to buy. To do this requires information from the market and a marketing strategy to put an organization in a position to achieve this objective. At this level the role of marketing can be seen to be:

Satisfying the needs of customers.

Drucker, an eminent writer on management, expands on this simple approach by suggesting that:

The purpose of Marketing is to know the consumer well enough to develop products that sell themselves.

Neither of these explanations directly give those not involved in marketing an indication of the activities in which marketing departments engage. The Institute of Marketing describes the process of marketing as:

The management process responsible for identifying, anticipating and satisfying customer requirements profitably.

The need for marketing.

Marketing is that function of management which manages the organization's relationships with its customers and potential customers. Figure 1.1 in Chapter 1 shows how marketing relates with other management functions. Marketing will

---

This chapter was co-written with Trevor Pye.

have significant impact not only on the success of the organization but also the direction the organization takes to achieve that success. The objective of marketing is to contribute to corporate objectives within the corporate strategy. To do this it must formulate a marketing strategy in conjunction with the other functions of management, particularly finance and manufacturing. There is no point selling products that cannot be made on time or at a profit. Marketing cannot operate in isolation. Marketing decisions influence the efficiency of manufacturing, and if not co-ordinated will lose profit.

Marketing is an 'intelligence' function, monitoring the market environment and enabling an organization to be aware of economic trends and competitors' products. The primary function of marketing, therefore, is managing all aspects of the relationship with customers.

While marketing is concerned with manufacturer–customer relations, it is the same relationship as that between supplier–manufacturer since to the customer the manufacturer is the supplier, and to the supplier the manufacturer is the customer. The supplier–manufacturer relationship is developed in Chapter 16, Purchasing. The marketing concept can also be applied within an organization. Modern management approaches in manufacturing and other organizations are towards delegating responsibility to small operating units. These may be called strategic business units or profit centres but the idea is the same. Each unit relates with other units in the organization as suppliers and customers. In marketing, this concept is known as the internal market. In Britain an internal market has been created within the National Health Service.

Marketing objectives

Marketing objectives would include maximizing not only sales but also the reputation and image of the organization. In order to sell products an organization must understand why people buy their products: the order-winning criteria. Products will only sell if people want them and prefer them to other similar products in the market. The organization must achieve competitive advantage over its competitors if it is to sell. This advantage may be in price but it may instead be in technological advance, or quality or the ability to sell 'off the shelf'. Marketing is not just selling. Selling is a reactive role, reacting to customer demand. Marketing is a proactive role, creating and increasing markets for an organization's products.

The concept of *total customer satisfaction* has developed as a modern approach to marketing designed to achieve success. It identifies all aspects of the manufacturer–customer relationship and ensures that customers are satisfied in all ways, i.e. price, quality, delivery and after-sales service.

Historical development of marketing

The need for marketing is a recent phenomenon and has come about by the increasing separation of producers from consumers. An analysis of the history of exchange explains how this separation has developed as markets change.

The members of a primitive society exist by self-production and self-sufficiency. The necessities for existence are provided by individuals supporting themselves and their families by fishing, hunting and farming. At this stage of market development no exchange is necessary due to the self-sufficient nature of the society. There is in essence no market and therefore no need for marketing.

As societies develop and grow, self-sufficiency starts to decrease. Those societies close to the sea or large lakes begin to realize that if they fished all year there are benefits to be gained. Instead of breaking off for part of the year, to till their plots of land, they can exchange their excess fish for farm produce grown by other people and be better off. This stage of market development involves a type of exchange known as decentralized exchange. A simpler way is to consider this process as a barter system. It involves a direct exchange between two people, each of whom wants the produce or product of the other. In this situation no 'money' is exchanged; the transaction takes place at the level of mutual and equal benefit. In fact money was not invented until long after the development of commerce based on barter and exchange. Here a simple market has started to exist but marketing is still unnecessary as the exchange is taking place directly between the individuals involved in the transaction, who are the end users of the products exchanged. The geographic area and product range are very limited and the consumer has adequate knowledge of all of the products and opportunity to meet other people with whom to exchange.

For larger communities with a greater range of products to exchange the barter system has clear limitations. As the number of exchanges increases, the time necessary to complete the transactions starts to affect the time available for the primary activity of production. A second complexity arises from the lack of money. Each exchange has to be considered individually to ensure the mutual benefit. With an increased product range the knowledge required to ensure this mutual benefit becomes difficult to manage, as each exchange has the potential to be at a different level.

The rise of the merchant or trader occurred to overcome this imbalance in the market process. In this situation merchants or groups of merchants began to act as middlemen to facilitate the flow of produce from producer to consumer. In this situation the market is literally the place where buyers and sellers gather to exchange their goods, such as the village square. This is a centralized exchange. From this develops the economist's notion of a market as a collection of buyers and sellers who are involved in the trading of specific products or ranges of products. This system of trading enables a larger range of products to be exchanged and over larger distances as merchants travel from village to village with a stock of products or produce.

As this form of trade developed there arose the need for common standards of weights and measures so purchasers knew how much of a product they were buying. These started with agricultural produce but extended to other products.

Laws were soon found to be needed to protect purchasers from merchants using weights, lengths or volumes less than the standard or those adulterating wine with water. The Sumerians and Egyptians had laws to punish such merchants.

The development of trade necessitated the invention of money, by the Lydians in the seventh century BC, as a common unit of exchange. To be of value in different countries, money had to be acceptable everywhere. It was for this reason that gold, silver and copper were used for money. This enabled products to be exchanged for money instead of other goods. It also expanded trade routes much further. Initially exchange was for a given weight of gold or silver; only later did emperors emboss their picture on metal discs and use different sized discs for different weights, i.e. different values. In this way coins were developed which consumers were able to use to make purchases. With the withdrawal of the Romans from Britain trade reduced and money ceased to be used. It was only reintroduced with the arrival of the Anglo-Saxons some centuries later.

The development of trade can be seen to have caused the development of writing as the details of trade were recorded. The development of accounts also arose from the need to record the financial details of transactions. (Figure 19.1 in Chapter 19 shows an account recorded 4000 years ago in an Egyptian tomb.)

The relationships between these historical models of exchange are shown in Figure 5.1.

CORE CONCEPTS OF MARKETING

Viewing the historical perspective leads to the definition of marketing used by Kotler and Anderson:

> Marketing is a social and managerial process by which individuals and groups obtain what they need and want, through creating and exchanging products and value with each other.

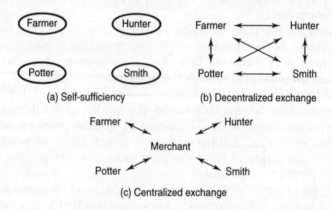

Figure 5.1 *Forms of exchange.*

This definition introduces the core concepts of the marketing process as shown in Figure 5.2.

Human needs and wants

The most basic concept underlying marketing is an understanding of the needs of the individual consumers. People need food, water, air, clothes and shelter if they are to survive even at a basic level. A human need is therefore a state of felt deprivation of some basic satisfaction. These needs are not created by society or by producers but are necessary for the fabric of human existence.

The American psychologist Abraham Maslow developed a *hierarchy of needs* which can be used as a model to understand human needs generally and is of value in the marketing environment. The elements of the hierarchy of needs are shown in Figure 5.3.

Figure 5.2 *Core concepts of marketing.*

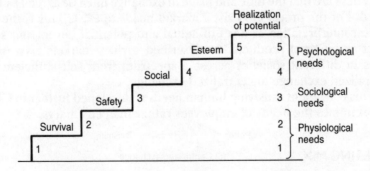

Figure 5.3 *Maslow's hierarchy of human needs.*

Whereas human beings need food, drink, clothes and shelter they tend to buy burgers, cola, trainers and live in the most expensive properties they can afford.

In marketing terms these are wants rather than needs, and describe the form needs take when shaped by culture and personality. This implies that while needs may be general, wants can be specific to individuals or groups.

For organizations to be able to sell products a demand needs to exist, or the organizations need to create a demand. Demands are therefore wants for specific products that are backed up by an ability and a willingness to buy them. Many people would wish to own a Jaguar or Rolls-Royce car, but only a few have the resources to purchase the product. Manufacturing organizations must therefore measure and anticipate demand for their products.

NEED SATISFACTION

Needs are therefore satisfied by products. Products are defined as anything that can satisfy needs or wants. In the broadest sense this includes people, places, organizations, activities and ideas in addition to the more obvious goods and services.

If products satisfy needs, then this implies that an exchange is taking place. For the consumer the felt deprivation has been satisfied and for the manufacturer the product is sold. Exchange is this act of obtaining something desired by offering something in return. The transaction involving the exchange of money for a product is only a part of the total exchange that takes place.

For this exchange to take place assumptions of contract need to apply at the basic level. The exchange must create value and leave both parties better off. For both parties to be better off a transaction needs to take place. This is normally accomplished by exchanging money for products, although from the 1970s onwards non-monetary transactions have increased both at an individual and organizational level.

Transactions must involve at least two conditions for the exchange to be of value. These are that the time and place of exchange must be agreed by the parties involved. For the process to exist a market needs to exist. This is the market in its widest interpretation as a set of actual and potential buyers and sellers of a product or range of products. As described earlier, markets exist in different degrees in different societies across a spectrum from self-sufficiency through decentralized exchange to centralized exchange.

This concept of satisfying human needs is developed further in Chapter 24, which examines the needs of employees rather than customers.

MARKETING MIX

Understanding the core concepts of marketing gives rise to the idea that successful marketing will identify a unique set of products for a unique set of consumers. This concept is known as the *marketing mix* and creates a definition of marketing as:

> Getting the right goods and service, to the right consumers, at the right price, with the right communications.

This concept is often described as 'the 4Ps' of marketing from the concepts of Product, Price, Promotion and Place as shown in Figure 5.4.

Product

Every organization has to make decisions about the products it intends to sell. How large should the range of products be? What features should the product have? Why do customers buy the product? What prices should be charged? To whom should they sell?

Design of the product

The design of a product is clearly very significant to its marketability. It influences customers. For some products customers buy because of technological innovation rather than price.

Name of the product

The name of a product is important as it creates an image for the product. A good name will encourage buying. The name has to be suitable in all countries in which the product is being marketed. A manufacturer of dog food produces two different brands. One tin has a label showing a ordinary dog and the other shows a poodle. These are aimed at different parts of the market. It has been suggested that the contents of the tins are identical but the tin with the poodle on it can be sold for more money because wealthier people will pay more if the product portrays a wealthy image. This is an example of product differentiation – different versions of basically the same product matched to different sectors or segments of the market. Product differentiation is considered in more detail in Chapter 8 in the context of product design.

Figure 5.4 *The marketing mix.*

Packaging and labelling

The design of the packaging and labelling of the product is part of the design of the product. This can attract a potential customer since many purchases are made on the basis of the packaging and the information on it rather than on the product itself. Within Europe labels are now multilingual. This enables a manufacturer to produce only a single product and single label for sales across Europe. Otherwise several labels would be required and products would be identified by country of sale and have to be stored separately. It also enables jet-setting business people to find a product they are familiar with in any European country they may visit.

All of these aspects of a product must be considered since they have a profound influence on the long-term success or failure of the organization.

Market strategies

Igor Ansoff defined four main strategies based on the opportunities for new products and new markets. These are shown in the matrix in Figure 5.5.

- *Market penetration* Market penetration means that the organization seeks to increase sales for its present products in its present markets through more aggressive promotion and distribution.
- *Market development* Market development implies that the organization must seek increased sales by launching its existing products into new markets.
- *Product development* Product development requires the firm to introduce improved products into its existing markets.
- *Product diversification* Diversification involves introducing both new products and new markets.

Types of product

The consumer is not, however, interested in organizational strategies, but in satisfaction of needs and wants. Products must then evolve in terms of design,

Product / Market	Existing	New
Existing	Market penetration	Product development
New	Market extension	Product diversification

Figure 5.5 *The Ansoff matrix.*

presentation, packaging, brand image, status and reliability. Products are therefore a mixture of features.

The product is not always a tangible entity; banks and insurance companies refer to their services as products. Kotler defines a product as:

A bundle of physical, service and symbolic particulars expected to yield satisfactions or benefits to the consumer.

Products are often referred to under four main categories: durables, consumables, intangibles and industrials, as listed below:

- *Durables* Durables are tangible goods purchased occasionally and normally consumed over long periods of time, e.g. cars, televisions or washing machines.
- *Consumables* Consumables are normally repeat purchases of items which are consumed or used in the short term, e.g. food, drink and household cleaning products.
- *Intangibles* Intangible products such as insurance policies, banking and entertainment are classified as services as no direct physical benefit or satisfaction is offered.
- *Industrials* The industrial products classification is used to distinguish between those products bought by the domestic consumer, who is the end user, and those bought by industrial organizations, which are not end users. These products are part of a supplier – manufacturer–customer chain which may have several links. This classification is not as well documented as consumer markets, but considering that some estimates indicate 80 per cent of all purchases are by industrial companies buying components, it is an important sector.

Satisfying customer wants

Since the sole purpose of a product is to provide satisfaction for consumers, every marketing organization must identify what customers want and recognize that this is a dynamic situation. Primarily this is a result of continual change in consumer's needs and their environment. Their incomes, lifestyles, customs, fashions and aspirations are dynamic not static. Therefore marketing policies must reflect that dynamism.

Some of the reasons for consumers demanding new satisfactions are as follows:

- Rising incomes and expectations
- Increasing education and sophistication
- Changes in social habits and customs
- Changing fashions
- Technology
- Economic factors

Product design takes into account the features in a product that have been found to be those which customers want. In making a purchasing decision potential customers will compare products on the basis of features as well as price. Consumer magazines such as *Which?* often tabulate the features of a range of similar products to provide consumers with the information they need to make a purchasing choice. Table 5.1 is an example of a preference matrix for purchasing a television.

Table 5.1   Preference matrix for buying a television

	Model							
	A	B	C	D	E	F	G	H
Size (cm)	48	48	48	48	51	51	51	51
Flat screen				●			●	●
Channels (no.)	20	30	30	40	30	40	40	50
Auto-tuning			●	●			●	●
Teletext		●	●	●	●	●	●	●
Fastext				●				●
Remote control		●	●	●	●	●	●	●
On-screen display			●	●			●	●
NICAM stereo				●			●	●
SCART socket			●	●			●	●
Headphone socket	●		●	●				●
With stand				●		●	●	
Price (£)	219	279	349	379	349	379	429	479

It is important for a manufacturer or provider of a service to be aware of the features that consumers want and the features in competitors' products. The preference matrix enables this data to be presented in a way that is useful for an organization to make such comparisons.

Potential customers may also use the concept of the preference matrix. They will exclude consideration of products beyond a price range and will assess the features in all available products. A customer is likely to place personal priorities on the features. Some will be considered essential, others indifferent and some even positively not wanted. With this approach some products will be eliminated and a choice finally made, probably on the basis of 'value for money' – a combination of price and features.

Product lifecycle

Due to this dynamic situation all organizations must accept that all products have a limited life. White comments that change is the only permanent element in the complex problems that face business managers in a developed economy. Change and limited product life can be expressed in terms of the product lifecycle as shown in Figure 5.6. This concept shows that sales of a product will vary depending upon the time the product has been on the market.

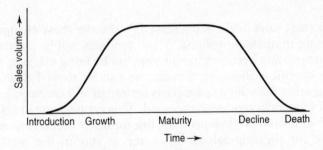

Figure 5.6 *A product lifecycle.*

The various stages of the product lifecycle are described below in a way that mirrors human life:

- Development
- Introduction
- Growth
- Maturity
- Decline
- Death

DEVELOPMENT
The development stage takes place prior to the product reaching the marketplace. During this stage market research must be undertaken, the product designed, prototypes built and the manufacturing system developed. At this stage costs will be very high and revenue will be nil.

INTRODUCTION
The introduction stage sees the launch of the product onto the market. Initial growth will be slow because the product is unknown and potential customers will be cautious of buying a new product, especially if it is from a company they have not heard of, since it will not yet have formed an image or a reputation.

GROWTH
During the growth stage the product reaches general acceptance and sales grow rapidly. As development costs are recovered and unit costs decrease with increased production volume, the organization begins to become profitable. As the market is growing rapidly, competitors are encouraged to enter the market.

MATURITY
As the product reaches the maturity stage the rate of change of demand begins to slow down and sales increase more slowly. Due to economies of scale and stable high volumes, high profits can be maintained. The competition is countered by non-price variables and the weaker competitors will leave the market.

DECLINE

At the decline stage sales begin to decrease. To counter these changes prices are cut in an attempt to maintain volume. Many products will be phased out at this stage as the competition becomes too intense, but by using selective retail outlets and selecting specific customers, the decline can be slowed down. For some products the market for selling spare parts for repair will continue long after the manufacture of the main product has ceased. This creates a small market, for the parts rather than for the original product, but for many manufacturers it will not be worthwhile for the original manufacturer to stay in the market. Smaller specialist organizations may enter the market at this stage to satisfy the small but continuing customer demand for the product. Manufacturers may licence subcontractors to make the product at this stage. In this way they remove it from their own manufacturing system but are able to offer the product to those customers who still have a need for the product.

DEATH

At some stage the product will cease being produced. It will have come to the end of its life.

It is not easy to predict in advance what the life of a new product is likely to be. It varies from one industry to another. It is high in engineering and low in fashion, but in all industries it is becoming shorter with the development of modern engineering technology. Some products had lifecycles of fifty years or more, many domestic items for example. The mechanical watch was available for many decades. New technology brought the LED (light-emitting diode) digital watch to the market in 1970s. The product lasted only a few years before being replaced by the LCD (liquid crystal display) digital watch, which in turn is being replaced by a consumer (or is it a manufacturer?) preference for the traditional analog, rotary display.

The importance of the product lifecycle concept is that it provides the marketing department with information to plan and forecast future demand enabling the organization to develop a marketing strategy. It reminds the organization of the following facts:

- Products have a limited life.
- While costs and revenues are not consistent throughout the life of a product, they can be compared with other products at the same stage.
- Products require different marketing initiatives at each stage of their product lifecycle.

The consequence of the product lifecycle, with its variable profit from individual products over their lifecycle, is that organizations need a portfolio of products at differing stages of product lifecycle if they are to retain reasonably consistent investment costs and profit levels. The organization has to plan to have a succession of new products coming through the pipeline as shown in Figure 5.7.

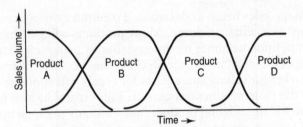

Figure 5.7 *Overlapping product lifecycles.*

Drucker has drawn attention to the need to keep all products under review to ensure that a only a small proportion of them are at the end of their lifecycle. He describes the following six categories:

● Tomorrow's breadwinners – new products or today's breadwinners modified and improved.
● Today's breadwinners – the innovations of yesterday.
● Turnaround products – those capable of becoming net contributors if something drastic is done.
● Yesterday's breadwinners – generally products with high volume, but badly fragmented into 'specials' or small orders.
● The 'also rans' – the high hopes of yesterday which did not work out well but were not total failures.
● The failures!!! They never got off the ground.

SWOT analysis

From the product lifecycle concept and Drucker's analysis of product categories, it follows that all products must be kept under review to assess their contribution to total profit. The decision to drop products from the portfolio is a crucial decision in the life of a product and the profit of the organization.

Products that are no longer making a positive contribution to the firm will consume revenue that could more profitably be used to develop and launch new products or modify existing products.

One method of assessing the health of a product is to carry out a SWOT (Strengths, Weaknesses, Opportunities and Threats) analysis of the product.

The results of a SWOT analysis allow the organization to assess how competitive each product is by comparing the product's strengths and weaknesses with the competition. The market can also be assessed to look for new opportunities that may exist for modified or new products. The threats to products can also be judged by assessing future competition and the economic situation.

The results can be developed into the matrix shown in Figure 5.8.

Growth Share	Low	High
Low	Dogs	Problem children
High	Cash cows	Stars

Figure 5.8 *The Boston matrix.*

The matrix was developed by the Boston Consulting Group and positions products into four categories:

- *Stars*   Products with both a high market share and the potential for further growth but which will require additional expenditure.
- *Cash cows*   Products with high market share but little potential for further growth. These products are 'milked' to provide cash for the development of other products.
- *Problem children* (or question marks)   Products about which difficult decisions need to be made. Are they dropped from the portfolio or is extra cash spent on them in an attempt to turn them into stars?
- *Dogs*   Products with both low market share and little potential for growth. They are clear candidates for elimination from the product range.

Price

To many casual observers the price element of the marketing mix consists only of the cash element. For the marketing practitioners the notion of pricing has wider implications.

The consumer may not solely consider products on their price alone; the idea of 'total price', 'value for money' and the strength of the need will also be uppermost in their minds. The total offer being proposed by the seller needs to consider the quality of the product, the packaging, the method of distribution, payment methods and the after-sales service facilities.

The organization therefore has many options and variables which it can consider when deciding what price to place on its products. The influences relating to total price are under the control of the organization itself and it can decide the emphasis it places on each of the elements.

Price determination

In a monopoly the price can be determined within the organization. The cost of a product is determined and a desired profit is added to determine the price.

There is no incentive to reduce costs or profit because the customer is in a captive market and will pay. This is known as a 'cost-plus' approach and is expressed by equation (5.1):

$$\text{cost} + \text{profit} = \text{price} \tag{5.1}$$

In a competitive market price is determined outside the organization by the market. If price is fixed externally, management must deal with the remaining factors in the equation. A traditional view would accept costs as fixed, and as long as profit was satisfactory no one worried about the costs. Cost was seen as the difference between price and profit, as seen in equation (5.2):

$$\text{price} - \text{profit} = \text{cost} \tag{5.2}$$

A modern view would see that by improving productivity costs can be reduced and profit increased. In order to achieve a desired profit, productivity must increase to enable costs to fall. Profit is seen as the difference between price and costs.

$$\text{price} - \text{cost} = \text{profit} \tag{5.3}$$

The constraints of the market, on the other hand, are not so easily controlled by the organization and will depend on both the strength of the organization and the product in individual markets and the existence of competitors in the market. While the type of market will affect the slope and shape of the product demand curve, the same basic relationship between volume and price will always exist. By varying the price, the firm will vary the profit margin per item of the product sold. A higher price may increase the profit margin per unit sold, but decrease the volume of the sales. Likewise a reduction in price may lower the profit per unit but increase the sales volume for the product. The sales revenue is the price multiplied by the sales volume. These relationships are shown in Figure 5.9.

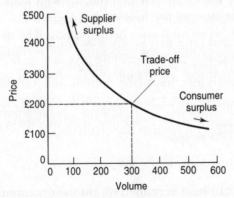

Figure 5.9 *The demand curve.*

On the demand curve in Figure 5.9 the market price for a quantity of 300 can be seen to be £200. Above this price the supplier will not be able to sell all the products brought to the market; below this price more consumers will demand the product than the supplier is willing to bring to the market. The sales revenue is the area of the dotted rectangle on Figure 5.9. As price is changed, so volume changes (or vice versa). An organization may wish to maximize its revenue and determine a price and volume at which this occurs. It should be noted, however, that the volume for maximum profit is not necessarily the same as the volume which maximizes revenue.

Depending upon the market conditions and the product's stage in its lifecycle, the organization may adopt differing pricing policies. With a new product which has a technological advantage over the competitors, the organization may use a 'skimming' policy. In this case the product enters the market at a high price, to enable high profits to be initially taken, while the product is a market leader. These high profits will, however, encourage the competition to enter the market, with follow-me products, to enjoy their share of the large profits. The 'second to market' manufacturer may adopt a different pricing policy. In this situation the second organization has to encourage the consumer to purchase what might be seen as a copy. The price may therefore be set artificially low in an attempt to seize market share; price is being used to penetrate an existing market.

For organizations selling specialized 'made-to-order' products or services, price determination is more complex. For a one-off product, cost determination is predictive and has to be based on estimation. Price determination will be by negotiation with the customer. The result will depend on their relative bargaining power.

Promotion

Promotion is the term given to a collection of methods by which an organization attempts to communicate either directly or indirectly with a market. The aim of promotion may be one or more of the following:

- To create a demand for a new product or brand.
- To maintain or expand sales of existing products or brands.
- To maintain or expand an organization's share of particular market segments.
- To create a favourable corporate image for the organization. Whatever the aim, the message must be consistent in its attempt to portray the right message to the right people.

Methods of promotion include the following:

- Personal selling. A direct face-to-face, two-way communication such as doorstep selling.

- Advertising. A communication to inform, persuade and reinforce the consumers' knowledge of the product.
- Promotional activity such as special offers, discounts, competitions, incentives and trade-in offers.

Some form of advertising is likely to be necessary for all products and services. The possibilities range from television, newspapers and magazines to billboards and leaflet distribution. Decisions are needed on the amount and placing of advertisements. Placing should be targeted to anticipated readership. Advertising by driving schools is often placed in buses on the grounds that bus passengers are those who do not yet drive. The frustration of sitting on a bus in traffic congestion may encourage such people to learn to drive a car. The advertisement is placed there to suggest such a course of action. Some food manufacturers pay a supermarket to display their products at a prominent place.

The range of methods available to the marketing organization in the area of promotion is often referred to by the acronym AIDA.

Arrest	Attention
Generate	Interest
Create	Desire
Prompt Buying	Action

Law in marketing

Clearly organizations need to be honest in their advertising and promotion of their products. The law protects customers in many ways. In Britain the law of sales is governed by the Sale of Goods Act of 1956. More recently the Fair Trading Act of 1977 has extended protection. Sellers are required to describe their products fairly. Sellers of food may have their premises inspected for hygiene, and may be fined or forced to close if these are not up to standard. Trading standards officers monitor sales to enforce weights and measures legislation. The law also gives rights to customers who are dissatisfied with a product and in cases where the product fails within a short time. Manufacturers often give guarantees of reliability for their products which are enforceable in law.

In some countries it is a legal requirement to state the country of origin of a product as part of the product information. Manufacturers must be honest in this description of their products. Some years ago, it has been reported, some towns in Japan were renamed Usa and Uk so that products made there for export to the West could be described as being made in the USA or UK in order to improve sales. It has been suggested that a town in Japan making cutlery was renamed Sheffield and labelled its products 'made in Sheffield'.

Public relations

Apart from promoting their products, organizations promote the organization itself. This is the function of public relations. In a large organization this would be a separate department. Public relations activities include sponsoring sporting activities and getting favourable media coverage that keeps the name and the image of the organization foremost in people's minds.

Place

Even when the product characteristics, price elements and promotional methods have been decided, the product still needs to get to the customer. Two factors must be considered: location and availability. These will be dictated largely by the type of product, although the use of technology in the distribution network may enhance the product features.

Location of point of sale

Generally the consumer may not be prepared to travel very far for a simple purchase such as a loaf of bread. However, if that purchasing action can be encapsulated in a more grandiose experience, such as a Sunday market or an out-of-town hypermarket, then consumer behaviour can be altered.

Distribution

The options available to an organization in deciding their distribution network are based on the types of marketing channels shown in Figure 5.10.

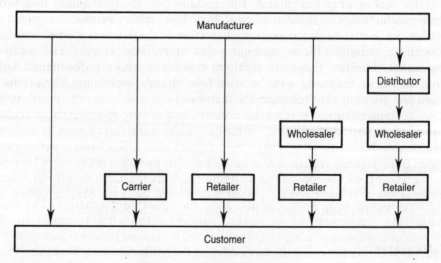

Figure 5.10 *Marketing channels.*

The direct flow of goods from the manufacturer to the consumer, using, for example, their in-house transport, is a one-channel distribution network. The introduction of a retailer, perhaps a subsidiary company, or a freight carrier or the Post Office introduces an extra channel and makes this a two-channel network. The further introduction of a wholesaler and or a wholesaler plus a distributor gives a three- and four-channel system respectively.

Figure 5.10 needs to be modified to include importers and exporters if the organization buys or sells overseas.

Distribution is a vital part of the marketing process. It ensures that the product is available to the consumer 'when' it is required and 'where' it is required. The cost of providing this level of service must not, however, outweigh the sales and profit generated. An organization must decide whether to do its own distribution or to subcontract this part of the process to another organization. Distribution is developed further in Chapter 15.

MARKET SEGMENTATION

To gain the full benefits from the marketing mix, the ingredients must be mixed in a specific way to match individual markets. Segmentation of the market enables the organization to differentiate between the needs of the various customer groups that collectively comprise the total market for a product. The product can then be designed, packaged, promoted and distributed to match the needs of these different groups. The needs of different groups create what are termed 'niches'. These are groups with an identified need. They may not be very large in size but it may be only one manufacturer which has identified the niche and so can operate profitably without much competition in this niche.

In this attempt to match the product to the consumer two broad segmentation types can be considered. Some researchers attempt to segment the market by looking at 'consumer characteristics'. These can be based on geographic, demographic or psychographic factors. Other researchers try to form 'consumer responses' to specific products in order to determine the benefits being sought. These segments can be based on brand loyalty, usage patterns or questions about quality compared with price.

To understand how segmentation can help an organization the marketing mix needs to be examined.

Demographic segmentation

Geographic factors such as climate clearly can affect the market, but differences between regions or districts can also have an impact on the market. The size of the urban conurbation will, for instance, affect issues such as transport. In other areas organizations use a regional image to sell individual products with both Bass and Allied Breweries using the term 'Men of the Midlands' and 'Yorkshire's Best' respectively to promote their best bitter beers.

Demographic factors including, age, sex, family size, family lifecycle position, income or occupation are examined to determine consumer wants. These factors are widely used as empirical evidence, and can be easily measured to test the links between buying patterns and demographic influences. Specific products have an appeal to differing age groups. The manufacturers of fashion clothing and compact discs, for example, need to determine their main consumer groups. Age probably gives the clearest segmentation. Products such as cosmetics and toiletries are clearly aimed at a market determined by sex, but cigarettes and motor vehicles can also be marketed on the same basis. Many medium-sized cabriolet cars are directly aimed at career women.

For products such as baby foods and school accessories the market may be enlarged if large families can be targeted effectively. An insurance company which attempted to sell pension plans to first-year university students would sell less policies than if they targeted the academic staff. For products such as insurance policies and saving plans it is the age profile of the consumers that is critical. Income segmentation attempts to generate an image of exclusiveness. Suntory, the Japanese whisky distiller, attempts this by associating the product with golf tournaments. Games consoles and personal computers share a common technology and many components but are marketed by the use of the product: an accountant or engineer may purchase a personal computer to assist their professional work but then also use it as a games machine.

Social factors independent of income are also considered to affect the patterns of buyer behaviour. The suppliers of products therefore need to be able to identify their target population by categories other than income. The Joint Industry Committee for National Readership Surveys (JICNARS) attempts to group consumers by the occupation of the head of the household. From this classification comes the concept of social class or grade:

A	upper middle class	3% of the population
B	middle class	15% of the population
C1	lower middle class	23% of the population
C2	skilled working class	27% of the population
D	working class	18% of the population
E	people on benefit	14% of the population

The proportions in each category vary from one society to another and also vary over time for a given society. From this classification comes advertising campaigns which can be targeted through specific national newspapers and magazines depending on the product and target consumer.

The JICNARS classification has its detractors and its limitations and has been supplanted by the ACORN (A Classification Of Residential Neighbourhoods) classification for many marketing decisions. This attempts to group consumers by residential characteristics, the type of property and the area they live in. With an analysis of thirty neighbourhood types it can enable the producer to target their consumers more accurately.

Since neighbourhoods are identified by a postcode, mail promotion can be targeted by knowing the type of household in each postcode area.

In recognizing that people with more money are more likely to spend more, some groups have been identified as being suitable for targeting with expensive products. These include the following:

Yuppies – Young Upwardly Mobile Professional People
Dinkies – Dual Income No Kids

Market segmentation needs to be reviewed periodically to ensure the market conditions have not changed. Henry Ford believed that the only segmentation was by price. During the late 1930s General Motors realized that the market could be resegmented, based on income and preference groups. With this strategy they replaced Ford as the dominant manufacturer of cars. General Motors have in turn failed to recognize a new segmentation based on size and economy, and have lost market share to Volkswagen and the Japanese manufacturers.

Hierarchical segmentation

One way to examine new options is to segment the market in a hierarchical way using several different segmentation factors. This enables the organization to thoroughly examine the market in order to determine the most successful option. Figure 5.11 shows how the market for tea drinkers could be segmented on a hierarchical basis.

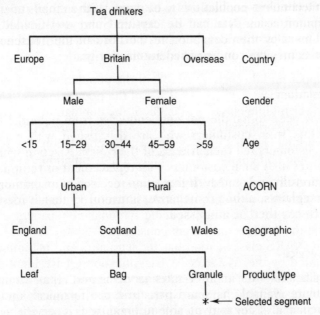

Figure 5.11 *A hierarchical market segmentation model.*

With all hierarchical segmentation models the marketer's knowledge of the product and the market must be used to determine the order in which segmentation factors are used in the hierarchy. For the tea drinkers in Figure 5.11 the cascade sequence used in the hierarchy is: country, gender, age, ACORN, geography and product type. At each level choices have to be made as to which segment or segments are selected to determine the segments of the total market in which the organization wishes to operate. For other products differing cascade sequences using different segmentation factors would be necessary.

SALES MANAGEMENT

As indicated at the beginning of this chapter marketing is much more than just selling. Nevertheless selling is a part of marketing and needs managing. Depending on the type of industry, an organization may have a salesforce who need training, motivating and paying. Organizations may offer an after-sales servicing facility.

Information technology in sales

The salesforce operates remotely from the head office of the organization. Information about sales made need to be transmitted back to the office. Information technology enables this to be done. With a mobile telephone, fax and laptop computer, sales data can be captured and downloaded to the office computer. This sales order data becomes a major data input to the manufacturing function. Sales invoices can be raised automatically.

Customer relations

Organizations need to handle not only enquiries from potential customers but also complaints from customers who are dissatisfied with a product. Since dissatisfied customers tell their friends of their dissatisfaction with a product, it is important to satisfy such customers with replacement or refund and to handle such cases carefully and quickly. Information received by an organization through customer complaints, although negative, is useful in that it identifies product failures which can then be addressed.

After-sales service

Some organizations offer an after-sales servicing and repair facility. They have the components available as spare parts and the technical knowledge of the product to repair it. They may be able to organize this service more effectively than other organizations. However, many manufacturers would not want to be involved in this area and would be content to sell spare parts to others or even to license the manufacture of spare parts to other organizations. This avoids the manufacturer's factory from having to deal with the manufacture of small quantities of possibly obsolete components.

Task sheet 5 Marketing management

'Confident' case study

GS Chemicals plc manufacture and sell a range of cosmetics and pharmaceuticals. Their drugs and cosmetics have a high profile in the market in small chemists shops and occupy very profitable market positions at the high-quality end of both markets. The company has gained an excellent reputation for research and development and has made several well-publicized technological advances. It has a policy of not testing its products on animals.

The company has decided that to rely on its existing market is not sufficient to increase its market profile or to increase its profits. The problem for the company was how to diversify its products without major investment in research and development. Furthermore, the company wished to gain a higher profile in the area of supermarket sales.

The company's marketing department has identified an opportunity for a new range of toiletries containing natural ingredients. The R&D department has already been working on a new series of natural toiletries. After trials of their new formulation they claim to have developed a new deodorant based on natural ingredients not used by anyone else and clinically proven to work effectively for longer periods than most other brand deodorants.

With a high degree of optimism for the new range of toiletries, senior management gave approval for the marketing department to plan the product launch, including selecting a brand name, packaging design, selection of distribution channels and the development of an advertising campaign.

The brand name chosen was 'Confident' which, it was agreed, gave the right image for the product. Both male and female deodorants would be sold under the same brand name, creating 'his' and 'her' versions of the product which would encourage intergender purchases. The packaging was impressive and based on light pastel shades. The product would be distributed through the existing network of wholesalers to avoid increasing distribution costs.

The marketing department recommended a premium price for the products since this was consistent with the quality image it wished to convey. Consumers would pay more for a more effective product than those offered by competitors.

The company allocated a large budget for the advertising campaign with well-known personalities on television saying that they used the products. The media campaign was planned for a three-month period. A public relations company was hired to generate publicity for the products in all the major up-market magazines. The campaign seemed to be a success. The supermarkets which had agreed to sell the products were well stocked and the media campaign created a significant product profile.

Sales figures were, however, very disappointing. Fourteen weeks after the launch retailers were demanding that GS Chemicals took back the unsold products and that they were refunded their expenditure.

The company hired a market research company to interview a large sample of consumers. They reported the following responses:

The deodorant I usually use works, I see no point in changing it.

I've seen the adverts on the telly but I haven't seen it on sale in my supermarket.

My deodorant smells nice, 'Confident' has a horrible smell. I couldn't use it.

I want a man's deodorant, 'Confident' looks like a woman's.   (male respondent)

I want a feminine deodorant, not one that's the same as a man's.   (female respondent)

I'm not prepared to pay that much for my deodorant; it can't be that much better.

This environmentally friendly stuff is a gimmick. How do I know it's natural?

I want to smell nice all day, any deodorant will do that.

In order to rescue the 'Confident' brand from oblivion and establish the product as a market leader, GS Chemicals has hired you as marketing consultants. Advise the Managing Director how you would rescue this product.

5.1   Identify any mistakes made by GS Chemicals in the launch of the 'Confident' products.

5.2   Advise on the importance of segmentation, targeting and positioning, which, if used, could help to rescue the product and gain a profitable market position.

CHAPTER 6

MARKET RESEARCH

WHAT IS MARKET RESEARCH?

Market research can be defined as follows:

> The gathering, recording and analyzing of all the facts about problems relating to the transfer and sale of goods or services from producer to the consumer.

Over the last fifty years the need for accurate information concerning an organization's products and their performance in the marketplace has become more important. Competition has increased dramatically as markets moved from a local perspective towards a national outlook and then added an international dimension. For many products, such as motor cars, televisions, audio equipment, and white goods (washing machines and refrigerators), organizations are now operating in a global market.

With increased product choice and disposable incomes, consumers become more discerning in their buying behaviour. Products are purchased on the basis of their styling, colour and features rather than their basic functions. This move to buying based on wants rather than needs makes it more difficult to quantify potential demand because there is greater uncertainty and fewer data available. While organizations need to be aware of the potential customer's wants, they must not be blind to the underlying needs that cause the felt deprivation as this is the real drive to buy.

Railway networks have been responsible in the past for this type of 'marketing myopia', believing that the consumer needed rail travel, when in fact the need was to travel. Alternative forms of transport – aircraft in the United States, motorways in Britain – could provide that need. The move to purchasing on the basis of wants rather than needs means that it is the total product which is being sold rather than the actual product. In these situations organizations are faced with a move towards non-price competition rather than competition based on price alone.

These three factors have highlighted the need for accurate, meaningful and

---

This chapter was co-written with Trevor Pye.

95

timely information relating to the product and the market. The process is shown in Figure 6.1.

TYPES OF MARKET RESEARCH

Data gathering is the front end of the process, where facts are obtained about the market or the product. The process can be considered in various ways: reactive or proactive; desk research or field research; use of primary or secondary data. Primary data is that acquired directly; secondary data is data that already exists. The relationship between different forms of market research is shown in Figure 6.2.

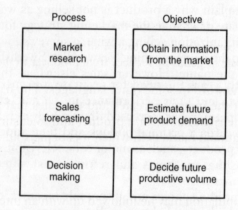

Figure 6.1 *The market research process.*

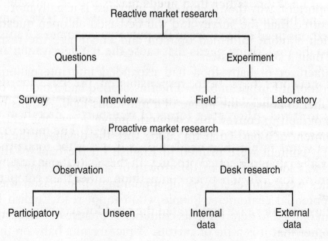

Figure 6.2 *Types of market research.*

Reactive market research

Market surveys

A survey is usually well structured and asks closed questions. These are questions structured to elicit a yes or no answer. They direct the respondent to give precise answers. The questions can be determined by reviewing previous questionnaires. Surveys collect data about perceptions of the product or of similar products in the market. They may ask what features are wanted in the product, or what image is created by a particular product. It is usual for a market survey specialist organization to be contracted to undertake the survey. Results are then analyzed and presented to the commissioning organization. Manufacturing organizations can receive valuable feedback from the market about their own and their competitors' products. This may explain why a product is not selling as well as expected and can be used to formulate or change the marketing strategy for a product.

In a case of an organization selling a wine, a survey showed that poor sales in Sweden were due to the name meaning 'seaweed' in Swedish. Sales there were boosted by changing the name. However, sales elsewhere then fell due to the name being an obscene word in some other languages. The name of a product is significant in terms of the image it creates and the sales that result. It is clearly important to create the right name for a product and to ensure that the name is right in each of the national markets in which it is being sold.

Interviewing

Some of the survey population may be followed up with an interview if the survey indicates that they are likely to be a prospective customer or if more information is required. An interview can reveal the reasons why consumers buy a product and, just as important, why they do not. The interview is usually more interactive and less structured than the survey and will be based on open questions which attempt to elicit a more informed opinion and encourage the respondent to provide more than a simple answer.

A manufacturer of baby food had expanded into international markets beyond its traditional European base. The tins of food had a picture of a healthy baby on the label to convey the image of a nutritious food. All tins of food had the same photograph to convey this corporate image across the whole range of foods. The content of a particular tin, along with the details of the ingredients, calories and fat content were printed in English on the back of the tin. The manufacturer was looking forward to growth of sales by selling in developing countries. Sales were, however, very disappointing. The reason was found by interviewing potential customers. People who cannot read English look at the picture on a tin of food to determine its contents. A picture of carrots on the label tells the customer that it is a tin of carrots. A picture of a baby on the label . . . well, no wonder no one bought any!

The main difficulty with either questionnaires or surveys is the question of bias. The design of the questions themselves may imply a required answer or force a particular viewpoint. The style and layout of the questionnaire may also encourage or discourage particular groups of potential consumers to ignore it. The problems associated with non-response can mathematically distort the results.

With the less formal format of the interview the individual prejudices of both interviewer and/or the interviewee can affect the outcome. Clearly the interests of the parties can have an impact. For example, asking a railway enthusiast for a view about the cutting of rail services is unlikely to give the impartial view required.

With the questionnaire or survey the control that market researchers have over the consumer is very limited. A further drawback is that the market researchers are only asking about behavioural patterns, not actually observing the behaviour itself. To overcome these difficulties experiments need to be conducted.

Experiments and market testing

Experiments are a research process in which one or more of the variables are manipulated under conditions that permit the collection of data which show the effects, if any, of the variables.

For the experiments to be successful a professional approach needs to be adopted. This includes the defining of objectives (what is the experiment trying to achieve?), establishing the controls (how will the variables be monitored?), and deciding the test markets and the number of them, (what is a representative segment of the market and how many observations are required to provide statistical reliability?). The number of observations required will have a major impact on the length of the market test. A long test will increase the cost, while too short a test may risk invalidating the results.

Two types of experiments are available to the market researchers, field and laboratory, with varying degrees of control over the test.

Field experiments are conducted in the natural market environment. In these situations the experimenter has very little control over the variables. To gain control of the variables, experiments can be carried out in a controlled laboratory environment with the experimenters directing the participants. This control reduces the chance of extraneous results, but may introduce an element of bias and artificiality which can negate the advantages of the control.

Proactive market research

Observation

Experiments are both costly and time consuming, but in many cases simple observation can provide the results required. With participatory observation the consumer is observed by the market researcher and then the behaviour is followed up by a questionnaire. This double-checking can prevent misconceptions by questioning the behaviour observed.

At a more detached level the observations can be non-participatory where the subject is viewed in a remote and unobtrusive manner. With a manual observation the market researcher may miscount, fail to see or misinterpret the behaviour being watched. The application of video camera security surveillance can be used to observe at arm's length how the consumer behaves at the point of sale. These observations can be used to judge the effectiveness of promotional material (does the customer see but ignore the product?), packaging (does the customer pick up the product but then return it to the shelf?), and the total product (does the customer pick up, examine and then purchase the product?).

The methods of market research discussed so far are based on an interaction between the consumer and the marketplace. These types of research are suitable only in situations where that interaction can be observed. They are therefore very suitable for consumer markets but not so suitable for markets where the buyer is remote or, more importantly, not an individual.

Desk research

For industrial marketing a different approach is required. This is often referred to as desk research, as it can be carried out from the marketer's desk by evaluating secondary data that already exists. This form of analysis can be based on internal or external data.

Internal data is quick and cheap to prepare since it is the information that companies already have to keep to run their organization. Information such as sales reports and sales figures are available from the marketing department. These can be compiled by product, sales territories, customer and function. Information from the manufacturing department concerning production volumes and delivery lead times is also necessary to complete a marketing information system. The design department also needs to contribute to this information system by providing data on new products, modifications and technological developments.

External data is required to give a view of the market in its entirety. Many organizations are in the business of compiling data and then making it available or selling it to others. These range from government departments, trade organizations and think-tanks to private and public companies engaged in market research as their primary activity.

SALES FORECASTING

An essential part of marketing data needed for planning in all parts of an organization is some estimate of the future volume of demand for individual products. Financial, manufacturing and manpower planning all require forecasts of future sales so that resources can be planned and organized to match the forecast demand.

Data for sales forecasts can be derived from three sources:

- Mathematical analysis of past data (extrapolation or causal analysis).
- Knowledge gathered from the market about consumer wants and competitors products (primary market research).
- Information on in-house developments and activities, such as product innovation and marketing promotions (secondary market research).

To be accurate and of use to the organization, forecasting needs to take account of all the expertise in the firm and is likely to be based on all of these sources of data.

Analysis of past data

Data consisting of a set of figures for the past demand of a product is the basis for the mathematical/statistical methods of generating a forecast. Such forecasts are only as good as the quality of the data used in the calculations. Mathematical forecasts make assumptions from the data provided, and management judgement must be used to interpret the information supplied and, if necessary, to modify the mathematical forecast to take account of additional information. The information needs to be challenged because the basic assumption made by mathematical forecasting is that because something happened in the past it is a good indicator of what will happen in the future. A forecast, therefore, is a management judgement based on all of the data available, of which a mathematically based forecast is only the first stage.

Patterns of past data

All of the techniques discussed here will analyze past demand data up to time, t, and make a forecast for a time, $t + 1$. In the techniques explained below, D_t is the demand at time t and F_{t+1} is the forecast for time $t + 1$ which is made at time t. As time progresses, forecasts are made at each period for the following period using new data available for the most recent period. Over time, therefore, a set of forecasts will be made, one at each time period. The period may be a day, a week, a month or a year, depending on the level of organization at which the forecast is made. (The higher up the organization, the longer the planning period and hence the periods for which forecasts will be made.) A visual representation of the patterns of past data is shown in the diagrams accompanying the techniques referred to. The pattern of past data is the main basis on which a forecasting technique is selected. The technique must match the data. If it does not the results will be inaccurate and misleading.

LEVEL DATA

In this situation the previous level of demand varies randomly over time but its general pattern is neither decreasing nor increasing. The most suitable method for dealing with this type of past data is to calculate the average and to use this as the forecast. Visually, a horizontal line of best fit is drawn through the data and projected forward to the next time period. This projection then becomes the forecast.

With a lengthy history of demand all the past data could be included. However, it would be more realistic to reject the very old data as this is less relevant as the basis for estimating for the future. In practice, therefore, the number of time periods (n) to be used in a forecast is selected and fixed for a set of forecasts over a period of time. The value of n is likely to be between 3 and 12 and depends on the amount of variation in the data and the attempt to base the forecast on more recent data. As time moves forward new data is added and old data removed. The average is recalculated and therefore moves forward with time as new forecasts are made. This type of forecast is therefore described as a moving average forecast. For a three-period moving average forecast made at time t the forecast is as follows:

$$F_{t+1} = (D_{t-2} + D_{t-1} + D_t)/3 \qquad (6.1)$$

A moving average using n values of data in the average is known as an n-period moving average. This technique is shown in Figure 6.3.

More complex moving average techniques are available. These differentially weight the data in the average in order to improve response and stability. Weighting can be linear or exponential. The details of these techniques are beyond the scope of this book.

TREND DATA

With this type of data the level of demand still varies randomly over time but also displays an increase or decrease, i.e. a trend, over a period of time. In this case a

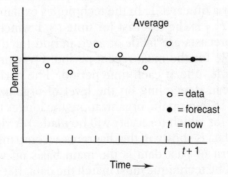

Figure 6.3 *A moving average forecast.*

moving average is not suitable since it would give silly results. For this type of data a sloping line can be drawn through the data and again projected forward to the next time period to derive the forecast. This is known as a trend line and could be straight or curved. The formula for a straight trend line is as follows:

$$F_{t+1} = a + b(t+1)$$ (6.2)

where a is the demand when $t = 0$, b is the slope of the trend line and t is the number of time periods of data.

The number of periods, n, of past data to be used must be considered as above. There are several techniques for trend forecasting, the mathematical details of which are not explained here. Figure 6.4 shows the basis of these techniques. Forecasts can be made graphically without the need for mathematics. In practice, however, data would be analyzed by computer.

CYCLIC OR SEASONAL DATA

The characteristic of a cyclic demand pattern is a random variation over time plus the indication of a cyclic variation in which demand over the period oscillates between higher and lower values. A trend may also be present. For some products, such as ice-cream, soft drinks or sports equipment, the demand varies with the time of year, and therefore this type of demand is referred to as seasonal variation. For some products the period of the cycle is not a year but could be several years. Some products, such as chocolates for Christmas or Easter, have a demand pattern best described as a 'spike' since demand occurs for only a brief period within the year. This type of data can be analyzed using cyclic techniques.

For this type of data the shape of the variation over the cycle is measured and the shape projected forward to give a set of forecasts for the whole of the next cycle. The shape of the demand during a cycle is measured by ratioing the demand for each period to the average or baseline measured over the cycle. These are called the seasonal factors for each period within the cycle.

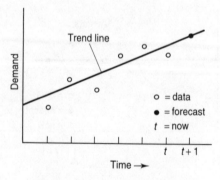

Figure 6.4 *A trend-based forecast.*

This technique requires data for several past cycles in order to determine both the shape of the cyclic variation and whether the overall pattern, ignoring the cycle, is level or has a trend. This can be determined by calculating the average over a whole cycle. Successive averages may show a trend. This is determined by calculating a set of averages over whole cycles. These are known as baseline averages. This technique eliminates the cyclic effect and enables a trend, if any, to be seen as a change in the baseline for successive cycles. The shape of the seasonal variation over a cycle is determined by averaging the seasonal factors for the same period in each cycle. This average seasonal variation is built onto a baseline for the next cycle projected forward from the baselines for each cycle of the data. This baseline projection may be level or show a trend. This process generates a forecast for each period of the next cycle. Figure 6.5 shows the basis of the cyclic forecasting technique. In this figure the variation is seasonal and data is shown for quarters of a year.

Once a set of forecasts has been made for a cycle, the forecast can be compared with actual data as it occurs. Because of the combination of seasonal and random effects it is not easy to detect any change that may occur in the trend of the data. To do this it is possible to remove the seasonal effect and to plot the baseline corresponding to actual data. This process is known as deseasonalization and is done by dividing the actual data by the projected seasonal factor for the period. This will display random variation but will also show a trend if one is present. Alternatively a moving average including data for a complete cycle can be calculated. This technique is used with economic data where, for example, unemployment or inflation is measured. News reports may refer to actual and deseasonalized data. The term 'underlying trend' is used to refer to any trend present in deseasonalized data.

DESEASONALIZATION OF DATA
The availability of deseasonalized data enables politicians to choose either actual or deseasonalized data according to the impression they wish to create. For

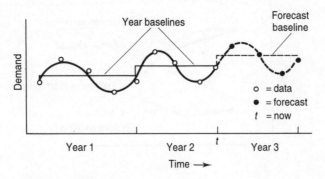

Figure 6.5 *A seasonal forecast.*

example, unemployment may have actually fallen from one month to another. Politicians in power will publicize this. If, however, a fall is expected at the particular period in the cycle and a fall actually occurs which is less than expected, then this might be a random effect or it could indicate that the underlying baseline has risen. Opposition politicians will publicize this view of the data. Thus opposite views can be formed according to the way the data is interpreted. Figure 6.6 shows how this situation can arise.

In Figure 6.6 the data has fallen to its most recent value. However, the data is at that part of the cycle where it is expected to fall. By plotting the deseasonalized data it can be seen that the baseline rises for the most recent data (because the fall in the real data was less than expected).

Both interpretations of this data are correct but they are different because they are measuring different aspects of the data. Actual data is more relevant in the short term but deseasonalized data is more relevant as an indicator of longer-term trends.

Comparative or causal analysis

There are some product markets in which an alternative technique can be used. This is where the demand for an individual product can be compared with, and shown to be caused by, factors external to the organization. Such factors could include the demand for another product, the price of another product or the general level of activity or unemployment in the economy. The technique cannot be used for all products, only those were the link between the two variables can be seen to be related, i.e. only those products where the demand for one product can be seen to be caused by changes in the external factor. These are referred to as causal relationships and are usually analyzed mathematically or graphically as shown in Figures 6.7–6.9.

For the relationship to be of any value in forecasting there needs to be a

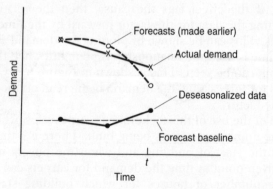

Figure 6.6 *Deseasonalization of seasonal data.*

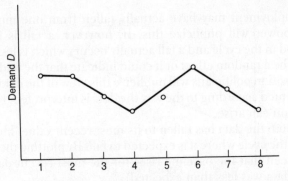

Figure 6.7   *Variation of demand over time.*

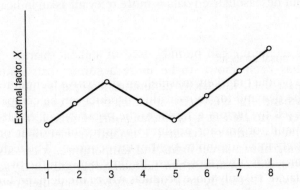

Figure 6.8   *Variation of external factor over time.*

time delay between the cause and effect. In this way data is already available for the causative factor and this can be used to forecast demand in the future. In Figures 6.7 and 6.8 demand does not appear to be closely related to the external factor. However, if the effect lags the cause, then the relationship must be examined by shifting the data for the factor forward by the time lag, in this case three time periods. This can be shown by plotting the demand against the factor. When the same time is used the relationship is poor, but when the time is shifted a close relationship can be seen. This is shown in Figure 6.9. Once established, this graph is used for forecasting, the demand being read off the graph according to the value of the factor.

An example of the use of this technique is carpet sales, which are related to and caused by the number of houses being built. There is a time lag of, say, six months between the start of building and when the buyers move in and buy carpets. At any given point in time the demand for carpets can be forecasted by basing it on the number of houses for which building started six months previously, a figure that is already known and published.

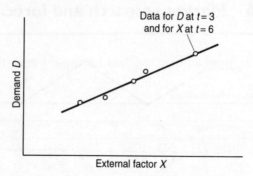

Note: data for *D* at time *t* is plotted against data for *X* at time *t* + 3

Figure 6.9 *Relationship between demand and factor.*

While this technique can be only used in suitable markets, where it has been used, it has been shown to be more accurate than simply projecting demand from past data. Electricity generating companies need to be able to forecast the demand for electricity since electricity cannot be stored. The amount of power generated at a particular point of time has to be planned to match the expected volume of demand. Demand for electricity varies over the day due to working hours, street lights and temperature. It can also vary over a period of minutes, for example when popular programmes or advertisements appear on television. Electricity generating organizations therefore forecast the demand for power in the short term by consulting the weather forecast and the *Radio Times*.

INFORMATION TECHNOLOGY IN MARKETING

Marketing involves the collection and processing of marketing data and making decisions on the basis of this data. Modern information technology can store and analyze marketing data. In market research, answers to questionnaires can be optically read or input directly into laptop computers or hand-held data capture units for downloading into office computers. Data on past sales can be analyzed, as mentioned above, to provide forecasts of future demand. Given the mobility of salespersons, it may be more effective for them to operate from home. With modern information technology people can work from home and remain in touch with the organization. Mobile phones, fax machines and modems to link computers at home to the office are modern information technologies which enable organizations to have some employees out in the marketplace, where they are more effective, without the delays which occurred if the employee only called into the office once a week.

Task sheet 6 Market research and forecasting

Table 6.1 provides the product demand data for these forecasting tasks.

Table 6.1 Data for demand forecasting

	Month											
	1	2	3	4	5	6	7	8	9	10	11	12
Data 1	336	373	333	373	351	332	370	343	330	320	326	339
Data 2	206	200	219	248	260	263	280	274	295	279	311	331
Data 3	344	387	502	581	676	794	1067	1292	1110	775	568	452
Year 2	331	376	518	605	697	850	1052	1284	1148	799	563	490
Data 4	205	339	421	495	467	410	307	206	115	95	105	137
Index	762	797	782	749	696	651	603	589	598	617	642	681

6.1 Plot data 1 and decide whether it is suitable for the use of the moving average technique of forecasting.

For tasks 6.2–6.4 imagine that you have not yet got all the data but only the data up to the point at which you make a forecast.

6.2 Calculate the three-period moving average forecast for periods 4–13 and then plot them. Compare the forecast with the actual figures (those in the data set).

6.3 Calculate the six-period moving average forecast for periods 7–13 and then plot them. Compare the forecast with the actual figures (those in the data set).

6.4 What effect does changing the number of periods in the moving average have on the forecasts?

6.5 Plot data 2 for periods 1–12.

6.6 Graphically estimate the trend line.

6.7 Forecast the demand for period 13.

6.8 Plot data 3.

6.9 Graphically estimate the baseline average for each year.

6.10 Forecast the demand for year 3.

6.11 Plot data 4 and index against time.

6.12 Identify the time lag between the index and the data.

6.13 Plot the data against the index, taking account of the time lag.

6.14 Estimate the equation relating demand to index.

6.15 Forecast the demand for periods 13 and 14.

PART 3

PRODUCT MANAGEMENT

LEARNING OBJECTIVES

3.1   Define 'innovation' and distinguish it from 'invention'.

3.2   Explain the need and significance of innovation for an organization in the long term.

3.3   Give reasons why an organization should develope an innovation strategy.

3.4   Explain the need for clear objectives in innovation.

3.5   List the strategic decisions needing to be taken as part of an innovation strategy.

3.6   Explain the three strategies for innovation.

3.7   Explain the possible organizational changes that an organization may need to make to develop an innovation strategy.

3.8   Explain the purpose of a design brief.

3.9   List the four stages of design.

3.10 Explain the concepts of design.

3.11 Explain the need for 'total design' teams.

3.12 Detail the design strategy options and show how they are derived from customer-based criteria.

3.13 Explain the concept of a product design lifecycle.

3.14 Detail the techniques of design analysis and explain their purpose.

3.15 Explain the procedures and benefits of protecting intellectual property.

3.16 Be aware of the implications of product liability legislation.

3.17 Define quality, reliability and conformance.

3.18 Describe the procedures for quality control by inspection.

3.19 Recognize the benefits of building-in quality assurance.

3.20 Identify the eleven stages of quality assurance.

3.21 Be aware of quality standards such as BS 5750.

3.22 Explain the purpose and benefits of 'quality circles'.

3.23 Describe the philosophy of 'total quality management'.

CHAPTER 7

INNOVATION

INNOVATION IN MANUFACTURING ORGANIZATIONS

What is innovation?

The dictionary definition of innovation is 'something newly introduced such as a new method or device', or as a verb 'the act of innovating'. Innovation therefore involves creativity, the generation of ideas and the transformation of ideas into reality. Creativity is not an easy activity to organize. Innovations do not arise daily and innovative tasks are not routine – they involve detailed analysis and much testing. Equipment can be expensive and the time taken to create a new product can be long and uncertain. For this reason it is important that innovation be effectively managed. In industry the terms product development, research and development, or simply R&D are often used to cover the functions of innovation and design. In Figure 1.1 in Chapter 1 the term 'research' is used to cover these functions. This shows how this function interacts with other functions in the organization. Innovation is concerned with ideas and concepts. Design is concerned with developing the detail of implementing the concept and is covered in Chapter 8.

The difference between discovery and innovation is that discovery finds some previously unknown principle, whereas innovation is the creation of a new marketable product. There is a similar difference between science and engineering. Scientists discover what is already there but engineers create what was never there before. Engineers apply the laws of science discovered by scientists to the development of new products.

> *Engineering is the appliance of science.*

Scientists develop and prove ideas; engineers translate ideas into reality. An industrial R&D function needs both scientists and engineers.

Inventions are new product ideas, but they are not innovations until they

---

This chapter was co-written with Robert Mansfield.

become marketable products. Many more items are invented than ever become successful products in the marketplace. Some inventions are good but are poorly handled. Other inventions have no commercial value and do not progress beyond the drawing board. Britain has had a good reputation for invention but a poor one for innovation, leaving countries like Japan and the United States to profit from British inventions. The television, video recorder and microwave oven were all invented in Britain but none are made by British companies.

The need for innovation

For any organization success depends on its competitive advantage. In many cases competitive advantage comes not from price but from technological innovation. This cannot be guaranteed long term: in time any advantage will be eroded. The latest technology can be duplicated by competitors; system changes can be duplicated by other organizations; product technologies become obsolete; patents expire – the status quo cannot be relied upon.

In a dynamic environment with technological advances occurring at an accelerating pace, rapidly changing patterns of needs, increasing international competition and decreasing lifecycles of products, the ability of an industrial organization to survive depends on its innovative abilities, on its capacity to anticipate new user needs, to recognize technological opportunities, and to develop, launch and market new or radically improved products.

The innovation function cannot be separated from the other functions of management. Figure 1.1 in Chapter 1 shows how the innovation or design function fits in the organization. Innovation is very expensive and must compete for funds with other functions. Innovation must work closely with the marketing function, which will be aware of market wants. The aims of an organization's marketing function must be to increase the share of existing markets, establish new outlets on the home market, sell existing and new products overseas, resist competition, replace products approaching obsolescence, and guard against threats from new technologies. Innovation and marketing together must develop plans for the introduction of new products. Innovation must also interact with manufacturing to determine the likely costs of manufacture.

External links for innovation

The innovation function must have links outside the organization. Staff need to ensure that they are aware of the latest technological developments and scientific discoveries. The organization must ensure that its staff are fully qualified and continuing to develop their technological skills. One way to achieve this is through a link with a university. The interaction between staff in industry and in a university is mutually beneficial. The cost of innovation to the manufacturing organization can be reduced through collaboration with a university since government funding is available for research done at universities. This enables

R&D work to be done at the university which the organization could not afford on its own. One disadvantage to organizations of such collaboration is that developments made must be made public since university staff improve their ego and promotion prospects by publishing academic papers.

Manufacturing organizations wish to keep many developments confidential until they get to a stage at which they can be patented. For this reason some organizations carry out their own research in order to achieve a lead on their competitors. Only large organizations can afford the money and the risk associated with doing their own research. Millions of pounds can be spent on innovation without a useful output. The defence industry has developed many ideas which never came to fruition. In these cases, however, it is the taxpayers' money that is wasted.

Corporate innovation strategy

The importance of innovation for the success of the organization, and the need for it to be fully integrated with the other functions of management shows the need for an organization to develop an innovation strategy or product development strategy. Innovation must interact with marketing and justify its proposals and expenditure to finance.

Innovation often implies a discontinuity from current practice and experience sufficiently great to require an examination of its possible effects on the organization's strategies, structure, skill requirements and attitudes.

Many organizations are preoccupied with their current manufacturing operations without considering the future. Product creation is as crucial as productivity and must be at the centre of a organization's corporate strategy. The ultimate responsibility for innovation and product design must lie with the managing director. Only if the innovation/design function reports at a high level within the organization structure will it have the status and power to be effective and to contribute to the effectiveness of the organization.

To be effective an organization must first decide what its core business is and what its aims and objectives are. To be able to do this, senior executives need an appreciation of the innovation process and the technological capabilities of their organization. They also need to decide which markets, market position and prime competitive advantage the organization is seeking. It is important to determine the correct combination of growth and profitability to satisfy short-term funding needs and also to achieve long-term growth. This process must be continuous. Objectives need to be regularly reviewed and amended in response to market and competitor developments.

One of the main strategic decisions for an organization is whether to specialize or diversify its product range. Specialist organizations can become more efficient by concentrating on one area of skill and reduce the cost and range of machines, tooling, etc. However, this can lead to disaster if technology or the market changes. Diversification spreads the risk but can lead to problems if management try to operate in unfamiliar sectors. The advantage of diversification is the reduction of risk to the organization from the collapse of any one market sector.

Another strategic decision is the level or extent of product innovation:

- Basic (technical breakthrough)
- Incremental (improvement innovations)
- Adapted adoptions (improvements on copies)
- Pure adoptions (straight copying)

The next strategic decision is the class of product innovation:

- Products new to the market (product leadership).
- Products new for the company (market share).
- Improved company products (market expansion).

A further important strategic decision is whether to be an innovation leader or a follower.

Technology push

An organization adopting a proactive strategy of 'basic' innovation and 'product leadership' is following a strategy of high risk and possibly high rewards. By considerable investment in technological drive and breakthrough they are producing products new to the market for a demand that does not yet exist. The development of new technology enables the creation of the product. The product is being pushed into the market in the hope that it will create its own demand. Large profits can be made before competitors can catch up since initially the innovative organization is the only one with the product in the marketplace. By the time competitors enter the market the development costs will have been recovered, at least to some extent. Prices can be reduced at this point in order to undercut the competitors. Examples of the use this form of innovation are the introduction of microwave ovens, mobile telephones, video cameras and recorders, nickel-cadmium batteries and many other electronic based products.

Technological innovation can be an order-winning criterion and some organizations wish to be technology leaders. Others may prefer to be followers, investing little in innovation and then moving into markets created by others. The investment risk is low but so is the return.

Market pull

Where an organization detects, by market research, the existence of an unsatisfied demand it can respond by adopting any of the other strategic levels and classes of product innovation. Products are pulled into the external world by market demands. This strategy is reactive and therefore less risky. However, the rewards are often less because competitors will also have detected the opportunity. An organization must be able to respond rapidly before its competitors can enter the market, forcing down the price. Changing fashion can also reduce the demand before products reach the market. Examples of this form of innovation include disposable nappies. The change in mobile telephones from analog to digital is a response to the demand from the market to prevent eavesdropping on such telephones.

THE PROCESS OF INNOVATION

Innovation is not a routine day-by-day activity; it is a creative process that requires inspiration. The way of thinking about routine activities is not adequate for creative activities, and different ways of thinking are required. The process of innovation requires innovative thinking. Systemic thinking, introduced in Chapter 3, is a framework within which to develop styles of thinking for innovation.

Creative thinking

Lateral thinking

Edward de Bono has written many books in the field of thinking and in particular creative thinking. His book *Lateral Thinking* develops a style of thinking that expands beyond the normal deductive process and provides a base for creative thinking. His books *Six Thinking Hats* and *Six Action Shoes* develop styles of thinking and action that match the different situations which managers find in organizations. These forms of thinking are of value not only in developing skills needed in managing the creative processes of innovation, but also as innovative styles of management for the other functional areas of management in an organization.

Delphi technique

The Delphi technique is based on deriving a consensus view from the views of a group of people who are acknowledged authorities in the field. The group do not discuss, debate or vote as a group, but their views as individuals are sought by a co-ordinator or facilitator. They do not need to meet; the process can be carried out by writing to them. It would be possible to take the view with the greatest support at this stage but it is unlikely that this view will have the support of the majority. The Delphi technique is iterative and goes to further stages using a process not unlike that used in multi-stage voting processes.

The authorities are told at this stage that some poorly supported ideas are to be dropped and are told the ideas which got support. The amount of support is not given since that may lead to a 'bandwagon' effect. Views are sought again with this further information. The process is continued until an idea carries majority support. The technique carries no guarantee that the idea derived is the best, only that it is likely to be the best because it is based on the consensus of a group of experts in the field.

Brainstorming

Brainstorming is a technique for use at an early stage of a creative process. It involves a group of people brought together to solve a problem. Ideas are raised by individuals, in no formal order, and written or drawn on a whiteboard where all can see them. The process is synergetic because ideas from one member of the group will stimulate other ideas in other members of the group which would not have occurred to them had they been thinking about the problem of their own. Brainstorming is a way of making the logical jumps which are necessary to get people out of thinking traps caused by their educational experience and the organizational environment they work in.

All ideas are recorded, including those that appear to be silly. If they are indeed silly they can be rejected at a later stage, but in the meantime silly ideas may inspire other people in the group and trigger other ideas that may not be silly. This may prove to be the route to an effective innovation and one that would not have occurred if silly ideas had been rejected immediately.

The Japanese engineer Shingo has related an example which illustrates the value of brainstorming. The process of die-casting involves injecting liquid metal into a die cavity made by two die halves which are then separated to extract the then solid component. A unwanted protrusion of metal 'always' occurs on the component where the two halves of the die come together. Brainstorming looked at ways of dealing with this protrusion. It could be removed by grinding, milling or hand filing. They were trapped into looking at cures. No one looked at the cause or investigated methods by which it could be prevented rather than cured. Soon it was recognized that protrusions were caused by the pressure in the die cavity caused by forcing metal into the cavity, which already contained air. This pressure forced the die halves apart, allowing metal to bulge out of the cavity. Solutions then looked at increasing the pressure holding the dies together. Eventually the idea arose of evacuating the air from the cavity to reduce the pressure. Vacuum die-cast components no longer have the protrusions which were once thought to be inevitable. The problem was prevented.

ORGANIZATION OF INNOVATION

Skills needed for innovation

The processes of innovation requires creative thinking and hence needs the employment of creative thinkers. Being a process of change, positive rather than negative feedback is required.

Ironically, good innovators are rarely good managers and so promoting good innovators to be managers can often be unsuccessful. In Japan only young people are employed in innovation since young people are often more creative. After the age of thirty innovators are moved into other functions where their innovative skills are valuable, such as manufacturing.

For a company new to innovation, part of the corporate strategy may be to bring about a change in the culture of the organization and restructure the links between departments. Successful innovation management demands integration of the innovation and design processes with the organization's other functions from manufacturing through purchasing to finance. This requires a more flexible approach than found in the traditional bureaucratic organizational structure with its rigid specialization and hierarchy.

Creative thinkers are not easy to organize. Creative thinking takes time to develop, and needs a suitable organizational culture for it to be effective. The nature of their work requires that creative thinkers work on their own initiative. The duration of tasks or time to get a successful result cannot be measured or predicted with any accuracy. Such employees are therefore left on their own. There is still, however, a need to ensure that their time is used effectively and that their cost is justified by the results they achieve.

Organization structure for innovation

There is a need for strong management of the innovation process by the setting of clear objectives and then ensuring that resources are provided to enable these to be achieved. The head of the innovation function must operate at a sufficiently senior level within the organization in order to have the independence and authority necessary for success. Innovation can best be organized as a set of projects. These give some structure and direction to the process without stifling initiative.

An innovation project which is properly managed from start to finish is far more likely to succeed than one where the initial objectives are vague and where management responsibilities are unclear. On average, 80 per cent of the costs of a product over its total manufacturing lifecycle have been committed by the end of the innovation or concept stage of the design process. If this is not managed properly a lot of money can be lost, and the organization can lose its reputation and competitive advantage in the marketplace.

Research and development managers, for their part, must ensure that the design engineer, the manufacturing engineer, the accountant and the marketing department all work together to produce a product that is technically sound, easily and cheaply manufactured and, most importantly, is what the customer wants.

Project organization for innovation

The organization of innovation needs to be different from that for routine activities, but an organization needs effective management of innovation to ensure that money is not wasted. Project management is considered to be the most appropriate form of organization for the innovation function. Project management is developed in Chapter 14.

Task or project groups seek to bring together the right people at the right level in the organization with the appropriate resources and to let them get on with it. This is a form of matrix organization referred to in Chapter 4. It is extremely adaptable and able to respond rapidly to changes in the market or the environment. It is a common form of organization for the innovation function since the nature of the work is project based, each project having a limited duration before the team goes on to another project. Individual members of the project team are brought in or moved out to other projects according to the phase of the project. This provides a flexible team needed for effective innovation. An organization involved in innovation may have several new product design projects ongoing at a particular point in time.

Funding for innovation

Corporate strategy must consider the capital requirements and selection of sources of finance. Innovation is expensive. Funds are required for market research, project evaluation, R&D effort, acquisition of technology, provision of necessary manufacturing facilities and initial marketing activities. The use of an organization's own capital reserves can lead to cash-flow problems. The issue of further shares to finance R&D will rarely be successful since shareholders will be hesitant to inject further money for the 'risky' expenditure that is involved. Loans and so-called 'venture capital' are easier to obtain but cost more and carry the risk that the funds may be withdrawn if delays or setbacks occur.

Senior management must create the right climate for innovation by providing sufficient resources to match the product strategy, the opportunity for pure research and appreciation of work done. Individuals employed in research must be given latitude to follow their own ideas and incentives for creativity; they also need a sympathetic management attitude to the inevitable mistakes and false trails.

Senior management must ensure that up-to-date research information about market requirements is made available to the design team, that the collaborative information and evaluation links between the design team and other parts of the organization are operating properly, that the organizational strategies and procedures for managing the design team process are adequate, and that there is a sincere and visible corporate commitment to high standards of product design.

FROM INNOVATION TO DESIGN

The major function of innovation is the development of ideas for marketable products. Once the idea is accepted and approved within the organization the second stage is the detailed design of the product. The output from the innovation stage to the design stage is the *design brief*. This should provide an unambiguous reference guide to the design process. It will identify the market, the concept design and the financial resources required. As such, the design brief is critical to future organization performance and success.

The design brief

Market

- Evidence of actual or potential customer demand.
- Markets and market segments aimed at.
- Advantages over competing products.
- Market share targets against competition.
- Special or unique features of the product.

Specification

- Basic performance requirements.
- Target costs and selling price.
- Compatibility with existing products.
- Guidelines on appearance, style and image.
- Relevant legislation, standards, code of practice.
- Requirements for reliability and durability.
- Recommended materials, components, quality of finish.
- Requirements for ergonomics and safety.
- Use of standard components and assemblies.
- Production requirements and constraints.
- Packaging, maintenance and servicing requirements.

Time and cost

- Project budget, timetable and launch date.
- Potential for future evolution.

Resources needed for design

- In-house design skills.
- External design expertise.
- Equipment.
- Design management.

LAW IN INNOVATION AND DESIGNS

In recent years the term 'intellectual property' has been coined to include not only patents, trade marks and designs but also 'know-how' and trade secrets.

Patents

Letters patent have existed since the fifteenth century, the first known patent being granted in Florence in 1421. However, it was not until 1623 that they were enshrined in legislation in Britain by the State of Monopolies Act. This act prohibited monopolies but permitted, as an exception, the grant of a patent. A patent is a grant by the state of a monopoly right to the exclusive use of an invention for a maximum of twenty years in return for a disclosure to the public of the invention and how it was put into practice. Before the introduction of patents many inventions were kept secret and then lost on the death of the inventor. The patent library provides a resource of ideas available for anyone, including competitors, to examine. They cannot use the idea directly but can use it as a starting point for their own development, eventually rendering the original patent obsolete.

A patent is a property which, like any other business commodity, may be bought, sold, hired or licensed. If a patentee does not wish to exploit the patent personally, someone else can be licensed to exploit it on payment to the patentee.

A patent has a value that forms part of the assets of an organization. Virtually all machines, products and processes (and the individual parts of them) across the industrial spectrum are patentable provided they satisfy three criteria. They must be new, they must be inventive and they must be capable of industrial application. The word 'industrial' includes agricultural and horticultural.

Patents need to be prepared by an expert skilled in the archaic legalistic language and terminology required to give the degree of flexibility and protection required. Patents are expensive to obtain, and in all countries, except the United States and Canada, renewal fees have to be paid to keep the patent in force over a period of time. These vary greatly in amount but they are generally payable annually and increase with the life of the patent.

Patents are also expensive to defend through the legal system. The infringement of a patent is covered by civil law not criminal law. The claim for damages and an injunction to stop the infringer from doing it again must be brought to the High Court by the plaintiff at his or her own expense. The small inventor is therefore at a disadvantage against the multinational organization which can afford the best patent barristers and the protracted legal process. On a successful outcome the inventor may be granted costs against the defendant. However, the inventor may never receive them without yet another court case and the infringer may long since have gone bankrupt.

National patents granted by the British patent office are only effective in the United Kingdom. Patents must be taken out in every country where protection is required. However, procedures are developing within the European Union for patents that will be valid throughout the Union. A European Patent Office has been established in Munich but the European Patent Convention has not yet been ratified by all member states. European patents are harder and more expensive to obtain as the search for prior similar patents is extended to all participating countries.

An invention can be exploited without a patent provided no one else already has rights protecting that invention or a part of it. Such exploitation would prevent anyone, including the inventor, subsequently getting a valid patent for it, but would not prevent others from copying it. It is therefore vital not to disclose an invention to anyone or to make commercial use of it before an application is made for a patent. Prior disclosure could prevent the grant of a patent or could invalidate a subsequently granted patent. If disclosure is necessary, care should be taken to ensure it is in the strictest confidence. A signed confidentiality agreement is essential to prove that it was not the intention to put it in the public domain.

Registered designs and trade marks

Trade marks can be registered. This gives protection to an organization and prevents other organizations from using words or diagrams in an attempt to pass off their products as someone else's. This practice was not uncommon in the past, particularly by developing nations copying the goods and packaging of other nations. Such international activities are not easy to prevent, as there is no effective, enforceable international law.

Designs too can be registered to prevent copying. The pattern or ornamentation on wallpaper or other two-dimensional objects can be registered. For three-dimensional objects the particular design, including the shape and configuration, can be registered. There are, however, limitations to the protection which registration can provide.

In a case concerning car exhaust pipes, a car manufacturer sought to prevent competitors from making such items for sale in the replacement market. The competing product had a different design but in order to fit onto a particular car it had to have a particular shape. The manufacturer argued that this was the same shape as their original product and thus was protected. The decision, however, was that the product had to have that shape in order to fit and was not simply a copy of the shape. To have denied the competitor the right to make his product would have been an unfair restraint on competition.

Copyright

Computer programs may be called products by those who create them but they are not products for the purposes of patent law. They are, however, works of literature! They can therefore be protected by the law of copyright, which provides similar protection. Drawings of products and components are considered as works of art and can also be protected by the law of copyright.

INFORMATION TECHNOLOGY IN INNOVATION

Good innovation depends on good information. The availability of new materials or process and new trends in the marketplace need to be absorbed. Modern information technology can assist in keeping innovators abreast of modern developments that may give an organization a competitive advantage. The main form of IT of value in innovation is the CD-ROM.

Academic journals, periodicals and library catalogs are now becoming available on CD-ROM. These are automatically indexed and can be accessed by keyword searches. This offers an innovator rapid access to selected data without the need personally to scan all the periodicals. Patent library data is also available in this form, saving much time on patent searches.

Task sheet 7   Innovation

7.1   In groups of about five undertake a brainstorming exercise. Use some of the group to observe the process while others participate directly in the discussion. The topic can be something relevant to your current studies in which all the group are involved.

7.2   After the exercise, discuss with the observer how the group worked and how it achieved its solution.

If a suitable problem cannot be found, consider the following:

How can car drivers be continuously or intermittently informed of the speed limit on the road on which they are travelling?

This is a problem since it is unreasonable to expect a driver to remember the speed limit on the basis of a single sign at a point of change of the speed limit.

CHAPTER 8

DESIGN MANAGEMENT

DESIGN

What is design?

Design is the detail stage of taking a concept design from the innovation stage and converting it into a detailed design of a specific product that can be sold at a profit. The emphasis is on the specific shape, form, material, process and tooling requirements for manufacturing each component and, where required, assembly into a product. The interest here is on the design of the product. Chapter 11 considers the design of the manufacturing system to make the product. Innovation and design are in most cases in the same functional department within an organization, the alternative name, research and development, referring to these two stages of the total innovation design process. Figure 1.1 in Chapter 1 shows how the R&D department fits into the organization.

Design is a blend of engineering and art that is applied in electronics, mechanical engineering, architecture, construction, furniture, packaging, graphics and fashion.

Mechanical design engineers are concerned with how things work. They design things to achieve a function and retain that function during the working life of a product in use.

Manufacturing design engineers are concerned with how things are made. They are aware of the manufacturing processes available and their relative costs and advantages. They influence design so that parts can be made more efficiently and specify the tolerances within which parts must be made.

Industrial designers are concerned with appearance, aesthetics and styling. All design engineers must interact with each other to design products that work well, can be made at a profit and appear attractive to customers, especially in the consumer goods industry.

Graphic designers work with the *packaging designers* to prepare what is often the first selling point – the packaging.

This chapter was co-written with Robert Mansfield.

123

Many organizations are now breaking down the demarcation barriers between the various design disciplines by setting up total design teams to consider all aspects of the design. Designers should therefore have an appreciation of the work of different departments in an organization such as manufacturing engineering, product planning and marketing, and an understanding of the interactions between them. In this way symbiotic relationships can develop.

In-house designers must be kept up to date. Training in new skills must be provided where possible. Outside consultants, university research departments, trade research associations and specialist institutes can be employed. Suppliers are a valuable source of expertise in the application of their products within your design. Trade exhibitions provide opportunities to pick up ideas and to view competitors' designs. Market research (dedicated and general), customer surveys, consumer magazines and technical journals are all valuable sources of information.

Marketing input to design

Design must interact with marketing. Together they can develop detailed ideas of design and packaging that will attract customers. The real challenge in design is to know the customers needs and wants, and to understand the real order-winning factors that cause people to buy a particular product. Customers have different criteria at different stages, all of which need to be considered in the design of a product.

- *Before purchase* Potential customers compare manufacturers' specifications, advertised performance, test results, appearance and list price. They may have seen the product in use by a friend and asked for an opinion of it (brochure characteristics).
- *Purchase* At the point of sale customers will see the overall design and quality, special features, materials, colour and finish. First impressions of performance, and purchase price are paramount (showroom characteristics).

Having bought the product, the criteria that generate brand loyalty and recommendation are as follows:

- *Initial use* Actual performance, ease of use, safety, etc. (performance characteristics).
- *Long-term use* Reliability, ease of maintenance, durability and availability of service and spare parts (value characteristics).

Design output to manufacturing

The major output of the design process is a design specification for the product. This specifies in detail for each component the material to be used, the processes to be performed and the result required. This design data takes the form of drawings, parts lists and process data. Process data includes a list of operations and for each, the type of process, the required size, the limits within which the size must be and the time it should take to make. Further details of this data are included in Chapter 10 since they form the starting point for managing manufacturing.

Information technology in design

The process of design can be more efficient and better integrated with the rest of the organization if use is made of modern computer technology. Computer-aided design (CAD) is the key to this. Using CAD, the process of drawing can be done on-screen. Initially this may be time consuming, but an organization soon builds up a library of component designs that can quickly be modified as required. With CAD, designs can be displayed and considered aesthetically.

The real benefit of CAD goes far beyond computer-based drawing. Strength and mechanical characteristics can be analyzed using finite element analysis. Data about the shape of the component can be used to design the tools, jigs and fixtures and to programme the machines to make it. Parts lists and process lists can all be produced as a part of the design process and stored on computer. Operation time data for machine processes can be generated. This data can be passed to the manufacturing department or to a subcontractor by downloading directly to their computer. Linking CAD with computer-aided manufacturing (CAM) enables CAD data to be used to programme and operate the machines used to make the components. These machines may be at a different site from the design, or even in a different country.

DESIGN STRATEGY

Part of the function of design management is to decide on the design strategy for the organization. There are three main strategies for managing design and linking design with other functions of management.

Sequential design

Research → Development → Design → Manufacture

This is a straight-line, 'open-loop' process. It usually means that manufacturing costs are high due to excessive process times and number of operations because manufacturing was not considered at the design stage. Also, because the design was not tested, the product does not meet the market needs and may not sell very well. The timescale from innovation to the market is also very lengthy. In the recent past it took seven years for a British motor manufacturer to bring a new product to market; a Japanese manufacturer could achieve this in less than three years.

This traditional approach to product development is sequential. Each stage of work is completed before going to the next stage. This is not only very time consuming but it prevents feedback from or the involvement of the staff at different stages of the process. It is easy to see why it took seven years to develop a new car and eleven years for an aero-engine. This has led to greater success for those who can reduce this design lead time and get their new products to market faster. Figure 8.1 shows the timescale required for this method of product development.

Iterative design

$$\text{Design} \rightarrow \text{Test} \rightarrow \text{Redesign} \rightarrow \text{Test} \rightarrow \text{Redesign}$$

This is a 'closed-loop' process with constant feedback within the design process. This results in good, well-designed products, though these may be too late for the market because of the long lead times and too expensive because of continual designing, testing and redesigning. It used to take seven years to get a new car from the concept stage, through design to manufacturing and sales.

Simultaneous design

An alternative approach to design is a process of simultaneous design (see Figure 8.2). This approach is also known as *simultaneous engineering* or *concurrent engineering*. It differs from the previous strategies in that as soon as enough information is generated at one stage, the next stage starts without waiting for the

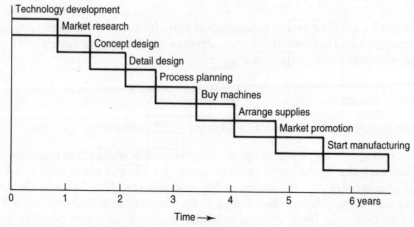

Figure 8.1   *Traditional product development.*

completion of the previous stage. This means that design activities proceed in parallel (concurrently) rather than in series (consecutively). In this way the timespan for introducing a new car is now down to five years in Europe and less in Japan. Simultaneous design also ensures that the different aspects of design are developed together rather than sequentially. In this way not only is the design throughput time significantly reduced, but a better design is achieved first time. This approach requires feedback and iteration between different functions, and so liaison between the different functions is of paramount importance. Information must be communicated to all who need it as soon as it is available. Computer-integrated manufacturing with a common database is of great benefit in this approach. By reducing the design throughput time an organization can make a rapid response to market needs and put new products onto the market ahead of competitors.

DESIGN MANAGEMENT

While it is the function of senior management to plan, motivate and review, and to create a structure and framework for the integration of the different functions of management, it is the function of design management to manage the concept embodiment and detail of the design that matches and contributes to corporate objectives.

Design must be effective technically and economically. Designers must be familiar with manufacturing and with costing. Design is normally managed as a set of projects, using a project management approach, as explained in Chapter 14, together with project appraisal, covered in Chapter 21. Costing is covered in Chapter 18. Design activities need to be efficient and well organized. Techniques developed in Chapters 12 and 13 can assist in doing this.

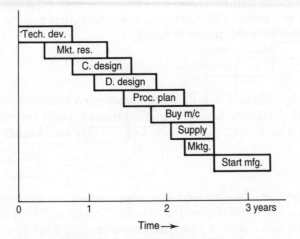

Figure 8.2 *Simultaneous engineering.*

DESIGN CONCEPTS

Simplification

This concept of design is based on the development of the simplest way in which a product can achieve its function. It views anything else as an unnecessary and costly complexity.

Standardization

This is a standard for a product, range of products or procedure which is agreed across an industry. This may be a standard of performance, testing, method of manufacture, composition or dimension. Examples include the now standard size of audio and video cassettes and the standard sizes of screw threads. The objective is to facilitate interchangeability between parts and reduce costs by reduction of stocks, tooling, materials, components and overheads. Both manufacturers and customers benefit from standardization. However, standardization can slow down beneficial design changes and the introduction of innovation. It reduces customer choice. The quest for standardization can lead to the use of unsuitable items or materials. Redundant stocks may be created during the transition period.

Variety reduction

This approach seeks to reduce variety on the grounds that excess variety costs money. A wide range of products and components increases the complexity of a manufacturing system and often results in small batches and lost production time due to excessive changeovers from the manufacture of one component to another. By using, where feasible, common components across a range of products the variety of components can be reduced. This saves time in design as well as in manufacture. It also reduces the variety of spare parts that need to be held by an organization involved in the repair industry.

Compatibility

Products made by different manufacturers may need to be capable of working together without alteration, e.g. computers, printers, audio and video cassettes. Compatibility is a form of standardization but one that goes further than normal.

Modularity

Modularity is the use of a standard size, shape or unit of measurement which enables different patterns of configuration to be constructed. For example, a brick is a modular unit and a wall is a modular construction. Different products can be made efficiently from standard modular components. A common example of this is hi-fi audio equipment in which separate modules are bought and fitted together to form a whole which is tailor made to the customer's own requirements.

Design for function

This part of the design ensures that a component functions as it is intended to, i.e. has the characteristics specified for it in the design specification. These may be mechanical, electrical or other engineering performance characteristics.

Design for appearance

Part of the reason that a product sells is its appearance. The aesthetic design must achieve this. Fashion and styling are part of this process.

Design for quality

The quality of a product must be considered at the design stage. Quality is often a significant order-winning criterion. Once a product has been designed and poor quality is found to result, it is too late and expensive to recover from the reputation and lost sales which may occur. Only by building in the requisite quality can organizations avoid such disasters. Quality is considered in more detail in Chapter 9.

Design for reliability

Reliability refers to the expected life of a component before it fails. The design of a product must build in its reliability. Reliability analysis is covered later in this chapter. This does not mean that all products have to be designed for the highest level of reliability or that all components in a product must have the same reliability. Such products would be expensive. They must be designed for the reliability appropriate to the market in which the product is to be sold.

Design for safety

The safety of the product in use must be considered at the design stage. Special care must be taken with products such as toys for young children, or any product that could be dangerous to children. Young children have a tendency to put things into their mouth. Eyes on teddy bears used to be made of glass and metal and were swallowed at an alarming rate, sometimes with fatal consequences. These days they are made of fabric so they are less likely to be chewed off. Child-proof car door locks and medicine bottles are examples of modern design for child safety.

Product liability legislation, referred to in Chapter 7, makes manufacturers liable for injury to users caused by their products. Design for safety is therefore important in preventing such problems.

Fail-safe design

A different aspect of design for safety is design of a product in such a way that when a failure occurs, the product cannot be operated or fails in a safe way. This is called fail-safe design. Modern trains are designed so they cannot be moved if any of the doors are open. Power-assisted brakes on goods vehicles usually operate with compressed air applying the force. Sometimes a part of the compressed air system fails. If the design was such that compressed air was required to put the brakes on, then the vehicle would be unable to stop in such a situation. This would lead to a dangerous situation. In practice, therefore, the compressed air is used to hold the brakes off the brake drum or disc against the force of a mechanical spring. If the air system fails, the brakes are automatically applied by the force of the spring. This is a fail-safe design.

Design for manufacture

This aspect of design is an attempt to ensure that the design of a component is not only one which achieves the function but also one which achieves this function in a way that can also be manufactured efficiently. Good design enables a component to be made using fewer processes and to be easily held in a simple, inexpensive fixture while being machined. Design tolerances must be specified – these are needed to achieve quality.

Design tolerances

Because process variation is inevitable, it is important for a design specification to state how much variation is permitted or tolerable within a component. This is known as the design tolerance. Since close, or tight, tolerances, with a narrow range of permitted variation, tend to cost more, tolerances should only be as close as they need to be to satisfy the function. For components that have to fit together, the tolerances on each must be such that they will in fact fit together. With loose tolerances on such components it is likely that not all pairs of components will fit together. This will cause difficulties, delays and extra costs in manufacture. For automated manufacture using robots for assembly it is essential that components that are to be fitted together do in fact fit together. Quality assurance procedures are needed to ensure that this happens. The relationship between design tolerance and process capability is covered in Chapter 9. Entire manufacturing systems can be stopped and damage done if a robot tries to assemble components that do not fit together.

Some products require very close tolerances. Aerospace components and the machines that process them are precision components and precision machines. For many products, however, less precise tolerances are acceptable.

It is an important part of the design process to specify suitable tolerances. This requires an understanding of the tolerances required by the customer and by the assembly processes and also of the variation inherent in a process. A process must be selected that is capable of achieving the required quality of output within the design tolerance. Machines must then be adequately maintained to ensure that they continue to provide this quality of output. This aspect is taken further in Chapter 9.

Design for maintenance

The product should be designed to be maintained efficiently. This aspect must be built-in at the design stage. Modern cars fit a lot of equipment into the engine area. During manufacture, many components are assembled onto the main body of the engine. This assembly is then lowered into the engine area. When the user comes to replace, for example, an alternator, the question arises as to whether the alternator can be removed and replaced without removing the whole engine. In some cases such replacements are not possible without removing the engine, an expensive process that requires lifting equipment.

Design for distribution

The design of a product must include some consideration of how the product is to be distributed. The product must withstand being transported, perhaps on the back of a lorry on a bumpy road. Packing in polystyrene can help, but the product itself must not be damaged by any transportation that can reasonably be expected.

Design for operating environment

Products must be designed to operate in the physical environment they are expected to operate in. This includes the extremes of temperature and humidity. For example, it might be better for handling equipment operating in a dirty factory in wet conditions to be designed to use pneumatic rather than electric power since the latter may be damaged by that environment.

Design for disposal

Disposal of products causes concern in a modern society in terms of pollution and litter caused by the disposal of toxic and non-biodegradable products. There is also concern over limited resources and over pollution caused by the energy required to make manufactured products. Energy and resource requirements can be reduced if products are recycled. Some countries have passed legislation requiring products to be recyclable. Some car manufacturers now claim that their products are 80 per cent recyclable. Achieving this requires thought at the design stage, particularly in choice of materials.

Optimal design

All of the aspects of a product explained above must be considered when the product is being designed. A balance must be drawn between all of these demands and a design produced which satisfies those features required in a product at minimum cost. This can be described as the optimal design, and is achieved by a balance between the cost of quality of the design to the manufacturer and its value to the consumer.

PRODUCT DIFFERENTIATION

This is a way of meeting the demand from different market segments with substantially the same product. It seeks to provide variety, but in a controlled and cost-effective manner. Motor car manufacturers make several different models of car using some common components, but for a given model there is a range of differentiated products that use a large proportion of common components. Each product variant is positioned in a particular niche or segment of the market. Examples of this are also seen in the car market. There are many versions of a car of the same model. These versions provide different features within the same model to cater for differing requirements at different prices. Manufacturing is efficient because the model as a whole is made in large volumes. Product differentiation provides a variety of products but without an excess variety of components.

Methods of achieving this are as follows:

- Cost reduction with lower technical performance but with special capability. For example, personal use, compactness, portability/pocket-sizing or add-on to existing systems.
- Quality at a given level of cost. For example, appearance, finish, general handling, user friendliness, reliability, lifecycle cost.
- Improved value at a given cost level. For example, variety in finishes, add-ons, customer group specificity, quality of user experience, user lifestyle, generic tailoring.
- Improved technical performance at a given cost level. For example, power, speed, extra functions.
- Enhanced performance and value, at an enhanced price. For example, ruggedness, extra reliability, resource economy, built-in response to specific user needs, user lifestyle specific tailoring.

THE PRODUCT DESIGN LIFECYCLE

The emphasis of design changes over the life of a product and is linked to the marketing concept of a product lifecycle.

Introductory phase

The primary emphasis here is on technical/functional design, imaginative leaps and rapid change, competing design philosophies and strategies. No standards are in place and so each organization develops its own. Secrecy is likely at this stage as organizations seek to develop a product ahead of their competitors. At this stage of design some disasters are likely.

Growth phase

Now primary emphasis is on manufacturing processes, including new developments in manufacture. There is customer pressure for standardization but organizations seek flexibility to outdo emerging competitors.

Saturation phase

Technical change is still rapid but there is an increasing emphasis on human design. There is pressure on the cost of design and its effectiveness. Pressure for standardization increases.

Post-maturity phase

The emphasis is now on changes of 'model' and minor improvements of the basic product. These methods seek to extend the life of the product. Products become more differentiated, some remaining a basic product at a basic price, others providing features that some customers want and which add value to the product for the manufacturer. There is also an emphasis on user appeal and on inexpensive ways of achieving competitive advantage. There is some evidence of user redesign, e.g. users customizing a car to their own wants.

DESIGN ANALYSIS

Value engineering

This is an approach to design using the technique of value analysis, which seeks to discover the most economic way of performing the function of each part of a manufactured product. In this way the product should achieve its function at a minimum cost. It is an example of the application of a systems approach to product design. It is more a discipline and attitude of mind rather than a scientific formula. Management must play a large part in developing the programme and create the co-operation that is essential for its success.

Value engineering starts with the function, then considers the cost. It defines the function of each item or part, and then considers alternative ways of performing the same function. Each alternative is costed. If at this stage no alternative is significantly cheaper, then either more alternatives are sought or the analysis is complete. This approach is based on the identification of the added-value and non-added-value activities in the manufacture of a component at the design stage and basing the design on the elimination of the non-added value activities.

Product ergonomics

Product ergonomics is the systemic analysis of the psychological and physiological requirements for a product from a human point of view. It is used to ensure that products are designed to fit as well as possible to human capabilities and limitations. One of the aspects addressed by ergonomics is the safety of a product in use. Safety and comfort will be enhanced if the design of a product conforms to ergonomic principles.

Anthropometrics

People come in different shapes and sizes. An understanding not only of the range of size and weight of people but also the ranges of reach and strength are required in order to design a product, or range of sizes of product, to suit this variety in customers. Anthropometric data – a measure of this variety – is available. Anthropometric data is of course different for men and women and also differs between ethnic groups. The most obvious examples of the need for products to account for the size of users are cloth sizes and the adjustment of the seat position in a car to suit the different leg lengths of drivers and so enable people of all sizes to reach the foot control pedals. The differences of size between people of different ethnic groups may need to be taken into account in design. A military plane designed in Italy was used by NATO forces. Norwegian pilots could not get into it. Norwegians are on average taller than Italians. Similarly, East Asians are generally smaller than Europeans.

Some Japanese cars are not just small cars, they are cars for small people.

An analysis of the design of a medium-sized lathe discovered that the ideal size of person required to operate it effectively needed to have a height of 1.2 m and an armspan of 3 m. Orang-utans are of a suitable shape for such work but not humans! Modern machine design and electronic technology are overcoming some of these problems.

Where people have to work with poorly designed equipment they may work very hard but they will be very ineffective and productivity will be low. A mismatch between the design of a product and the people using it causes accidents and illnesses due to heavy lifting and bad posture, which can cause backache. Modern machine tool design has eliminated many of these problems, but in Britain most of these products are imported and many poorly designed machines are still in operation.

It is not only size that varies. Physical strength varies, and men are on average stronger than women. Product design must take account of the strength of users. As vehicle size increases, so the strength required to turn the steering wheel and to apply the brakes increases. For this reason in the past women rarely drove buses or heavy goods vehicles. Women drivers would tire their muscles more quickly and be less able to cope with an emergency due to having less strength. Modern goods vehicles and most large cars have power steering and power-assisted braking. This reduces the strength requirements for operation. Such vehicles reduce muscular effort and tiredness for drivers of both sexes and are safer for this reason. This form of design enables women to do jobs previously done only by men because of the high level of physical effort previously required.

Motion stereotypes

A further aspect of ergonomic design is that products should conform to conventional expectations of movement known as motion stereotypes. In a technological society people grow up with a set of conventions which then become programmed in the mind. For example, on an analog clock or watch the hands always rotate in the same direction, giving rise to the term 'clockwise' for the direction of rotation used on clocks. A tap in a wash-basin is always turned anticlockwise to turn it on and to increase the flow. This arises from the convention that, except in rare specialist circumstances, screw threads are always right-handed spirals. This means that a clockwise rotation causes a movement away from the operator and an anticlockwise rotation causes a movement towards the operator. In the case of a tap or valve the anticlockwise rotation lifts the tap washer off the seal and permits the flow of water or other fluid.

Interestingly, the convention for rotary knobs on electrical equipment is the opposite of that on hydraulic equipment. On electrical equipment knobs are turned clockwise to turn the equipment on and to increase the signal. Similarly, in all instrumentation based on a hand sweeping across a marked scale, the hand will move to the right or rotate clockwise as the signal is increased. Car speedometers always rotate clockwise with increasing speed.

The significance of these concepts for design is that these conventions are preprogrammed in the minds of users. Any equipment that uses a different convention is likely to be misused by the user, not out of carelessness but because of the expectation of a conformance with convention.

Most motor cars now have their controls for lights, indicators and windscreen wipers on the same side of the steering column as other manufacturers. Anyone who has changed from an old car in which this was not the case will find themselves putting on the wipers instead of the indicators because they are preprogrammed by the experience of their older car. This difficulty should not now occur as positioning is standardized and the new arrangement has become the convention. This standardized positioning of controls is now the same in both left- and right-hand drive vehicles. If the design suits right-hand drive cars, then it will not suit left-hand drive cars as the work load of the left and right hands will be different because of the opposite positions of the gear lever and hand brake.

In complex equipment such as aircraft there is a large array of instruments and controls for the pilot to use. Some aeroplane crashes have been attributed to non-conventional instrumentation. While pilots are extensively trained to operate each type of aircraft, if instruments vary, then, especially in times of crisis such as an engine malfunction, a pilot may move a control in a way that is thought to be correct but in fact is wrong and causes a greater crisis, leading to a crash. When aeroplanes crash the cause is categorized as mechanical failure or 'pilot error'. The latter can be subdivided into pilot carelessness and 'system error', meaning an error of the type described above in which the origin of the error lies in the design of the system and its failure to interface adequately with the user. This is not carelessness of the operator but faulty design.

Useful life analysis

This is a method for predicting the average period of time during which a component can be expected to operate under normal circumstances without wearing out. This is used to compare alternative design solutions and as a cost reduction tool. This technique is used to determine service intervals and to design components for required lifetimes. It is based on life tests of components from which their expected life, or mean time before failure can be estimated. This will clearly depend on the design of a component. The use of thicker materials will probably extend the life of a component but will use more material and cost more. The competitive economic pressures cause a move to thinner or different materials. The effect of this on the reliability of the component must be assessed.

In the past products were designed to last a long time – some domestic products would last for a hundred years or more. These days they last only a few years. This is due to the pressure to reduce prices. The cynic would say it is being done to ensure that customers have to buy new products more often and so create a larger market.

There is no point in having some components in a product with very long expected lives if many of the other components have a more limited life. The expected life of the product should be reflected in the design of the components. Not every component in a product, however, has to have the same life. This could lead to a non-optimal design at above minimum price. Many products are designed for an overall expected product life but with shorter lives for some of the components. These components are therefore expected to be replaced during the life of the product. In cars, for example, oil, spark plugs, distributor contact points and tyres are expected to wear out and be replaced regularly. It would not be feasible to design all components in a car to last, say, ten years.

Reliability analysis

Reliability analysis extends the concept of useful life analysis from the component to the product. The reliability of each component can be estimated, but what is the reliability of the product as a whole? This requires statistical analysis which is beyond the scope of this book but the concept is explained here. Reliability analysis is a method for predicting the ability of a product to perform a required function under stated conditions for a given period of time.

The reliability of a product can be visualized with a graph known from its shape as a 'bath-tub curve'. Figure 8.3 shows a bath-tub curve of reliability. This has some similarity to the product lifecycle (Figure 5.6 in Chapter 5), but in this case applies to an individual product. The life of an individual product cannot be precisely predicted in advance, but statistical analysis of a large population of products enables the probability of failure at a given age to be calculated. The concept of product reliability applies to the human population where it is referred to as a mortality curve.

Reliability is not a measure of the expected life but of the probability of something still working after a stated time. For example, a component may have an 80 per cent probability of still being working after five years. For a complex product or system comprising several components the reliability of the product will be less than the reliabilities of each of its components due to the combination of component reliabilities. The system reliability (R_s) is derived by multiplying the reliabilities of its components (e.g. R_a and R_b):

$$R_s = R_a \times R_b \tag{8.1}$$

This formula applies where a system fails if *either* of its components fail. Clearly the reliability of the system is less than that of any of its components. If both component reliabilities are 0.9, then the system reliability is 0.81.

To offset the effect of reducing reliability it is possible to build 'redundant' components into a product. These are not required in the normal operation of the product but enable the product to continue working when some of its

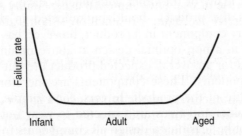

Figure 8.3   *Bath-tub reliability curve.*

components have failed. With additional components connected in parallel, the system fails only when all such components fail simultaneously. The formula for this case multiplies the probabilities of success, which are one minus the probabilities of failure:

$$(1 - R_s) = (1 - R_a)(1 - R_b) \tag{8.2}$$

In this way the reliability of the system or product can be increased. If both component reliabilities are 0.9, then the system reliability is 0.99. In the nuclear industry the consequences of failure are so horrendous that steps must be taken to ensure very high reliability of equipment. If a temperature-sensing device is used to monitor the temperature in a reactor, it is important that it is reliable, otherwise monitoring and hence control of reactor temperature may be lost. Therefore instead of relying on a single temperature sensor, several are used. If one fails or shows an abnormal temperature the others are there to continue providing a correct reading of temperature.

Maintenance analysis

This is a method which determines the consequences of different design alternatives on the future maintenance of the finished product. This analysis is intended to improve availability of the product by reducing lost time due to maintenance and service requirements.

Feedback from those involved with the product at every stage, especially after it leaves the factory, i.e. purchasers, wholesalers, retailers, service engineers and customers' guarantee claims, will provide useful information for the design of second-generation products.

THE BENEFITS OF GOOD DESIGN

Good design can contribute to competitive advantage in three main areas:

- *Cost* By reducing manufacturing costs and minimizing lifetime costs to user.
- *Product acceptability* By improving performance, uniqueness, reliability, appearance, ease of use, environmental acceptability and safety.
- *Service* Effective packaging, presentation, simplification of maintenance and repair.

In a study of the effect on return on investment of various strategies, Berliner and Brimson made the following discoveries:

- Improving manufacturing methods produced a 25 per cent improvement.
- Improving preproduction produced a 66 per cent improvement.
- Improving manufacturing strategy produced a 150 per cent improvement.

- Improving manufacturing methods produced a 200 per cent improvement.
- Improved design gave a 500 per cent improvement.

DESIGN STANDARDS

BS 7000

To ensure a proper consideration of all aspects of the management of design, the British Standards Institution has prepared BS 7000 – a checklist of points that management need to consider:

- Have the corporate objectives for the design function been properly defined and, thereafter, periodically reviewed?
- Are these corporate objectives understood by all involved and have they inspired enthusiasm?
- Is the company's product strategy compatible with its corporate objectives?
- Have sufficient resources been provided to match the product strategy?
- Are procedures in place to ensure that up-to-date information about market requirements is available to the design team?
- Are the collaborative, information and evaluation links between the design team and other parts of the business operating properly?
- Are the organizational policies and procedures for managing the design team process adequate?
- Is there a sincere and visible corporate commitment to high standards of product design?
- Are achievements and expenditure being monitored against time?
- Are results being properly evaluated and is this evaluation being communicated to all concerned?

Design in corporate objectives

Those responsible for design need to understand the following:

- The business in which the company is involved, including its technologies.
- The markets on which the company has to set its sights.
- The company's targets for the correct combination of growth and profitability to satisfy short-term funding needs and to achieve long-term growth.
- The market position and prime competitive advantages that the company is seeking.
- Objectives need regular review and amendment in response to market and competitor developments.

Long-term product strategies

- Are a company's products susceptible to being made obsolete by competitors' technologically innovative breakthroughs?
- How important are changing lifestyles or increased customer expectations?
- Is it likely that new standards or legal requirements will present opportunities or threats?
- Can alternative distribution or marketing methods be exploited?

LAW IN DESIGN

Recent product liability legislation in many countries has strengthened the rights of the public as users of products by enabling them to sue a manufacturer for damages caused by a fault in the design or manufacture of a product. This has a major implication for the design of products since safety in use must be assured in the design.

INFORMATION TECHNOLOGY IN DESIGN

The main form of modern IT in design is computer-aided design (CAD). At its most basic CAD is a graphics package that enables drawings to be created, stored and printed. In manufacturing this enables libraries of component drawings to be created so that new drawings can be created by modifying similar existing ones. Modern CAD, however, goes a lot further than line drawings. Components and products can be viewed on screen in pseudo-3D as 'solid' objects.

The analysis of the stresses and strains in a component can be computer modelled using CAD data together with software for finite element analysis (FEA). In this way the component can be designed and tested to ensure it achieves its function.

The real benefit, though, comes from the ability with modern IT to transfer CAD data to other areas of the organization for other uses. The information created in CAD about a component identifies its shape, size and where particular shaped features are. This is the same information needed to manufacture the product and for some processes needed also to make the tool or die. CAD data can therefore be transferred to processors which translate this data into a suitable form for these further uses. Where components or tool manufacture are subcontracted this data can be downloaded to suppliers at different sites for use in their manufacturing systems.

Expert systems are becoming available which assist designers in design problem solving. At a more basic level, databases of product information are available which give rapid access to the availability of components elsewhere which may be suitable as standard components for purchase.

Task sheet 8   Design

8.1 For a motor car, list the changes that have been made to the product over the last ten years. Which of these are due to:
- new materials?
- new technologies?
- improved safety?
- reduced pollution?

Do these changes increase the price or the value of the product?

Discuss the implications of these changes for the future of manufacturing in Britain.

8.2 Compare similar British and foreign-made products. How do they differ in:
- Their style and attractiveness?
- The number of components?
- The ease or difficulty of maintenance?

Assess which would be easier to make.

CHAPTER 9

QUALITY MANAGEMENT

INTRODUCTION TO QUALITY

Quality is likely to be one of the order-winning criteria of a manufactured product. It can also affect the productivity and profitability of a manufacturing organization. Quality is needed not only in the quality of the manufacturing of the product but also in all other activities of the organization. Quality does not just happen, it needs a quality strategy, quality planning and a manager responsible for the quality function with sufficient authority to ensure that quality is an inherent part of the culture of the organization. Non-manufacturing organizations also need quality; the same concepts apply to the services they offer or the tasks they perform. The principles outlined here are therefore of value in all organizations.

What is quality?

Quality can be defined in many ways. It is concerned with how well a product or service achieves its purpose. It is therefore formally defined as 'fitness for purpose' or as 'conformance to specified characteristics', these characteristics having been laid down in the design specification for the product. How long a product lasts before it breaks down is often thought of as a measure of good quality, but 'reliability' would be a better word to describe this feature of a product. Reliability is important, but is just one aspect of quality. The concept of reliability was referred to in Chapter 8.

A Rolls-Royce is thought of as a 'quality' car, but this uses the word in the sense of possessing 'luxury' features. It is equally important that cheaper, mass-produced cars are of good quality even though they clearly do not possess the luxury features of a Rolls-Royce.

Quality, then, is neither luxury nor reliability. This confirms the definition given above that quality is conformance to specified characteristics. Quality must be capable of being measured.

The need for quality

Good quality often enables an organization to win orders in the market. Conversely, poor quality will not only lose orders but will damage reputation and lose later orders. It has been said in Chapter 5 that one dissatisfied customer will tell at least twenty other people how poor your product is, so the loss of future orders could be substantial.

Poor quality means that components due to be assembled together do not fit properly, giving a shoddy product or causing additional work, and cost, to make them fit. This is of particular importance in automated assembly where the components just have to fit together properly. If they do not fit, then the system will literally 'crash'. In manual assembly if components do not fit operators will try other components or send components back for modification. This causes extra costs and delays, and reduces the productivity of the system. When components are scrapped it is not only the material cost that is lost but also the cost of all the work done on the component (the added value) up to the point at which it is scrapped.

Poor quality leads to a reputation for poor quality, so the costs of poor quality are not just those of immediate waste or rectification but also the loss of future sales.

Poor quality costs money. Good quality may cost money too, but in most cases it is considered that the costs of poor quality exceed those of good quality. Hence it was said by Crosby in his book of this title that 'Quality is free' in the sense that having adequate quality systems costs less than not having them. Crosby is also credited with the concept of 'zero-defect' manufacture, which is based on the view that it is cheaper to prevent defects than to rectify them.

Quality cannot be isolated from other functions of management. Quality is about products, and hence relates to design, to manufacturing, to marketing and to finance. Quality is something which links all of these functions together.

At its most basic quality means doing things right – making a product or performing a task the way it was designed and planned to be done. In an ideal world this would be easy, but in the real world many things can go wrong between plan and execution – between design and manufacture.

CAUSES OF POOR QUALITY

In order to achieve good quality it is worthwhile understanding how systems can fail to deliver good quality. In a manufacturing or other system which is repetitively producing the same component or performing the same task, the output should be the same each time. The component should be the same size and have all the characteristics stated in the design specification. In practice there will be deviations and variations, the causes of which need to be understood if they are to be controlled or eliminated.

Deviation

Deviations take the form of gross differences from the specification – operations that have not been performed or performed so badly that it is obvious that something is wrong. For example, holes drilled where they should not have been, or not drilled where they should have been.

Deviations may occur because of clerical or human errors. A necessary operation was not requested or was forgotten. It can also be due to a breakage or malfunction of the machine or of the tool which performs the work on the material. This can occur in automated manufacturing where there is no human operator who would have seen such a failure and stopped the process in order to rectify the problem. Automated systems can, however, have automated condition monitoring systems to perform automatically the same control function as the human operator. The Japanese concept of *bakayoka* is based on automatic detection of abnormalities.

Deviations are 'sudden' changes in that components being produced suddenly become different from those produced just previously.

Variation

Variations are minor differences from the design specification. For example, holes have been drilled but are they exactly in the right position and exactly of the right size? Is the overall size of the component correct ('correct' meaning within the range of size stated in the design specification)? Variations are of two types: random and gradual change.

Random variations

Random variations are inevitable in any process. Minor differences in, for example, the hardness of a material cause variations in the forces used in a process. This can cause components to be produced in slightly different sizes. The mechanical linkages in machines can become worn, allowing a slackness in positioning of tools. This can also contribute to random variations in size. The nature of the process itself is also a cause since some processes are controllable to a greater extent than others. Such variations define the 'process capability' of a machine. Precision processes are those capable of achieving repeatability of output within very close limits. Less precise processes achieve output only within a broader range of limits. Some of these variations are inherent in the process but others are controllable. Good machine maintenance – replacing worn parts, for example – can reduce part of the random variation.

The magnitude of random variations of a process can be measured in two ways. The range of a set of values is the difference between the largest and smallest values. This is easy to measure but is not very useful as it is based on only two extreme values which may not be typical. The preferred alternative is to measure the standard deviation. This is a statistical measure of the extent of variation from an average. Its value is approximately one-sixth of the range.

Gradual variation

Gradual variations are also inevitable in processes in which tools are used to change the shape of a material. These occur as the result of the gradual wear of tools or dies due to friction: as tools wear they become smaller; the cavity within an injection tool becomes larger. As a result components gradually become larger. This type of variation is controlled by changing the tool before the component size goes outside the range specified. A precision component would require more frequent tool changes or a tool that wears less quickly, in both cases leading to higher costs.

Gradual variations can be monitored and seen if a graph is plotted of the size or other characteristic being measured against time or output units. An example of such a graph is shown in Figure 9.1.

Process variation and design tolerance

Process variation can cause components to be of unacceptable quality. The question is how variable is the process and how much variation can be tolerated in a component? A process must be selected that has the capability of achieving output within the range stated in the design specification. If quality is poor, as evidenced by a high reject rate, a failure of components to fit together or customer complaints, it may be due to the process variation being greater than the variation permitted in the design specification. Quality can be improved by reducing process variation. The Japanese engineer Taguchi has developed approaches based on identifying the causes of variation for particular cases and measuring their effect. By selecting a suitable process the amount of variation can be reduced. This seeks to prevent defects from being produced, being better than allowing them to be produced and then thrown away. Precision processes usually cost more than others because they need to operate within very close limits.

It is important for a design specification to state how much variation is permitted or tolerable within a component. This was covered in Chapter 8. For automated manufacture using robots for assembly the quality of components must be such that components that are to be fitted together do in fact fit together. A process must be selected that is capable of achieving the required quality of output within the design tolerance. A commonly used rule is that the process variation should be such that four standard deviations each side of the mean should be within the design tolerance. This gives a high confidence that random process variations will not exceed the limits of the design tolerance. Machines must then be adequately maintained and tools changed sufficiently often to ensure that they continue to maintain this quality of output. The process should be monitored for gradual variations and tools changed before it becomes possible for the process to produce output outside the design tolerance.

THE TRADITIONAL APPROACH TO QUALITY

The traditional approach focused on manufacturing quality only and was based on techniques of 'quality control'. It took the form of inspection of work after it had been processed. Unsatisfactory work was rejected and satisfactory work passed on to the next process. Inspection costs money and only detects faults after processing, when the damage has already been done. Often, inspection was done away from the process (in a clean room as opposed to a dirty factory). This injects a time delay between processing and inspection. If a fault is found that occurs in all items, then further faulty items will have been produced before the fault is detected. The longer the delay, the greater the number of further components that will have been made which may be faulty. If items are rejected, not only is the cost of material lost but also all of the value added to the item prior to its rejection.

Inspection

Inspection of all items (100 per cent inspection) is expensive, so often only a sample is inspected (sampling inspection). This reduces costs but increases the risk that defective components will not be detected.

Inspection does not improve quality – it merely filters out the poor quality in a process. It reduces or eliminates poor quality (i.e. defective components) entering the next process, but only by rejecting those items with defects or of a size or quality that does not conform to the design specification at the inspected stage. In repetitive manufacture the number of items coming out of the end of a set of processes may be less than the number for which raw material was put in. Ordering of material may have to take account of such losses. Thus raw material costs are higher in manufacturing systems in which losses occur.

Inspection is also carried out on components arriving from suppliers – to make sure they conform to design and quality. This is known as goods-inward inspection. This is an expensive process, necessitated by a mistrust of the quality of goods supplied.

With modern technology it is possible to build-in automated inspection into a process. By inspecting all components at the time of manufacture any deviation from standard is detected early and corrective action can be taken automatically. The problems associated with delayed inspection are avoided. In many cases deviations can be measured which are still within the specified tolerance, and defects are avoided by stopping the process before components go beyond the permitted range of size. If, however, the specified tolerances are narrower than the random variation inherent in a process, then a large proportion of components will be outside the tolerance and will have to be scrapped or rectified. Clearly this situation must be prevented by reducing process variation.

Process control charts

Data obtained, manually or automatically, at the time of processing can be plotted graphically in order to provide a visual record of the performance of a process. Critical dimensions or other characteristics are measured. A typical control chart is shown in Figure 9.1. The horizontal axis is either time or a cumulative output volume. The vertical axis is a measure of a critical characteristic of the component such as its length or diameter.

The objective value of the characteristic being measured is plotted as a horizontal line. An analysis of past output will indicate the amount of random variation. From this the standard deviation of the process from the mean (average) can be measured. Further horizontal lines can then be drawn on the graph at positions representing one, two and three standard deviations above and below the objective line. These values – the control limits – are derived from an analysis of the process. It is expected that 67 per cent of values will lie between ± 2 standard deviations, and hence within that part of the chart which lies between these control limits; 99.8 per cent will lie between ± 3 standard deviations.

As output is produced, the value of the chosen dimension is plotted. This will show both random and gradual variations and also any deviations. Figure 9.1 shows the expected value of a size or other characteristic of a component and also indicates the two and three standard deviation control limits. The data shows a gradual increase in the size of the component, but this increase is still within the control limits.

Some output can be expected to be beyond the two standard deviation control limits. These are known as 'warning limits'. Some output of this type is not a problem as it is expected, but successive values in this region indicate a problem that needs to be corrected. Virtually no output should occur beyond the three standard deviation limits. These are known as 'action limits'. If such values

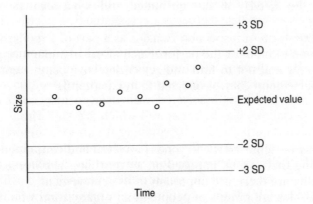

Figure 9.1 *A quality control chart.*

occur, this indicates that the process is out of control and action is required to correct it.

An empirical rule can be used to assess whether a process is performing normally or is out of control. If eight successive values are on the same side of the mean, four out of five are beyond one standard deviation, two out of three are beyond two standard deviations or just one value is beyond three standard deviations, then it must be assumed that there is an assignable cause for this and that this could not occur if the process were in control. Such outputs therefore indicate a process out of control.

Control charts can also be applied to the inspection of raw materials received from suppliers in order to confirm that the supplier is supplying materials which conform to the size and standard specified in the purchase contract. The modern approach will be based on working with suppliers to ensure that they assure the quality of their materials before they send them to the manufacturer.

THE MODERN APPROACH TO QUALITY

A modern approach takes a systemic view of quality and the interaction between different parts of an organization that are necessary to achieve it. The approach is based on 'quality assurance' and seeks to ensure that defective work is not produced. This is seen to be less expensive than making defective components and throwing them away. Quality assurance procedures cost time and money but will usually save more money than they cost. In this sense it can be said that 'quality is free'!

In this approach quality is 'built-in' at the design stage to ensure that a product will meet its quality requirements. This requires co-ordination between the design and manufacturing functions. This approach also seeks to ensure that the quality meets the customers' expectations. This requires co-ordination between design and marketing functions. Design tolerances must be specified according to the needs of the customer and what can be achieved by manufacturing.

While some form of inspection remains as a part of a modern approach to quality, it is done to monitor the performance of the manufacturing and quality assurance systems and not to find and reject defective components. This is an example of double-loop control referred to in Chapter 3.

Quality circles

Quality will only be achieved if everyone is involved and committed to quality. It is not something that can be imposed on an unwilling workforce. The human aspects of quality are therefore important to its achievement.

Quality circles are groups of people in an organization who meet to share experience and develop better quality in their work. This involves people in quality and commits them to it as a philosophy. Quality circles need a participative style of management for them to work. They only work when quality has been established as a common objective. Without this, employees will not be interested in being involved in achieving quality. The use of quality circles was developed in Japan by the American quality consultant W. Edwards Deming in the late 1940s. Initially they were concerned only with quality. As experience and confidence in this participative style of management grew, the agenda of quality circles extended to other areas. By the 1960s their interest extended to safety, another common objective. By the 1970s quality circles extended to cover all aspects of productivity. This was possible because in Japan productivity is a common objective. This expansion of the agenda shows that the quality philosophy extends beyond just the quality of manufacture to all aspects of an organization's activities.

Europe and America totally ignored these developments in Japan and did nothing about them until the late 1970s in America and the 1980s in Europe. In these thirty lost years Japanese quality and productivity increased significantly, allowing Japan to grow and dominate world manufacturing. European manufacturing in this period was unable to compete because it failed to develop modern approaches to quality and productivity.

In 1957 the British Productivity Council made a film, *Right First Time*, which advocated all of the procedures of modern quality management. It fell on the deaf ears of the neanderthal managers of the time. Japan used these ideas, Britain did not. The decline of British manufacturing over the last forty years is a result of this failure. Only now is British industry beginning to apply the lessons that it could have learned a long time ago.

Traceability

In the aerospace, automotive, military and food industries it has been found necessary to keep detailed records of manufacturing so that each product and component can be traced back to the day it was made, the machine it was made on and the batch of material from which it was made. This requires serial numbers on products so that they can be identified and records being kept for many years. These procedures obviously cost money, and so the extent to which they are done needs to be determined.

The need for traceability can be illustrated for components used in an aircraft. Aeroplane crashes are very expensive in terms of loss of life and damage to the reputation of the builder. If an aircraft crashes and its cause is found to be faulty material in a component, then all airlines will want to know – and passengers will want to be assured – that any other aeroplanes of similar model they are flying in do not have a similar fault in that component. They therefore want to know whether the components in their aircraft were made from the same batch of material. This can only be known by keeping records of serial numbers. This has to cover not only the equipment originally installed in a product but all subsequent replacements. This requires traceability at both product and component level.

Other industries do not require traceability to this extent, but some form of traceability is increasingly being required in other industries. Notices of recall of cars and other products for modification are based on an identification of a fault in a component and require some form of traceability. The use of serial numbers or batch numbers on products is an indication that traceability is being used at least at product level. With foods the 'best before date' is not only an indicator to the consumer of the shelf-life of the product, but is also a batch identifier for traceability purposes.

QUALITY ORGANIZATION AND MANAGEMENT

The importance of quality requires the development of a corporate quality strategy and the appointment of a senior manager with responsibility for quality. The quality function needs to be independent of manufacturing and to report at a sufficiently high level in the organization structure to carry the authority to ensure that quality is achieved. This independence of quality is needed to prevent quality problems from being 'swept under the carpet', i.e. hidden in the hope that no one notices. Quality covers all aspects of an organization's activities. Only the managing director has the authority over all areas and functions, but responsibility for quality is usually delegated to a quality director. The quality director advises the managing director on the quality systems, procedures and documentation needed.

Figure 1.1 in Chapter 1 shows in outline where quality fits in the organization. It needs to interact with manufacturing but also with marketing. The human aspects of quality mean that human resource management needs to be involved in quality. Quality, quite literally, must involve every person in the organization.

Organizations need to audit their quality systems to ensure that quality is always achieved. They need to do this to be able to assure their customers that their products will always meet the specification. With ISO 9000 (see below) and other quality standards it is becoming more common for customers to visit a manufacturer to do their own audit on the manufacturer's quality systems.

The achievement of quality requires the active involvement of everyone and the design of the following systems:

- Appropriate design systems
- Appropriate manufacturing systems
- Appropriate management systems

A modern, systemic approach to quality expands beyond a product and its manufacture to the whole of an organization's activities. There is a need for quality in office procedures, customer and supplier relationships and all other management systems. Quality assurance is needed at all stages of the whole process from design and marketing, through purchasing, manufacture and distribution to the use, maintenance and eventual disposal by the customer. This approach to quality is also needed in non-manufacturing organizations.

The British Standards Institute has developed standards for quality assurance (see Figure 9.2) which embrace this breadth. There are a range of analytical techniques for analyzing tasks and processes which are beyond the scope of this book. *Quality function deployment* is the term now used for a systemic approach to quality planning and management using a variety of analytical techniques.

In order to assure quality within the organization, it is important to assure the quality of materials and services entering the organization. Quality assurance of supply is therefore vitally important. This aspect is covered in Chapter 16, Purchasing.

Total quality management

The activities of an organization are only as good as the people who do them. To achieve quality requires the development of attitudes conducive to quality and commitment from senior management. The ethos and culture of an organization must be developed so it is appropriate for the development of quality.

'Right first time – Right every time' is the basis of this approach to quality. The term TQM – Total Quality Management – embraces this modern approach to quality.

Crosby, Deming, Juran and Taguchi are credited with developing a modern approach to quality management. They have produced lists of steps or stages in achieving quality which are too detailed to include here and are left for a further

Figure 9.2 *Eleven stages of quality assurance.*

study of quality. They are all, however, based on the concept of a systemic approach, a participative style of management with commitment from directors and the involvement of all employees.

Quality is so important that it cannot be left to an obscure department which inspects incoming goods and processed components. Quality involves all the people in an organization acting in an integrated manner. Quality assurance procedures are crucial to the survival and profitability of an organization. A *quality manual* is a means of documenting quality procedures.

The achievement of quality requires the development of a quality culture within the organization and the establishment of quality as a common objective. Commitment at all levels is needed and with it a quality strategy that involves all functional areas within an organization. An interdisciplinary *quality project team* (or quality improvement team) is one way of achieving a continuous improvement of quality. The concept of continuous improvement is one to which Japanese manufacturers are committed. The word *kaizen* is used to cover this concept. A further Japanese concept, *poke yoke*, developed by the Japanese engineer Shingo, is also of benefit in ensuring quality. This means designing components and procedures in such a way that errors become impossible. Assembly of components will be done wrongly if the design allows them to be done wrongly. If forms can be filled in wrongly, they will be. It is important when entering data onto a computer that errors are avoided. *Poke yoke* procedures need to be built-in to the design of the system.

BRITISH AND INTERNATIONAL STANDARDS

ISO 9000

ISO 9000 is an international standard for quality assurance. It applies primarily to manufacturing organizations but different versions are available for educational and other organizations. All organizations have a need to assure the quality of their products or services. ISO 9000 is a framework of procedures for assuring quality. This standard is now used in Europe (as EN 29000) and was developed from a British Standard BS 5750. Organizations that develop quality procedures for their activities can be assessed and certified as having achieved this standard in quality assurance.

A manufacturing organization buying components from a supplier which has been ISO 9000 certified should be able to buy with the confidence of knowing that the supplier has adequate quality assurance procedures. Their components should therefore conform with the specified design and quality. Inspection of incoming goods should not therefore be needed for such suppliers. In practice this does not always happen. The supply contract between the manufacturer and supplier must deal with non-conformance to specification. Therefore some form of sampling inspection will still be done, but it does not need to be as extensive or expensive as previously. The emphasis should be on defect prevention rather than defect rejection.

INFORMATION TECHNOLOGY IN QUALITY

The monitoring of the quality of manufacturing output requires the collection of a lot of data. Modern computer systems can assist in this process. Computer-aided quality assurance (CAQA) can store, analyze and present quality data. It can link to automated sensing and measuring equipment in manufacturing, to manufacturing, planning and control, and to warehouse, distribution and financial software.

Data for process control charts can be captured automatically and presented on a control chart. The process can be stopped when abnormalities occur or tools changed automatically as gradual variation indicates tool wear.

Quality reports for management can be produced which analyze or summarize output levels, and the frequency of any problems.

Traceability requires the holding of data over a long period of time. It necessitates the linking of a serial number with a date and machine used for manufacture. For the aircraft industry, it also involves updating this data as components are replaced. This is not easy since replacement can take place anywhere in the world. Such records are needed over the life of each component.

Task sheet 9   Quality management

9.1   Consider a domestic product which you have owned for some time and which on some occasion has failed.

How old was the product?

Was the failure expected?

Was the product repaired?

Were service facilities available locally?

Did the product conform to your expectations?

9.2   For a motor car:

List the components which are likely to fail before the end of life of the product.

Are these components replaced when they fail or before they fail?

For which components is failure in service dangerous? How should this be dealt with?

9.3 If you are a student:

List the factors which make up good quality of education for you.

How can the quality of education be measured?

List any incidents or poor quality you have experienced. Try to identify their cause.

How can quality of education be assured?

What involvement do you get as a student in the academic quality assurance process?

What part does the academic level of study play in the quality of education?

MANUFACTURING MANAGEMENT

155

4.24 Construct a network diagram for project planning.

4.25 Determine project duration and the critical path.

4.26 Construct time-based network and resource schedules.

4.27 Build sectional budgets for a project.

CHAPTER 10

MANUFACTURING MANAGEMENT

THE ROLE OF MANUFACTURING MANAGEMENT

Manufacturing management is that function of management responsible for managing the primary activities of an organization – those concerned with conversion of an organization's inputs into its outputs. Although the particular term 'manufacturing' implies a manufacturing organization, the function is required in all organizations. It may be called operations, production or works management in an industrial organization; administrative or office management in an office-based organization. All organizations have and must identify, their primary activities – those that perform the primary function of the organization. These are present in hospitals, airlines, hotels, universities, banks, governments and military services. The central role of manufacturing or operations management is seen in Figure 1.1 in Chapter 1.

There will be differences of detail between different types of organizations but the common role of manufacturing or operations management is responsibility for managing the resources, known as the 4Ms:

Materials + Machines + Manpower + Money

used to perform the primary tasks. Management itself is a resource and so is time. Time is a non-renewable resource – once lost it cannot be recovered. The objective of manufacturing management is to manage these resources effectively and contribute significantly to the achievement of corporate objectives. An alternative perspective identifies the focus of the activities of manufacturing or operations management, known as the 4Ps, which can be seen to relate to the 4Ms above:

Products + Processes + People + Profit

The single word that best describes the objective of manufacturing management is productivity, meaning manufacturing or production effectiveness, which is the major theme of this book.

Manufacturing cannot be managed in isolation; it requires collaboration with other management functions to ensure that they are working well together.

157

Manufacturing management manages the central and 'internal' activities of an organization, whereas finance and marketing are more involved with 'external' relationships with shareholders, banks and customers. More closely related to manufacturing within the organization are design, quality, materials (purchasing) and human resource management (including personnel) departments.

For any organization to operate effectively, directors and managers must recognize that the organization is a system and develop a 'holistic' approach to the business and its activities. This involves recognizing and understanding the interactions between different functions that are necessary for the organization to be effective. All of the different management functions must be involved in setting the corporate strategy of the organization. Each function will then set its own functional objectives and develop a functional strategy to achieve those objectives. Chapter 2 referred to the process of strategy formation.

Consultation, agreement and procedures are necessary to ensure that these functional strategies are compatible with each other and that they work together synergistically to enable the organization as a whole to operate effectively. This requires the development and application of a systems approach. If functional strategies are so designed and are compatible with each other, then the different management functions will have common objectives and will be able to work together to contribute to and to achieve corporate objectives and to improve corporate performance.

In practice, each management function may pursue its own objectives in ways that conflict with and cause difficulties for other management functions. Each function acts in a way it believes is beneficial to the organization, but without a systems approach there is limited understanding, so the effects of decisions on other functions and on corporate performance is not seen. This leads inevitably to poor performance, low productivity and reduced profit, or even to the failure of the whole organization.

An example of this is when a marketing department offers prices or promises delivery dates that are unrealistic. These boost sales and hence their own objectives (and perhaps salaries or sales commissions). They do so believing that higher sales are beneficial to the organization and in order to appear to be doing their job successfully. If, however, the products are unprofitable at that price or manufacturing has to be disrupted to make them on time, then there will be a lot of extra work, a lot of hassle, and productivity and profit will fall.

The role of manufacturing management can therefore be seen as designing manufacturing systems and the management systems necessary for the effective operation of the manufacturing system.

FUNCTIONS OF MANUFACTURING MANAGEMENT

These functions are illustrated for a manufacturing organization but are required, in a suitable form, in all organizations.

Manufacturing systems strategy

Strategic decisions are needed on which activities or processes are to be done 'in-house', which components the organization will make and which they will buy or subcontract to other organizations; how the in-house activities are to be done and how many of each product are intended to be produced. This will determine the number and type of employees and machines required. As well as machines for the actual processing, further equipment for handling, transporting and storing needs to be specified. Tools (specific to the items being processed) also need to be specified, designed and acquired. The degree of automation and integration of the manufacturing system must be decided. This function of management interacts with design, marketing and finance. The objective is to specify the conceptual design of the manufacturing system to fit the corporate strategy. A detailed consideration of manufacturing systems strategy and manufacturing technology are beyond the scope of this book but some basic analytical techniques are covered in this chapter and in Chapter 11. The make–buy decision is an important strategic decision and is developed further in Chapter 16. Manufacturing strategy is developed later in this chapter.

Design of the manufacturing system

This function is concerned with the design and organization of a new or changing manufacturing system. It identifies what type of process is needed and how many machines are required; where are they to be sited within the factory; what work will be done, how it will be done and how long it will take; how many people are required to do it and how to make their jobs interesting and satisfying. This function of management interacts with quality, personnel and finance. The objective is the detail design of the manufacturing system to provide the required capacity efficiently and to cope with variation and uncertainty within the strategy framework set in the conceptual design. This book does not cover the technology of manufacturing processes or automation. Readers should refer to a book on manufacturing technology or manufacturing engineering for these aspects. This book does cover the strategic, organizational and managerial aspects of manufacturing system design in this and the next chapter. It examines the decisions that need to be made at each stage of the process of manufacturing system design and shows how these decisions need to be based on a systems view of an organization and integrated with marketing, finance and other management functions within an agreed corporate strategy.

Operation of the manufacturing system

This function manages the flow of material through an existing manufacturing system. It identifies when work will be done; which machine it will be done on and how much and when materials and bought components need to be bought. This function of management therefore interacts with both materials (purchasing) and marketing, and translates product demands into manufacturing tasks and into purchase requirements. It determines the scheduling of tasks, the loading of machines and manpower resources, and controls the input, output and levels of stocks of material at various stages of processing. The objective is to manage the day-to-day operation of the system to achieve required outputs with minimum inputs. This function is explained in Chapter 13.

Physical asset management

This function manages the 'physical' or hardware assets, i.e. the buildings, machines and other equipment. Its parts are sometimes called building services and plant engineering.

Building services manages the buildings, initially deciding how big a factory is needed, what facilities and services are needed in it and where it should be located. An important decision is whether to buy or rent a factory. The location of the activity is a strategic manufacturing decision. This is considered in Chapter 15 on logistics. Once acquired, the building will need maintenance. Properly maintained, a building and its site should be a non-depreciating asset and can increase in value over time.

Plant engineering manages the machines as assets. These machines need to be bought, installed, maintained and eventually replaced, so decisions are needed on when to buy machines and which particular machines to buy in order to provide the required capacity. It is not only the machines used directly in manufacturing that need to be considered. Conveyor belts, workbenches, stores racking, fork-lift trucks and other handling equipment also need to be bought and maintained. The objective is to provide the machines and other facilities at minimum cost but to ensure they operate effectively. A balance is needed between the cost of a machine and the cost of running it. This function would include energy management.

Maintenance crucially affects the availability and hence the utilization and capacity of the equipment. Maintenance therefore needs to be managed. Machines do not produce while being maintained. Maintenance can be done at night-time or weekends. The loss of productive time due to machine breakdowns can be reduced if machine parts are regularly replaced before they break down in a management system known as preventative maintenance.

People owning cars usually practice some mixture of preventative and breakdown maintenance. Items for which the expected life is known and short, such as engine oil, spark plugs and tyres, are likely to be replaced on a regular basis before they cause a breakdown. Other items such as the clutch can be replaced when noise or performance indicate a near end of life condition. Despite this preventative maintenance, breakdowns will still occur. A user must balance the probability of breakdown and its likely effect and cost against the cost of preventative maintenance. The principles outlined here for a car are equally applicable to manufacturing equipment.

The objective of physical asset management in manufacturing is to provide the maximum capacity and availability of the machines and other equipment at minimum capital and running cost. A modern technique called *total productive maintenance* takes a systems approach to the hardware assets of an organization. This approach analyzes costs at the system level and provides data for management decisions.

ORGANIZATION STRUCTURE FOR MANUFACTURING

Figure 4.1 in Chapter 4 showed an organization structure diagram for a manufacturing organization. It is not possible to show the structure of all the functions within an organization in a single diagram. It is relevant now to consider specifically how the manufacturing function could be organized. The principles would be the same for the primary function in a non-manufacturing organization and for other functions in an organization.

The manufacturing function of management comprises the four roles referred to above, and these are reflected in the organization structure. Manufacturing has to interact with all of the other functions in the organization, receiving and providing data, taking decisions and developing strategies together. It also manages the day-to-day manufacturing operations. A possible organization structure for manufacturing is shown in Figure 10.1.

The primary or 'line' management function manages the supervisors and operators doing the manufacturing. The term 'line' does not relate to a production line but to the conceptual line of responsibility and accountability in an organization structure which links the managing director at the top of the

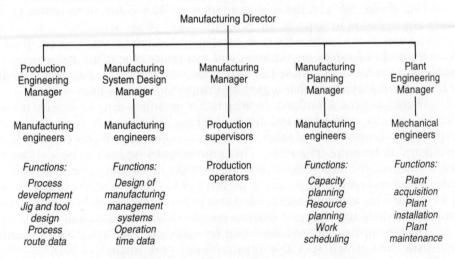

Figure 10.1 *Organization structure for manufacturing.*

organization structure to the operators at the bottom. All managers on this line are referred to as line managers.

In an organization of any size the total set of manufacturing activities will be split into several departments or sections. The different ways of doing this are considered in Chapter 12. The organization structure must match the physical split of the manufacturing activities. Each section will have a supervisor. In the old days this person would be called a foreman. In a modern factory the title may be 'group leader'. This person is the first-level line manager, directly managing the operators and the work they do.

The data handling, analytical, planning and maintenance functions are support functions, i.e. they support the primary function – the manufacturing itself.

As well as the organization being designed to manage the day-to-day activities, provision must be made for handling change. This could be done using project teams, people who plan and manage the introduction of new products or processes, or a reorganization of manufacturing. Managing the introduction of change is considered in Chapter 14.

MANUFACTURING STRATEGY

Part 1 of this book showed the importance of improving productivity in order to achieve competitive advantage. Britain, and other European countries, need to achieve high levels of productivity if they are to achieve competitive advantage and sell in world markets. This concept, of being as good or better than the best manufacturers in the world, is known as 'world-class manufacturing'. This book is orientated to the development of world-class manufacturing in Britain and Europe and identifies what is needed to achieve it.

The choice of a suitable organization structure for manufacturing is important if this is to happen. In all of the areas of an organization it is not sufficient merely to identify what tasks need to be done and to set-up departments to do them. Leadership, organization and management systems are needed to enable and to ensure that these tasks are done effectively, that best use is made of the resources used and that organizational performance is improved.

There has been a tendency in the past for manufacturing to do what it was told to do – by marketing and finance! Manufacturing merely reacted and responded to the demands of other management functions, which were somehow considered to be more important. This approach can be seen to be a failure – and a cause of the problems of low productivity in Britain. The approach is reductionist and parochial; it fails to perceive or to understand the organization as a system. This approach was applied also in managing people – a centralized, autocratic style of management that can only be described as neanderthal.

Modern approaches, and those that are successful, are based on a systemic or holistic view which sees that organizational performance is based on the interactions between the resources – money, materials, machines and manpower.

A manufacturing strategy needs to be developed in conjunction with other functional strategies to form a cohesive corporate strategy. In this framework manufacturing develops a proactive approach, is involved as an equal partner with the other functions of management in setting strategy. This is the only way of ensuring that the functional strategies are compatible, that the organization has a corporate strategy which enables each management function to operate effectively and in doing so contributes to the overall effectiveness of the organization as a whole.

New forms of organization are needed to replace the autocratic and hierarchial structures based on centralized decisions and employees being told what to do. New structures can create a motivated workforce by giving employees challenges and responsibilities for decisions. The role of manufacturing management in this new form of organization is proactive and dynamic, always seeking improvement in manufacturing performance. The Japanese use the word *kaizen* to describe this approach. This, after all, creates the competitive advantage which sells products and creates the wealth from which profit is generated and wages paid. Many existing manufacturing organizations, in which history constrains performance, cannot compete in a changing economic environment of increasing competition from overseas.

A modern approach, business process re-engineering (BPR), is based on developing a systems approach to understanding the organization and using within this framework a set of manufacturing engineering techniques such as quality assurance, value engineering and integrated real-time information systems. These techniques are applied not only to manufacturing, as in the past, but to all the activities of the organization. This approach seeks to change the culture of the organization and the attitudes, vision and motivation of all its employees. This approach is developed further in Chapter 25.

MANUFACTURING DATA

This section identifies some of the basic data in manufacturing and defines some of the terms used. Chapter 11 considers the techniques for analyzing manufacturing systems and measuring their performance. Data is derived from the analysis of the products being manufactured or intended to be manufactured and used to plan the capacity and operation of the manufacturing system. This analysis interacts with the design function. It should *not* merely take and operate on data provided by design but should be involved jointly with design to influence the design of the product to enable its efficient manufacture.

Product volume data

This identifies the expected volumes of production for each product. This data is usually derived by the marketing function using forecasting techniques covered in Chapter 6. Short-term volume data forms the base for scheduling of activities over a period of days or weeks. Long-term volume data estimates the expected volume for a product over the timescale from its introduction to its death. This timescale is referred to as the 'life' of a product, and is the time period over which it is being manufactured, *not* the life of the product bought by a customer. This data is needed to plan the capacity of the factory and hence its size and the number and type of processing facilities. The concept of a product lifecycle was introduced in Chapter 5 and shown in Figure 5.6.

The significance of the product lifecycle for manufacturing is that the design of a manufacturing system must be capable of dealing with low, varying and high volumes of output of the product at different times. This normally means using a different design of manufacturing system at different stages of the lifecycle. In this way the manufacturing system can be designed to suit the particular pattern of demand and thus be efficient. An inappropriate design of manufacturing system will lead to inefficient operation.

Product structure data

This identifies the components of a product and the assembly relationship between them. (For non-manufacturing it identifies the outputs and their constituent activities.) The data may be in the form of a parts list or diagram. A parts list lists the components in an assembled part. Some such components may be assembled parts containing components. In practice there can be several levels of assembly, known as subassemblies before the final assembly of the finished product. A parts list will list those parts used directly to make a subassembly. These parts are at one level below the subassembly.

The term 'structure' refers to the assembly structure of the product not its physical structure. The number of each component is known at each level of assembly. From this the total requirements of each component in the final product can be calculated by multiplication of the requirements along paths to the final product. This can be seen in Table 10.3 below. These data are derived from the bills of material shown in Table 10.2, multiplying along the paths shown in Figure 10.2. When more than one product is manufactured, product structure data must include the requirements for all products. In some cases products share components with other products.

When presented as a list of all the parts at each level which are contained in a product this data is known as a *bill of material*. This is a structured list of components and subassemblies at each level in a product. It starts with the finished product. Indentation can be used to show the components of a particular subassembly. The number of each component in an assembly is included and each component and assembly is usually given a part number. Table 10.1 shows a set of parts lists. These can be combined into Table 10.2, which shows bills of material for two products which share some common components.

Table 10.1 Parts lists for products and assemblies

Part 4		= Product X
part 1	subassembly 1	2
part 2	subassembly 2	2
Part 5		= Product Y
part 2	subassembly 2	2
part 3	subassembly 3	1
Part 1		used in part 4
part A	component A	1
part B	component B	2
Part 2		used in 4 and 5
part C	component C	1
part D	component D	2
Part 3		used in part 5
part E	component C	2
part F	component F	3

Table 10.2 Bills of material for two products

Product X					Product Y				
Part	Description			No.	Part	Description			No.
4	product X			1	5	product Y			1
1	subassembly		1	2	2	subassembly		2	2
	A	component	A	1		C	component	C	1
	B	component	B	2		D	component	D	2
2	subassembly		2	2	3	subassembly		3	1
	C	component	C	1		E	component	E	2
	D	component	D	2		F	component	F	3

The example in Tables 10.1 and 10.2 uses a simple part numbering system. In practice a logical numbering system is often used. Part numbering can be hierarchial. For example, a component numbered 12345 would be a component used in subassembly 12340, which is used in subassembly 12300, which is used in final assembly 12000, which is used in product 10000. Alternatively, a part number can be used to indicate the geometry of the part (flat, cylindrical, threaded, etc.) and the processes used to make it. The Brisch and Opitz coding systems, named after their originators, are based on this concept. These systems can be used for sorting components in a way that is useful in determining their process requirements.

A bill of material can be presented diagrammatically as a *product structure diagram*. The diagram shown in Figure 10.2 is for the same products as in Table 10.2.

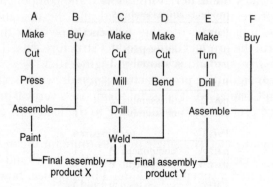

Figure 10.2 *Product structure diagram.*

From the product structure diagram the total requirements of each component in the product can be seen and calculated. These are needed so that a schedule of product requirements can be translated into a schedule of component requirements. This is known as 'product explosion' and is covered in more detail in Chapters 11 and 13. For the data in Figure 10.2 the total requirements are shown in Table 10.3.

Table 10.3 Total requirements of components in a product

Component	Total requirement	
	Product X	Product Y
A	2	0
B	4	0
C	2	3
D	4	6
E	0	2
F	0	3

Process route (or process planning) data

This method uses product structure data but also identifies the processes (or activities) required to make the components of a product or output. The facility and resource requirements are determined for each component or activity. Operation times are determined from work measurement data. The data may be in the form of a list or a diagram. This data is needed for all of the processes required in making each component and for assembling them to form an assembled product. Such detail is not needed for basic components which are bought but is needed for subcontracted manufacture as it forms the instruction to the subcontractor on how to manufacture the component.

Process route data is often in the form of lists of processes for each component and indicates the material from which the component is made, the machine or other processing facilities required and the operation times. For assembled products, a diagram is of value as it also shows how the components are assembled to form the product using product structure data. This diagram is termed an *outline process chart* and is shown in Figure 10.3.

In its basic form an outline process chart is not drawn to scale in time. For use in scheduling, in Chapter 13, it is useful to impose a vertical timescale on the diagram and to place processes on it according to the lead time *backwards* from the product.

It is conventional to specify a single process route for the manufacture of a component. This is logical on the grounds that the 'best' and most efficient process is specified, often a specific machine. This practice, however, prevents flexibility. The route is specified at the design stage but produces data which is used at the operation stage. At the operation stage people using the data have to assume that only a specific machine can be used if it is the only one specified. This can lead to machine overload and to delays even though, in practice, it may be possible for the process to be done on a different machine. This lack of knowledge can cause inefficiency. It is preferable, therefore, to specify alternative process routes at the time that the data is being generated. This may require more computer space, but it can lead to greater efficiency in operating the manufacturing system.

Work method and time data

This data identifies the methods by which work is done and the time it should take to do it. This method enables work to be designed so it is safe and efficient, and can be used to determine the amount of labour and machines required to provide a given capacity. Techniques for designing and measuring work are covered in Chapter 11. Some of the terms used in manufacturing are listed later in this chapter.

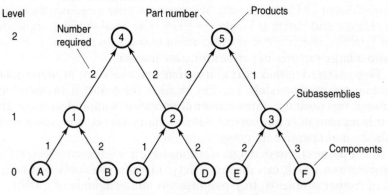

Figure 10.3   *Outline process chart.*

INFORMATION TECHNOLOGY IN MANUFACTURING MANAGEMENT

The complexity of manufacturing systems and the inaccuracies in their data were a cause for computing to be used in manufacturing much later than its use in, for example, accounting. In accounting and in the payroll the data has to be right whether or not the information is computerized and the information was generated in the past only once a week. Old and slow mainframe computers could deal with such tasks. Computerization was therefore a simple matter. In manufacturing, decisions are taken all the time, so real-time computing is needed. Different people with different functions need to access the data. Modern computers provide the speed, memory and access needed for manufacturing. The data referred to above can be held in databases which can be accessed at PCs through a local network. Database systems can be designed which store, sort and select relevant data and present the data required on-screen and in reports to a desired format.

This manufacturing database will contain all of the manufacturing data referred to in this and the following chapters and will communicate with data in marketing (sales forecasts), design (parts lists and process routes), purchasing (materials ordered), manufacturing planning (work schedules, capacity and maintenance), quality (rejects, if any), personnel (hours worked by employees) and accounting (manufacturing costs). Further details of the use of computers in manufacturing are given in later chapters in this part of the book.

LAW IN MANUFACTURING

Aspects of law specific to manufacturing are those concerned with the safety of the manufacturing processes and of the employees. Environmental aspects such as noise and pollution are also relevant.

In Britain legislation designed to protect employees from exposure to unsafe working practices has existed for many years. Early legislation included a Factories Act of 1833. There have been many similar acts since then, culminating in the Health and Safety at Work Act 1974. The need for such legislation is seen in the fatalities which occur at work, about 600 deaths per year in Britain. There are also a large number of serious injuries at work.

Basic safety legislation requires that machines are properly guarded and cannot be operated unsafely, i.e. they should not operate if an operator's hand is in a dangerous position. The number of operators without fingers or arms shows that this legislation is not perfect but the injury rate is decreasing with modern legislation and operator training.

It is a requirement that organizations have a safety committee with employee representatives which can discuss safety, identify problems and take action to prevent further problems. In Japan safety is on the agenda of quality circles.

Safety is enforced by factory inspectors who have the right to enter a factory at any time for inspection and must be called in after a serious accident. Inspectors have the right to issue two types of order for the enforcement of safety. An Improvement Notice requires the organization to make some improvement within a given time. For more serious cases a Prohibition Notice requires the organization to stop the machine immediately on the grounds of safety and not to use it until it has been made safe. Failure to conform can result in very large fines.

The 1974 Act applies not only to manufacturing but also to offices, universities, restaurants, shops, bus and train stations, construction sites and to agricultural activities – in fact to every place of work.

The Control of Substances Hazardous to Health (COSHH) Regulations 1988 require organizations to prevent hazards to employees and to society as a whole from potentially dangerous processes and materials. Toxic fumes, acids, hot liquids and some organic compounds are unhealthy to employees and unpleasant to residential neighbours. The cheapest solution of releasing pollutants into the air or into the local river are no longer environmentally acceptable. COSHH regulations require the safe use and disposal of such pollutants. Asbestos is an example of a substance now regarded as so dangerous that it should no longer be used. Handling the asbestos still in use requires precautions to avoid inhalation of carcinogenic asbestos dust.

JAPANESE MANUFACTURING METHODS

Historical background

Until the 1850s Japan was a closed society ruled by an hereditary military dictatorship. It remained feudal and agricultural. Industrialization started when Thomas Glover, a Scot, created the steel and shipbuilding industries in Japan. As a country Japan has few natural resources. It has to import most of its raw materials.

Following devastation in World War Two, the Japanese were quick to rebuild their industries. At that time Japan had a reputation for cheap, shoddy goods. Japanese companies blatantly copied the designs, packaging and logos of European and American companies. Japan has now moved beyond this stage in its industrial development but other developing East Asian countries are using this approach. In the late 1940s Deming, already mentioned in Chapter 9, set the Japanese on the path to productivity and quality improvement. He introduced quality circles, the active involvement of employees and the detailed analysis of manufacturing in search of continuous improvement. Japanese success in both quality and productivity is now obvious to the rest of the world. Some figures on Japanese productivity were quoted in Chapter 1.

Company structure in Japan is different from elsewhere and affects consumer buying behaviour. Most of the larger Japanese companies are known in Europe and America for selling cars and domestic electrical goods. It will be noticed that the same companies are in both of these markets, e.g. Mitsubishi makes cars and electrical goods. What is less well known is that these companies are also in shipbuilding, steel-making, coal-mining, petroleum, textiles and many other engineering industries. They also own banks, insurance companies, housing societies, travel agents, farms, supermarkets and petrol stations. Employees are thus able to buy everything they want from the company they work for, and are looked after from cradle to grave by the company. Employees therefore prefer to buy from their own company as a matter of loyalty and self-interest. This attitude is called *kokusanka* and is the reason why Japan imports very few industrial goods.

Modern Japanese manufacturing

Japanese industry has continued to develop new approaches to the continuous improvement of productivity. The Japanese word *kaizen* means continuous improvement. *Kaizen* is the basis of Japanese success in productivity improvement. Just-in-time (JIT) and *kanban* are Japanese ideas which, if used properly, can reduce throughput times and stock levels. The balanced outputs and synchronized workflow of a *nagara* production system are designed to achieve high productivity.

Japanese success in manufacturing is such that their approaches have been studied and some of them are now being copied in Britain. Caution is needed, however, in transporting Japanese ideas to Europe. Part of the reason for the success of the Japanese approach is the culture of the society in which it operates. Japanese methods do not necessarily work in Europe, which has a totally different culture. Japanese ideas can be used but need to be considered and modified to match the European culture. Ouchi, in his book *Theory Z*, showed how Japanese styles of management needed to be adapted, in this case to American culture. Social structure and low pay differentials in Japan also result in a low level of industrial conflict. Strikes are not, however, absent in Japan. International Labour Office figures have occasionally shown lost time due to strikes to be higher in Japan than in Britain, but more often the reverse is the case. Some figures for lost time due to strikes are shown in Chapter 23, Figure 23.1.

Higher productivity in Japan is not primarily due to differences in technology. It may be partly due to working 'harder' and longer hours but is mainly due to working 'smarter'. Better organization, better management, fewer levels of management and a greater delegation of responsibility for day-to-day decisions to employees are the key factors in Japanese manufacturing success. World-class manufacturing, to which European organizations aspire, is predominantly Japanese. Japanese levels of productivity, throughput times, stock levels, design, quality, reliability and maintainability are the benchmarks with which European organizations compare their own performance.

Japanese manufacturing managers do not sit in offices issuing orders. They do not have offices! They may have a desk somewhere in the factory. Their role is problem solving and supportive. They do this by being in the factory. Their style of management is known as *gemci genbutsu* – literally 'go and see'. This style of management is known in Europe as MBWA – Management By Walking About – and is similar to that advocated by Townsend in his 'upside-down organization' referred to in Chapter 4.

Over the last few decades Japanese companies have established factories in Britain and America. These factories operate with very high productivity. High productivity is therefore possible in Britain. It was always the workers who were blamed for low productivity in Britain. Japanese management has shown how productivity can be improved. It was the neanderthal style of British management which was the cause of low productivity in Britain.

At the end of this chapter a Japanese–English dictionary is provided which lists the words now used in Britain for the modern practices in manufacturing that have been imported from Japan.

DEFINITION OF TERMS USED IN MANUFACTURING

Available time
The time that a facility is available. This will be the working hours that the facility is manned and available for production. It would exclude preventative maintenance time but include breakdown and other delay times.

Capacity
The maximum output volume of a facility. For a single product it is working hours/ operation time.

Changeover time
The time to change a facility from producing one component to another (also called set-up time).

Cycle time
Time available for a repetitive task. Known in Japan as *Takt* time.

Down time
Time delay due to breakdown or maintenance.

Due date
Date by which a task must be completed. In-process due dates can be set for completion of intermediate stages of the work.

Facility
A machine, equipment or person or providing a process.

Idle time
Time for which a facility is not used because there is no work to do.

Lead time
Time from order to delivery. Applies to material orders from suppliers, sales orders by customers and to orders by one stage of manufacturing on another.

Loading time
Time to load and unload work onto and off a machine.

Operation time
Time for a task, operation or activity to be done manually or by machine.

Process time
Same as operation time, also called task time.

Productive time
Time period during which a facility is producing output. Usually less than available time.

Rest allowance
Time for an operator to recover from the fatigue of work (see Chapter 11).

Set-up time
Time to reset a machine for a different component (also called changeover time).

Standard minute
Unit of work. The volume of work that should be completed in one minute if the operator works at a standard rate.

Standard rate
The rate of work at which an operator completes a standard minute of work in a minute.

Standard time
Time it should take to do a task if the operator works at a standard rate of work.

***Takt* time**
Japanese term for the time available per unit of output. Measured as working hours/output volume.

Throughput time
Time for product to go through a work area. This is not the process time but includes also various waiting times and other delays. Sometimes called manufacturing lead time.

Utilization
The proportion of available time that a facility is working and producing.

Waiting time
Time waiting for an operation or for a move between operations. Excludes waiting for maintenance which is included in down-time.

Yield
The proportion of input material which is used in good output. This shows waste material (e.g. cutting components from a sheet of material) and defective components scrapped.

JAPANESE–ENGLISH DICTIONARY

Anki   Relief from the stress otherwise caused by fear of doing a task wrongly. Achieved by operators doing work designed using *poka yoke*.

Andon   A display of machine status on a board in a workshop. Yellow and red lights indicate a problem. Yellow means the machine is still working but a problem has been identified. Red means the machine has stopped because of a problem.

Bakayoka   Devices attached to a machine to detect abnormal operation. Based on in-process measurement of tool wear and condition monitoring. Used to achieve *poka yoke*.

Bucho   A head of a functional department.

Gemba   A supervisor, first-level manager.

Gemci genbutsu   A 'go and see' style of management.

Heijunka   Levelling the production schedule; providing a capacity that is the same for each process, producing the same volume of output each week and operating a market-pull, make to order system using *kanban* and just in time methods.

Jidhoka   Providing operators with the equipment, training and authority to be self-sufficient. This includes giving operators the authority to stop the line when a problem occurs and to display a light on the *andon* board.

Kacho   A head of a work section in a department.

Kakaricho   A head of a subsection within a section.

Kanban   A card or visible record which is used by a downstream process operator to order components from an upstream process area.

Kaizen   Continuous incremental improvement of all activities of the organization. Rates of future improvement are set as objectives and shown on an experience curve.

Keiretsu   Mutual shareholding among a group of companies in an industry. Creates mutual dependence and provides mutual support.

3 Ks   To be avoided:

Kiken   Dirt: cleanliness is important for pride in work.

Kitanai   Danger: safety is important for efficiency as well as for accident prevention.

Kitsui   Hard work: excessive hard work is unnecessary; greater output arises from working smarter not harder.

Kokusanka   National culture of preference for buying home-produced products. A form of nationalism, a Japanese national goal being self-reliance. This applies to manufacturers as well as the retail customer.

3 Ms   To be avoided:

Muda   Waste: to be avoided by eliminating the production of defective components.

Mura   Unevenness of output levels: to be avoided and so achieve *heijunka*.

Muri   Excess stock levels: to be avoided by using just-in-time scheduling methods.

Nagara   Literally means simultaneous operation on two or more tasks by an operator. However, the concept extends to a balanced zero-stock workflow through a synchronized manufacturing system.

Poka Yoke Error-proof design of components for assembly operations such that the only way to do the assembly is the right way, thereby ensuring zero defects. Uses jigs, fixtures and sensing devices (*bakayoka*) which prevent the task being done unless components are correctly positioned and correctly orientated.

Ringi The group decision-making process by consensus in which a middle manager (*kacho*) will circulate a proposal to other managers at levels above (*bucho*) and below (*kakaricho*) inviting approval.

Ringi-sho The document used in the *ringi* decision process. Managers sign the document to signify involvement and approval.

Soto Outside a person's group or sphere of influence. Contrast with *uchi*.

3 Ss to be achieved:

   Seiri Usefulness, contributing to effectiveness.

   Seiso Orderliness, leading to both safety and productivity.

   Seiton Cleanliness, for safety and pride in work.

Takt **time** Time available to make one unit of output.

Uchi Inside a person's sphere of influence; the interactions, consensus and norms which develop in a group. Contrast with *soto*.

KEY JAPANESE FIGURES IN MANUFACTURING

Ishikawa Originator of the 'fish-bone' diagram, sometimes called an Ishikawa diagram – a diagram with a pattern like fish bones used as a fault tree. It is a cause–effect diagram used in failure analysis.

Ohmae Business strategist with views on the global economy and the criteria for success.

Ohno Introduced *kanban*, *jidhoka*, zero defects and changeover time reduction into Toyota.

Ouchi Theory Z style of management (compared with Theories X and Y). Adapted the Japanese style of management to the American environment.

Shingo Involved with Ohno in *Kanban* and zero defects at Toyota but also with the reduction of changeover times to enable small batch sizes to be cost-effective. Press die changeover times in excess of eight hours were reduced to a single minute. Hence SMED, single minute exchange of dies.

Taguchi Application of sensitivity analysis to process variation to identify causes of variation. The objective being to reduce process variation to a value significantly less than the design tolerance and so ensure zero defects.

Task sheet 10   Manufacturing management

You plan to establish a small manufacturing business making a single product with yourself as manufacturing director responsible for designing and operating the manufacturing system and measuring its performance.

10.1   Consider the data needed by manufacturing management for each of these tasks.

10.2   Identify the information flows between manufacturing and other functions of management.

10.3   Consider any other sources of information.

10.4   Suggest some ways in which manufacturing performance could be measured. (Present your thoughts in the form of lists or tables.)

The data in Table 10.4 is a bill of material for a product. It includes the operation time for processing each component, subassembly and the final assembly of the product.

Table 10.4   Bill of material for a product

Part no.	Description				No. required	Operation time (min)
1000	Product 1 (final assembly)				1	3.0
	1100	subassembly 2			2	2.5
		1110	subassembly 4		2	1.5
			1111	component A	1	6.0
			1112	component B	3	–
		1120	subassembly 5		1	3.5
			1121	component C	2	4.0
			1122	component D	2	0.5
	1200	component E			4	–
	1300	subassembly 3			3	5.0
		1310	component F		1	1.0
		1320	component G		2	–
		1330	component H		3	2.0

10.5   Construct a product structure diagram.

10.6   Calculate the total number of each component in the product.

For an output volume of 500 units per 40-hour week:

10.7   Calculate the number of machines/facilities required.

10.8   Calculate the utilization of each machine/facility.

If labour is organized into two groups – (a) component processing and (b) assembly – and labour is flexible within these groups:

10.9   Calculate the labour requirements.

CHAPTER 11

MANUFACTURING SYSTEM DESIGN

MANUFACTURING SYSTEMS

A manufacturing system can be defined as a set of facilities that operate together to produce a set of components or products. A manufacturing system comprises hardware, software and human components. It will have objectives set in terms of output levels, quality and cost. The role of the manufacturing systems engineer is to design manufacturing systems which achieve these objectives.

The term manufacturing system can be applied at different levels of system. At its most basic level a manufacturing system could comprise a single machine producing a single component, but the term would normally apply to a set of machines linked together with handling equipment such as a conveyor or a robot. The system comprises not only the hardware but also the information systems and the human operators, if any.

At a higher level of system a workshop may comprise several basic manufacturing systems and a factory may comprise several such workshops. The factory itself, along with all its support functions, can be seen as a manufacturing system comprising smaller manufacturing systems as subsystems. Figure 1.1 in Chapter 1 reinforces this view of manufacturing as a system. The design of the whole manufacturing system is the subject of all of the chapters in this part of the book. This chapter addresses particularly the analytical techniques used to design basic manufacturing systems.

MANUFACTURING SYSTEMS ANALYSIS

Chapter 10 referred to the strategic manufacturing system decisions. This chapter considers the detail decisions and the analytical techniques used to derive the data on which such decisions are made. Chapter 12 goes on to consider the organization and physical layout of a factory. At this stage of the process of designing a manufacturing system, it is necessary to analyze in detail the tasks that need to be done in operating the system. Included in this analysis is not only the basic manufacturing tasks, which could be done by machines, operators or a

176

combination of operators and machines, but also the management tasks needed for the control of the machines and the flow of materials through them.

The analysis, specification and selection of manufacturing hardware is beyond the scope of this book. The analysis here will identify, quantify and decide among alternatives what tasks are to be done and how they are to be done. The analysis is at the most basic level of detail but nevertheless one that contributes to the effectiveness of the system as a whole. With any analysis, however, the cost of the analysis must be considered relative to the likely benefits. A vast army of analysts may cost more to the organization than the benefit they generate. This approach may appear to be reductionist – and if used without the framework of a systems approach it will be. It should be used within a systems approach, and if used this way will be effective.

The analysis considered at this stage, therefore, is of the work that is or will be done by machines and people. It is not a mechanistic analysis but must take account of people, their attitudes and motivation. Used appropriately it can create challenging jobs and achieve the motivation and commitment of staff. This is just as important to the productivity of an organization as the mechanical design of the work they do.

The main function of this analysis is therefore the design of work and the determination of the number and type of machines and employees required. The objective is to design work so it can be done efficiently, safely and cost-effectively.

METHOD STUDY

Method study analyzes work for efficiency and safety. Used in a reductionist manner it may look at work only from the organization's viewpoint and ignore the people doing it. This can lead to alienated employees, work that is unsafe, and jobs that are of short duration, repetitive, tedious and lacking in interest.

Used within a systems approach, method study will involve the people doing the work. They, after all, know most about it and will have ideas for ways of improving it. In the past employees were not allowed to think or to suggest improvement; only managers were supposed to do that. Method study, properly used, enables work to be designed so it is efficient and maximizes productivity. Adam Smith applied method study to the manufacture of pins and increased productivity by a factor of 2400. F. W. Taylor developed the approach and used it to increase productivity.

Method study uses a range of techniques to identify what work is done and a critical examination approach to existing designs to generate new ones. It is based on a six-stage methodology:

- *Select* work that is causing a bottleneck or likely to give a big improvement.
- *Record* how the work is currently done, quantifying time and distances moved.

- *Analyze* critically what is done, considering alternatives.
- *Develop* and define the most efficient way of doing the task.
- *Install* the new method with any training and equipment required.
- *Maintain* the effectiveness of the task in a changing environment.

Method study can be applied predictively to work design, as well as historically to existing work.

Several types of diagrams and charts can be used to record and visually relate the elements of work comprising a task. An *outline process chart* was introduced in Chapter 10 and shown in Figure 10.3. Here four further diagrams are introduced.

Flow process chart

This is a sequential list of the elements of work comprising the task, with elements classified, profiled and quantified. Tasks are classified as operations, inspections, moves, delays or storages (in a store). Figure 11.1 shows a flow process chart for the process of selling hamburgers.

Task: selling hamburgers

No.	Activity description	Oper. ○	Insp. □	Move ⇨	Delay D	Store ▽	Time (min)	Dist. (m)
1	Enquire customer needs						0.03	
2	Receive customer order						0.15	
3	Key order into till						0.10	
4	Put tray on counter						0.05	
5	Go to drinks machine						0.06	4.5
6	Fill cups with drink						0.18	
7	Bring drinks to counter						0.06	4.5
8	Put drinks on tray						0.03	
9	Go to burger rack						0.04	3
10	Pick up burgers						0.08	
11	Bring burgers to counter						0.04	3
12	Put burgers on tray						0.03	
13	Go to fries machine						0.06	4.5
14	Fill cartons with fries						0.18	
15	Bring fries to counter						0.06	4.5
16	Put fries on tray						0.03	
17	Check order complete						0.07	
18	Advise customer of cost						0.05	
19	Collect cash						0.10	
20	Give change						0.18	
21	Thank customer						0.02	
22								
23								
24								
	Total	14	1	6	0	0	1.50	24

Figure 11.1 *Flow process chart.*

Flow diagram

This indicates where the elements of a task are done on a plan of the work area, and hence shows where work flows around the workshop. This is useful for planning layout. Layout is covered in Chapter 12. Figure 11.2 shows a flow diagram for the process of selling hamburgers.

Gantt chart

This shows, for a group of people and/or machines working on a set of tasks, how their different tasks fit together in time. It is usually the case that a task cannot be started until a previous one has been completed. A Gantt chart – named after its originator Henri Gantt – shows these logical dependencies (also known as precedences). It shows how one facility may be waiting for another and how each facility is utilized over a period of time – days, weeks or months. Each task is shown as a block and drawn to scale in time. This chart is used for scheduling and loading facilities (Chapter 13), and for scheduling projects (Chapter 14). Figure 11.3 shows a Gantt chart.

Multiple activity chart

This chart is similar to a Gantt chart but is used to show the interactions and interferences between the work of several operators who are dependent on each other. It examines short-cycle repetitive work and determines the utilization of

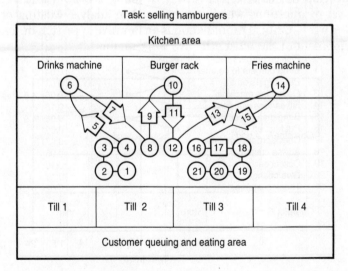

Task: selling hamburgers

Figure 11.2 *Flow diagram.*

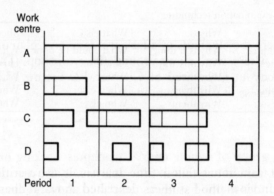

Figure 11.3 *Gantt chart for machine loading.*

operators and machines working in a group. It enables group work to be designed so as to be effective and operate with high productivity. Figure 11.4 shows a multiple activity chart.

Further diagramming techniques are used in other chapters in this part of the book. Network diagrams are used for line balancing in Chapter 12 and for project scheduling in Chapter 14.

Critical examination

A 'critical examination' technique can be used to analyze work. It is based on a series of structured questions that lead from what is done to what is best. It enables a user to develop the most effective way of doing a task. Table 11.1 shows a structured set of questions which can be used to analyze existing or proposed ways of doing work. Critical examination is an iterative approach designed to lead to the development of the most effective work method.

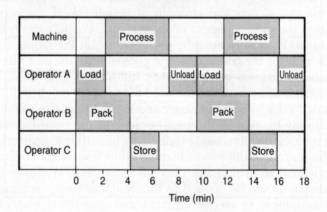

Figure 11.4 *Multiple activity chart.*

Table 11.1 Critical examination technique

Question	What	Why	What else	Should do
Purpose	What is done?	Why is it done?	What could be done?	What should be done?
Means	How done?	Why this way?	How else?	How should?
Place	Where done?	Why there?	Where else?	Where should?
Sequence	When done?	Why then?	When else?	When should?
Person	Who does it?	Why them?	Who else?	Who should?

Value engineering

Value engineering, sometimes called value analysis, is also a critical examination technique and includes method study as described above. It does, however, have a broader view and analyzes the design of a product in conjunction with the methods needed to do the tasks required to make the parts in the product.

Value engineering therefore uses the questions in Table 11.1, but only after having asked questions, in the same format, about the components in the product, the material used and the processes needed. The objective of value analysis and method study is to analyze and design products and work that are efficient by eliminating unnecessary and non-value-adding activities. The concept of value analysis was introduced in Chapter 8.

Ergonomics

Ergonomics is a systemic approach to work and product design based on understanding person–machine interfaces. It enables the design of work and products that 'fit' people (with method study, people had to 'fit' the work!). Product ergonomics was referred to in Chapter 8; here the emphasis is on applying the concepts of ergonomics to the analysis and design of work. Work design, as ergonomics may be called when applied to the design of work, identifies the type of work best done by people and by machines and uses anthropometric data about the shape, size and strength of people. Work for people must be designed within the physical limits of their ability. People vary in their physical capacity to do work, so work must be designed to suit the range of people likely to be doing it, not just the national 'average' person. In the past there was a need for a lot of heavy manual work in many industries. This caused illness and accidents. Modern equipment has removed the need for much of this type of work and has enabled work to become easier and safer. It has also enabled women to do jobs done only by men in the past because of the need for muscle power. Ergonomics is based on increasing productivity by working 'smarter', not working harder. In this way the reduction of manual effort should not lead to a reduction of output but instead to an increase. This was shown by F. W. Taylor in the 1880s.

Ergonomics should be applied to the design of workplaces, specifying the position of equipment to enable efficient and safe operation. The height of workbenches, desks and seats must be suitable or backache and other illnesses caused by bad working posture will occur. Operators on assembly lines may be doing tasks with short cycle times repetitively during the working day. This can lead to the overuse of some muscles, particularly those of the wrist.

Consider the checkout operator in a supermarket. Picking up and moving items over a laser barcode reader is an example of a short-cycle repetitive task that is done thousands of times in a working day. Loads may not be high but the continuous use of the same muscles can cause repetitive strain injury (RSI) and tenosynovitis.

In many jobs the working posture puts strain on the body, particularly the spine. Dentists not only spend a lot of time looking into patients' mouths, they often stand up and bend over in order to do so, especially when working on a patient's upper teeth. This can cause strain to the neck and lower back. Over the years dentists chairs have reclined patients further and further back until they are now almost horizontal. In this way dentists can avoid bending and reduce backache problems in later life.

These medical effects of poor work design not only cause ill health but cause far more lost production time than industrial disputes. Improvements in both health and productivity will arise from properly designed work.

As well as physical load, the mental and information-processing load involved in work must be assessed. If information load is excessive, employees will make errors. Inspecting small items passing on a conveyor belt is not effective as after a short period of time inspectors fail to see the defects they are supposed to be looking for. Air traffic controllers process information about the position of several aircraft and must guide them to safe take off and landing. Information overload can cause information to be forgotten or calculations to be made in error. Such errors are often referred to as human error, and many aeroplane crashes have been caused this way, but these are in fact system errors. The work system can cause such errors. The solution lies in suitable system design which avoids overload. Modern technology has reduced information overload, but work patterns may need to be changed, with operators changing to other tasks or having rest periods between repetition of the task. Rest allowances are needed to allow for both physical and mental overload.

The concept of motion stereotypes, referred to in Chapter 8, applies in the design of workplaces to the design of control devices and visual display instrumentation. Equipment must be designed to conform with these expectations otherwise errors will occur in performing tasks, especially if operators are stressed due to dealing with problems. Many industrial accidents have been attributed to non-conformance with motion stereotypes.

A further area of analysis in the ergonomics of work is the physical aspects of the environment in which the work is done. This recognizes the adverse effects of unsuitable temperature, humidity and light levels, and the adverse health effects of noise, dust and fumes. Working environments must be designed to avoid these problems. In doing so employee absenteeism due to ill health can be reduced. The greater productivity arising from reduced absenteeism more than pays for environmental improvements.

JOB DESIGN

Job design examines the sociological and psychological aspects of work and looks beyond the set of tasks to the whole job done by an individual. The aim is to recognize the need for a proactive approach to the design of jobs, one that creates the climate for positive attitudes, motivation and commitment. This requires delegation of responsibility for decision making and for control of work done and output. In the past, this aspect of management was ignored. This attitude led to dull, boring and monotonous jobs in which employees got a clear message that they were not important and should switch off their brains at work. Managers were there to take all the decisions; operators were treated almost as slaves. The result was an alienated workforce doing as little as necessary to keep their job but putting in no effort and no initiative in making suggestions for improvement even where improvements were obvious to the operators. The effect was low productivity, low profit and, in some cases, bankruptcy of the organization. More detail on the design of jobs is covered in Chapter 25.

WORK MEASUREMENT

Work measurement – which was developed by F. W. Taylor in the 1880s – involves measuring the time content of work and determines the *standard time* for a task. Work measurement is based on a six-stage methodology:

- *Select* the task to be studied.
- *Record* the elements making up the task.
- *Measure* the times for elements.
- *Analyze* occasionally occurring elements.
- *Define* the standard method and standard time.
- *Install* the method and train the operator to use the time.

An element of work is a small part (usually less than 1 minute) of a task. In all techniques of work measurement, analysis is at an elemental level. This enables the work method to be identified. Differences in task time may occur because different elements of work are included in successive repetitions of the task. A single measure of the time for the task as a whole would prevent the identification of this assignable cause of variation. Measurement at an elemental level enables occasionally occurring elements to be identified and to be quantified in terms of both duration and frequency of occurrence.

For machine work the standard time can be calculated knowing the speed of the machine and the task required. For manual work, this can be measured observationally or predictively. Since people can choose their own speed or rate of work, the time to do a task will be a variable, varying inversely with the speed of work. In order to obtain usable data, a 'standard rate' of working must be defined. A standard time for a task is then the time required for a qualified and motivated worker to perform a specified task using a specified method if the worker works at a standard rate of work and takes the appropriate rest period. A standard time includes allowances for occasional elements of work, delay and for rest pauses between tasks.

There are several different techniques of work measurement. The choice between them depends on the cycle time and frequency of use of the task since these determine the cost and benefit of the technique. Figure 11.5 shows the range of work measurement techniques available.

Observational techniques

Observational techniques of work measurement require an operator to perform a task in order that it can be measured using a stopwatch. *Time study* examines the time for a task; a *production study* examines a whole set of tasks in a period of time in order to determine how time is spent. Measurement is at an elemental level as described above.

Since the rate of work may not be standard, an assessment of work rate is made, which is then used to adjust the observed time to the standard time (the time it would have been if the operator had been working at the standard rate). This approach eliminates the variation of task time due to the variation in work rate and derives a time for the standard rate of work.

A task must be observed many times in order to identify the appropriate method and to obtain accurate results, given the random variation inherent in any form of measurement.

These techniques can only be used while a task is being done and can cause a lot of industrial conflict since operators may seek to obtain a larger time than actually required. Time study is time consuming and is only used for repetitive manual tasks.

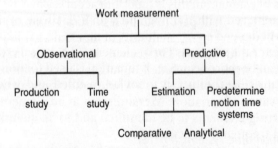

Figure 11.5 *Work measurement techniques.*

Assessment of rate of work

Observational techniques of work measurement require an observer to 'rate' the observed work. This process establishes the operator's rate of working relative to the *standard rate* of work. This is done subjectively but the result is expressed numerically using a rating scale. BS 3138 defines the rating scale in use in Britain. The standard rate is arbitrarily defined as 100. Slower rates of work would be assessed using numbers less than 100, and work faster than standard would use numbers greater than 100. In practical terms the scale could be said to range from 50 to 150 since a work rate outside this range would be difficult to achieve and even more difficult to measure with any accuracy. Rating is not a scientific process but a subjective assessment of the speed and effort of work. Observers need training to achieve reasonable accuracy and consistency. Table 11.2 shows the British Standard rating scale. For the standard task of walking, the rating scale can be related to walking speed.

Table 11.2 British Standard rating scale

Rate	Description of effort	Walking speed (m.p.h.)
150	Excessive effort, increased heart rate	6
125	Fast, dexterous and skilful	5
100	Brisk, businesslike, motivated	4
75	Slow, unhurried, unmotivated	3
50	Clumsy, fumbling, very slow	2

Predictive techniques

Predictive techniques of work measurement enable times to be derived before a task is performed and hence can be used for planning purposes and managerial decisions. These techniques are based on the following approaches.

Estimation

Estimates of task times can be made using estimators with experience of the work and of the process of estimation: this is quick but accuracy may be low. Acceptable accuracy can be achieved with training and with a database of times for similar tasks. Accuracy will depend on the similarity of new tasks to old ones. Estimation may be of the time for the whole task or of elements. Analytical estimation is based on elements and can be more accurate. Estimation is used for non-repetitive tasks where more expensive techniques cannot be justified and where 'reasonable' accuracy is adequate.

Predetermined motion time systems (PMTS)

These techniques analyze tasks down to the level of the motions of the body needed to do the task. At this level elemental times have been predetermined and are available as tables of data. A widely used technique is *methods-time measurement* (MTM), which was developed in the 1950s by observing work on cine-film. Analysis generated times for reaching, grasping, moving and positioning items using the hands. Times were also derived for leg, head, eye and body motions. Such times are 'universal' in the sense that they can be built up to determine the time for any task, since any task comprises basic motions of the body. The technique inherently includes method study since the method has to be analyzed in order to attach time to the basic work elements in the task.

At this basic level the technique is time consuming. The application/task time ratio is 350, i.e. a task lasting one hour will take 350 hours to be derived. By building times for commonly occurring sequences of elements, application can be made faster. Tables of data of this type will be orientated to particular types of work. There are tables, for example, for maintenance work and others for clerical work. Thinking time cannot be measured, so work involving mental tasks is not suitable for these techniques.

These techniques can be very accurate but are very time consuming and hence very expensive. They are used only for repetitive short-cycle manual or clerical tasks where accuracy is needed and the cost can be justified.

Synthesis

Where an organization maintains a library of task and element times derived from previous measurement, it is possible to use pre-existing times for these elements. An organization may also build up its own data for commonly occurring sequences of work elements. In this way the time for a new task can be generated without direct measurement. Such a time is said to be synthesized from pre-existing data.

Rest allowances

In addition to measuring the basic time for a task, it is necessary to take account of the necessary delay between repetitions of the task. The fatiguing nature of work, referred to above, is such that operators need some time for rest between tasks. The amount required depends on the physical and mental load of the tasks being done. All operators require some time away from their work during the working day for 'T' breaks and 'P' breaks. This is normally taken as a 5- or 10-minute break during the morning and a longer period for lunch. Such formal breaks stop production, so as an alternative work can be organized in groups to allow operators to take such breaks without the machines stopping.

Fatiguing jobs require longer breaks or moving operators to different tasks. Rest allowances must include breaks for recovery from mental as well as manual load.

While it is practice to subtract break times from the working time, it is more convenient to add a rest allowance to each task as a percentage of the task time.

In this way operators 'credit' themselves with an element of break time for every task they do, the amount depending on which task they do and how many times they do it. They do not take a small break after every task but a longer break after the completion of several repetitions of a task. The standard time for a task includes rest allowance. It is therefore a realistic time to use as the basis of calculation of capacity and daily output since all the delays for rest are taken account of in the standard time.

STANDARD TIMES

The standard time for a task is the time it is expected to take. This implies a standard rate of work and includes occasional elements of work and a rest allowance. In this form, then, the output expected in a day can be calculated, ignoring breaks since they have been included in the standard time (see capacity measurement below).

Standards times for tasks are needed and used for the following:

- Determining manpower and machine requirements.
- Scheduling of activities and loading of resources.
- Comparing different methods of work.
- Measuring utilization, performance and productivity.
- Performance-related payment systems.
- Costing of manpower and machines.

Standard costs are referred to in Chapter 20. For the cost of labour and of machine use the standard cost is derived from the standard time in the following way:

standard cost = standard time × cost rate

PRODUCTIVITY MEASUREMENT

Productivity can be measured at different levels of system (task, individual, group, department, factory or company) and in different units (time, physical output or money). There is no single measure of productivity; different measures measure different things. It is important to recognize what is actually being measured in each of the possible measures.

Time-based measures of productivity

Performance

$$\text{performance} = \frac{\text{standard time} \times 100}{\text{actual time for task}} \tag{11.1}$$

Performance (or efficiency) is a task-level, time-based measure which measures the speed of work relative to standard performance. Performance measures how efficiently a system is working while it is working. Standard performance (which is defined at 100) is the standard rate of work and is the performance achieved by an operator actually taking the standard time to do the task. When a method changes, the standard time is reduced. This is appropriate as the time expected to be taken for the task is reduced. However, in so doing, method improvements are excluded from this measure of productivity.

Utilization

$$\text{utilization} = \frac{\text{productive time}}{\text{available time}} \qquad (11.2)$$

Utilization measures the proportion or percentage of time available which is actively used. Utilization is likely to be less than 100 per cent due to down-time and idle time.

Effectiveness

$$\text{effectiveness} = \text{performance} \times \text{utilization} \qquad (11.3)$$

Effectiveness is a combination of performance and utilization, measuring the output over a given time compared with what it could have been. It includes the effects of both the speed of working and the proportion of the available time worked.

Physical or output measures of productivity

A physical measure of productivity can be made at the level of the factory by measuring the number of products (output units) made and ratioing this to the total number of employees or man-hours used. Different values would result according to whether all employees were included in the measure or only the direct factory operators.

$$\text{productivity} = \frac{\text{output units}}{\text{employee}} \qquad (11.4)$$

or

$$\text{productivity} = \frac{\text{output units}}{\text{man-hours}} \qquad (11.5)$$

These measures can be made for organizations producing a single product, or a range of similar products for which a single measure is possible. For example, in the motor industry it is possible to measure the number of cars produced per employee per year. Figures of this type for the motorcycle industry were quoted in Chapter 1.

Where an organization produces a range of different products it is not possible to add the outputs of the different products to produce a measure of output. In such cases it is necessary to convert the output of each product into a total work content. The outputs of each product are then measured in the same units and so can be added to form a measure of output for the factory as a whole. The total work content in a product can be called its target hours and would include hours for office as well as factory employees. The factory output can then be measured as the sum, for all products, of the target hours per unit of a product multiplied by the output level of that product. The input measure is the total man-hours of all employees.

$$\text{productivity} = \frac{\text{sum of (target hours/unit} \times \text{units/week)}}{\text{total man-hours/week}} \tag{11.6}$$

Physical measures define productivity broadly and include such things as improved methods. The target hour per unit would not be changed with changing method. In this way the measure reflects the increase in productivity due to improvements in methods.

Financial measures of productivity

Financial measures of productivity are also possible. These are measured at the level of the organization as a financial system. These are discussed in Chapter 20 but some examples are given below so that they may be compared with the physical and time-based measures given above.

$$\frac{\text{added value}}{\text{wages + salaries}} \tag{11.7}$$

$$\frac{\text{profit}}{\text{assets}} \tag{11.8}$$

$$\frac{\text{sales}}{\text{assets}} \tag{11.9}$$

Productivity measures compared

Each of these measures is measuring something different; each has its own value for particular purposes. Task-level measures of productivity define productivity narrowly, focusing on the level of the direct operators and measuring their speed of work. This measurement can only be made if all the work done has been measured.

Plant-level measures define productivity more broadly and include the improvements in output that result from better methods, better work scheduling and reductions in lost time.

Organization-level financial measures are even broader since they include other aspects of organization that contribute to improved sales, added value or profit. Further financial ratios of performance and productivity are covered in Chapter 20.

All of these measures can be used as the basis of measuring and controlling productivity. Any of them can be used as the basis of performance-related pay, a topic covered in Chapter 25.

CALCULATION OF MACHINE AND LABOUR REQUIREMENTS

At the planning stage of designing a manufacturing system, it is necessary to calculate the number of machines and amount of labour required for a given intended output of the system. Output is expressed as the volume of the set of products to be produced over a period of time. The time available per unit of output can be calculated as follows:

$$\text{time available per unit} = \frac{\text{working time}}{\text{output volume}} \tag{11.10}$$

The cycle time is the time required for a repetitive task. If the cycle time is less than the time available, then only one facility is needed to achieve the required output; if it is bigger, then more than one facility is needed.

The machine requirements can be calculated knowing the standard time for each operation necessary to produce the product. For a single process the number of machines required is given by the following equation:

$$\text{machines} = \frac{\text{output volume/week} \times \text{standard time}}{\text{working time/week}} \tag{11.11}$$

This calculation will normally produce a non-whole number, which is then rounded up to the next whole number to show the number of machines required. For example a result of 0.4 would show a need for 0.4 of a machine. Clearly one machine would have to be provided. The value 0.4 would then show its utilization, 40 per cent.

A result of 1.4 would show a utilization of 1.4, i.e. two machines are required with a utilization of 70 per cent. A result of 1.1 would mean either two machines, each utilized 55 per cent or working 10 per cent more hours than used in the calculation and so reducing the requirement to one machine.

When the number of machines is rounded up, the capacity provided will be greater than the demand on which the requirement was based. The capacity of these machines can be calculated using the formula for capacity given below.

Where several components are produced on the same machine, these can be added to determine the total machine requirement. These calculations can be done for several components involving several processes using the calculations as illustrated below for the products referred to in Tables 10.1–10.3 and Figure 10.2. Table 11.3 takes the total requirements of components in products from Table 10.3 and multiples them by the product requirements in months to derive the component requirements by month. The calculations here are based on forecasts of monthly output at three points a year apart. This will go on to show the future machine requirements to meet these forecast manufacturing volumes.

Table 11.3   Calculation of component requirements

Compo-nent	Product X	Y		Product	Jan 1997	Jan 1998	Jan 1999		Compo-nent	Jan 1997	Jan 1998	Jan 1999
A	2	0		X	1000	1200	1400		A	2000	2400	2800
B	4	0	×	Y	400	500	600	=	B	4000	4800	5600
C	2	3							C	3200	3900	4600
D	4	6		Product demand per month					D	6400	7800	9200
E	0	2							E	800	1000	1200
F	0	3							F	1200	1500	1800

The detail of the calculations is as follows:

$$\text{no. of A in Jan 1997} = \text{no. of A in X} \times \text{no. of X in Jan 1997}$$
$$+ \text{no. of A in Y} \times \text{no. of Y in Jan 1997} \quad (11.12)$$

Table 11.4 takes the calculations one stage further using operation time data for the components on three processes. This derives the total workload on each process for each month. The calculation follows the same approach as that illustrated above in equation (11.12).

Table 11.4   Calculation of process time requirements

Compo-nent	Jan 1997	Jan 1998	Jan 1999		Oper-ation time process	Component A	B	C	D	E	F		Time reqd	Jan 1997	Jan 1998	Jan 1999
A	2000	2400	2800		I	1.2	0.5	0.6	2.4	3.0	0		I	24080	29260	34600
B	4000	4800	5600	×	II	2.0	1.5	0.8	0	3.0	2.0	=	II	17360	21120	24880
C	3200	3900	4600		III	0	0	0	1.0	0	1.4		III	7600	9900	11720
D	6400	7800	9200													
E	800	1000	1200		Operation times in minutes								Processing time (min)			
F	1200	1500	1800													

If the factory works 8 hours per day, 5 days per week and 50 weeks per year, then the working time available is as follows:

$$8 \times 5 \times 50 \times 60/12 = 10{,}000 \text{ minutes per month} \quad (11.13)$$

The number of machines required for each process is shown in Table 11.5. The raw data is a set of non-whole numbers. These are then rounded up to show the number of machines required. Notice that the number of machines required increases over the three years to accommodate the forecast growth in output. Since the number of machines thus provided is greater than the theoretical number, the utilization is less than one and is shown in Table 11.5.

Table 11.5 Calculation of machine requirements

Machines reqd	Jan 1997	Jan 1998	Jan 1999	Machines reqd	Jan 1997	Jan 1998	Jan 1999	Machine utilization	Jan 1997	Jan 1998	Jan 1999
I	2.41	2.93	3.466	I	3	3	4	I	0.803	0.977	0.867
II	1.74	2.11	2.49	II	2	3	3	II	0.870	0.703	0.830
III	0.76	0.99	1.17	III	1	1	2	III	0.760	0.990	0.585

The calculation of labour requirements is the same as above if there is one operator per machine. However, it will be different if labour is flexible and moves between different machines or does work not involving machines. These calculations also have to take account of the intended location of machines and people in a factory. Aggregation (addition) of requirements can only be done if different items of work can in fact be done on the same machine or by the same person.

CAPACITY MEASUREMENT

Machines, other facilities and manpower resources are made available to provide an organization with capacity to perform tasks. Capacity can be defined as the maximum possible output in a given time period. Capacity needs to be measured in order to know the limit to which a resource can be loaded with work within a given time period. It is also used as a maximum output against which to compare actual output in an alternative measure of effectiveness.

Capacity is easy to measure only in the simplest of situations. For the manufacture of a single product on a single machine:

$$\text{capacity (units/week)} = \frac{\text{available time per week}}{\text{standard time per unit}} \tag{11.14}$$

The available time is the number of hours worked. This can be extended to cover several machines of the same type, the capacity being that of the set of machines. This formula is also appropriate for a single product in a product (line) layout, but the capacity of the line is that of its 'bottleneck' process, i.e. the process with the minimum capacity. Line balancing is a process of balancing the capacity of an assembly line and is covered in Chapter 12.

For a process layout with multiple products, in which the product 'mix' varies (mix = proportion of each product) it is possible only to measure capacity in terms of standard hours of facilities available. There is no guarantee that X standard hours of work can be processed through a workshop with a capacity of

X standard hours since in such a measure no regard is made to the type of resource available or required. Separate capacity measures would be needed for each process.

INFORMATION TECHNOLOGY IN MANUFACTURING SYSTEM DESIGN

Computer-aided design/computer-aided manufacture (CAD/CAM)

Computer-aided design (CAD) was referred to in Chapter 8. The CAD database contains much of the data needed in manufacturing. CAD/CAM links CAD with manufacturing in a manufacturing database containing process route data. Many modern manufacturing processes can be computer controlled. Computer numerical control (CNC) machines are programmed for a sequence of movements necessary to complete a manufacturing task. The process instruction data needed is that defining the shape of the component and the processes required on it. By transferring design data from CAD, manufacturing data can be created efficiently. A library of CNC programmes can be stored in computer memory and downloaded to the machine when required.

The areas of application of IT in manufacturing are shown in Figure 11.6. Application in layout planning is covered in Chapter 12 and in work scheduling in Chapter 13.

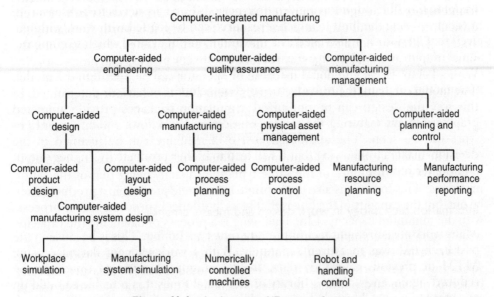

Figure 11.6 *Applications of IT in manufacturing.*

Automated manufacturing systems are controlled by a central or 'host' computer. This controls not only the operation of the machines but also the movement of materials on automated handling equipment such as conveyors or robots. Components can be identified by an automated vision device or by a barcode. The position of a component can be detected by optical or mechanical devices. Components can be automatically inspected. Tool wear can be monitored and tools changed automatically. The host computer receives information from these various devices and actuates the operation of the machines and other equipment. With this extent of computer control it is possible to design unmanned manufacturing systems. These systems may be expensive, and the cost must always be balanced with the cost of labour. Even so, such computer-controlled systems cannot cope with every eventuality. In some organizations the complex situations are dealt with by humans during the day and the factory programmed to run unstaffed with non-complex components during the night when humans are less inclined to work. Since there are no humans present, lights in the factory are not required, saving electrical costs in what is called a 'lights out' operation. If the system cannot deal with a problem at night it switches itself off and waits for the humans to arrive in the morning.

Simulation of manufacturing systems

Simulation is a powerful computer application for use in manufacturing system design. The process routes, task times for operations at each process and the output rates required can be used to build a computer model of the manufacturing system. This enables the behaviour of the system to be modelled and problems found before the design is committed to reality. Since errors, once made, are very difficult to rectify and can cause inefficiency for years, it is worth simulating the system. Different configurations of the system can be tested. For example, the simulation of an initial design may reveal a bottleneck process that limits capacity. With a keystroke an additional machine or operator can be built into the model. The model can 'run' the manufacturing system and its behaviour determined. In this way the design can be optimized. Simulation packages provide animated graphic displays featuring the factory on-screen. Workflows and any queues of work can be seen. The additional cost of a machine can be justified to the decision-makers by showing that it will lead to higher productivity, higher output and higher revenue.

Information technology in work design and measurement

Many work measurement techniques are time consuming. This is because in the past each task was measured separately. With a work element database using MTM or previous timestudy data, the generation of standard times can be reduced significantly. With a library of elemental times it is possible to build up and store times for 'blocks' of commonly occurring sequences of elements. Such systems lose their universality but enable a much faster generation of standard times. Without a database much work measurement would be prohibitively expensive.

The design of work for human operators can be assisted by SAMMIE, a computer system which simulates the human frame and enables the workplace and its equipment to be designed for optimal and effective operation.

Task sheet 11   Manufacturing system design

A workshop comprises 12 work areas (A–L, labelled sequentially in three rows of four) each 5 m square and separated by gangways 1 m wide. Material is delivered to the goods inward stores (A) where it is unloaded, inspected and stored. It is then moved in turn to areas D, J and L. In each area the work is left on a pallet awaiting processing and similarly awaiting transport to the next process. Finally it is moved to the despatch area (I) where it awaits inspection and when inspected is stored and eventually despatched.

11.1 Complete a process chart. Invent some times for the various activities and measure the distance the work moves within the workshop.

11.2 Produce a flow diagram for the above activities.

11.3 List the stages you would need to go through to measure the time for a simple manual task using a stop-watch. Consider the following:

Will you measure the whole task or break it into elements?

How many times will you measure it?

What happens if the operator uses a different method?

How do you deal with elements of work which occur occasionally?

How do you deal with the effect of a non-standard rate of work?

How do you account for rest pauses between repetitions of the task?

11.4 Visualize and list the basic motions of the hands necessary for replacing the fuse in a 13-amp electric plug. Assume that the plug, screwdriver and spare fuse are on a table in front of you. List the actions of the left and right hand separately.

CHAPTER 12

MANUFACTURING ORGANIZATION

OBJECTIVES OF MANUFACTURING ORGANIZATION

When products, processes and volumes have been determined, it is necessary to consider the locations within a factory or work area of the machines and other facilities. This analysis determines the space requirements for a manufacturing system and how it should be organized or 'laid out' physically. This stage of manufacturing system design requires analysis at four levels:

Level 4 Location of the factory geographically
Level 3 Location of departments in a factory
Level 2 Location of machines within a department
Level 1 Location of materials and equipment at a workplace

The decision of where the factory should be located geographically, level 4, including the location of storage facilities, is considered in Chapter 15, logistics. This chapter considers first decisions at level 3, how the total manufacturing task can be split into separate sections or 'departments', and then level 2, the layout of facilities within these sections. The analytical techniques as the same but they are applied at different levels of system. The detailed layout of a workplace, level 1, uses the techniques of method study and ergonomics explained in Chapter 11.

A proposed factory is likely to comprise many machines and may be planned to produce a range of different products. There are several different ways in which a set of machines can be arranged, and some ways may be more efficient than others. Analysis centres on 'matching' the layout of the factory to the products being manufactured and the processes used. An efficient layout enables manufacturing to be done more easily with reduced throughput times and costs. An inefficient layout constrains manufacturing effectiveness and is likely to do so for a long time since layouts are not easy to change.

Given that manufacturing volumes and the products themselves are likely to change over time as a result of product lifecycles, there is a probability that a once-efficient layout will become inefficient because it will no longer be producing what it was designed to produce. It was mentioned in Chapter 10 that a factory layout also needs to change to suit the different volumes at different stages of a product lifecycle and to cope with variable volume at the growth and decline stages.

It is important therefore to build-in some flexibility into the factory layout. This is not easy to achieve and will be limited in effect. Building-in flexibility costs more money than dedicated machines but may be beneficial if the manufacturing system is then able to deal with a range rather than a single product. Every so often a re-layout will be needed to match the layout to the product range in order to improve performance.

TYPES OF LAYOUT

There are five basic types of layout which split the total set of manufacturing activities into sections. This is the first stage of factory layout, level 3, the detailed layout within a section, level 2, is considered later in this chapter.

Fixed position layout

This type of layout is used for manufacturing very large assembled products on a one-off or very small volume basis. The product is assembled at a fixed position, with operators and components coming to this position for assembly. Examples include the manufacture of ships and aircraft (at least their final stages of assembly). The constructions of buildings, road and bridges are organized similarly because of the type of product involved. Figure 12.1 shows a fixed position layout.

Process layout

This type of layout, also known as job layout, locates similar processes together in the same section or department, different processes being in different locations or in different departments within the factory. Each section performs only one or a few of the whole set of process required on a product. There are likely to be several machines of the same type processing several different products. The volume of each product may vary but the variation of the total volume is likely to

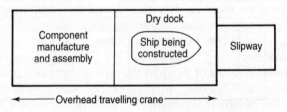

Figure 12.1   *Fixed position layout.*

be less. Variable volume of a product is therefore not a problem with process layout.

The sections or departments thus formed operate on the basis of providing a service facility. They do the tasks they can with the process they have. These sections receive work from other sections and, when completed within their section, pass the work on to the next section. Throughput times are very long as a product has to pass through several separate sections.

Process layout enables a range of products with different process routes to be handled. Work is simply sent to the appropriate section for the next process on a given product. With such variable routes internal transport must be flexible and may use forklift trucks. Employees in a particular section focus on their own processes and do not 'see' the manufacture of the whole product and are not aware of the timescale for completion of the product through all of its processes. This lack of knowledge leads to insular attitudes and an 'over the wall' syndrome where no one cares what happens to a component once it has gone on to the next section. The barriers between sections are not only physical but psychological.

Process layout is very flexible in that the same facilities can be used for a variety of products. The facility can therefore cope (to some extent) with changing volumes and products.

Process layout is used for small and changing product volumes and by subcontactors offering a service based on a particular process. Such organizations do not have any products of their own. The work may be a one-off, a single or small number of batches. If the work exists only for a short time period, then the layout cannot be oriented to a particular product but instead to an ability to process a wide range of jobs that change frequently.

Process layout is also used in larger organizations that have their own products at the growth and decline stages of the product lifecycle. In these cases the layout provides flexibility in handling variable volumes. Figure 12.2 shows a process layout.

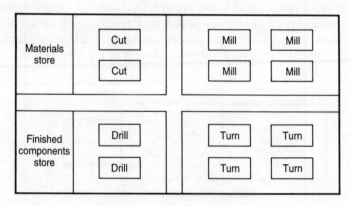

Figure 12.2 *Process layout.*

Cell layout

This type of layout locates together all of the processes required for a set of products. Cells should be relatively small and will usually deal with a small number of related products that share some common processes. Process routes do not have to be the same.

Cellular layout enables short throughput time and good control since the whole set of processes for a product is done within the cell. Cellular layout is flexible and can cope with a changing mix of products without much loss of efficiency. Cellular manufacturing is the basis of modern manufacturing. Figure 12.3 shows a cell layout.

Product layout

This type of layout, also known as line layout, locates together all of the processes for a product. In its pure form a line will deal with only a single product on a single process route, but some flexibility can be introduced to enable a line to deal with a small range of similar products so long as they have the same process route. Given a single process route in which all items being processed follow the same sequence of processes, the machines are placed in an order to suit this sequence. It is then possible to automate the handling and transfer of items between processes by means of a conveyor belt, robots or similar devices.

Product or line layout is often, but does not have to be, in the form of a straight line of machines. An alternative is a U-shaped layout, which is in effect two straight lines joined at one end. This fits better into a near square workshop and has both its start and finish at the same end of the workshop for loading and unloading from and to stores.

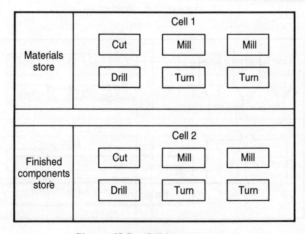

Figure 12.3 *Cell layout.*

Product layout is very efficient for the volume for which it is designed. It is tailor-made to a specific volume of a specific product. It may contain tailor-made machines for processes which are specialized for the product. These can be very efficient for the job they do. Lines do, however, lack flexibility, though they may be able to cope with some product variation. Lines are inefficient if volume falls and often cannot be used to manufacture different products.

Product layout is used for high and stable volumes of manufacture, such as occur in the mature phase of a product lifecycle. In these cases this type of layout is very efficient and processes work with minimum throughput time. Figure 12.4 shows a product layout.

Flow layout

This type of layout applies to those processes in which the material being processed flows continuously during the working period, often 24 hours a day, 7 days a week. The product is usually liquid or granular so its flow is often through pipes. Flow layout is used for oil refineries, chemical processing, and steel and

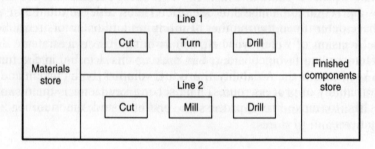

Figure 12.4 *Product layout.*

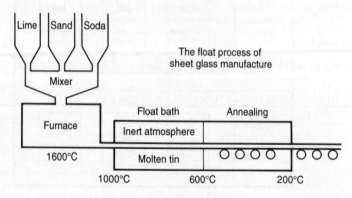

Figure 12.5 *Flow layout.*

glass manufacture. The layout is efficient when operating at its design volume but cannot be used for anything else. It cannot easily operate at reduced volume since stopping and restarting the system is expensive. Where, for example, furnaces are used, these would be heated for 168 hours per week even if they were used for manufacturing for only a day shift (say 40 hours per week). These plants therefore usually operate on three shifts per day for 7 days per week. Figure 12.5 shows a flow layout for glass manufacture. Molten glass moves continuously. If it moves at 6 m per minute it will produce 3000 km of glass in a year.

LAYOUT GENERATION

Data needed for layout planning

The starting point for layout planning is the product volume, product structure and process route data mentioned in Chapter 10. Since layout is not easy to change, layout planning is a medium- to long-term planning task. Some forecast of future product data is therefore required, not just current data.

Strategic manufacturing decisions, referred to in Chapter 11, will have determined which components are to be made in-house and hence which processes are required. Analysis of operation times, also mentioned in Chapter 11, will have determined the number of machines and operators required.

The decision of which basic type of layout to use is a strategic decision. Several factors need to be considered in making this decision and generating a layout. A key factor is the variability of product volumes, both current and future, and the variability of process routes. The second key factor is the distance that material has to move from process to process, and the number of such transportations.

Applicability of different layout types

Table 12.1 shows the characteristics of manufacturing systems and applicability of the different types of layout. This assists in selecting a layout type.

Layout and workflow

The planning of workflow is covered in Chapter 13 but it is relevant here to see

Table 12.1   Manufacturing characteristics and layout type

Layout type	Fixed	Job	Cell	Line	Flow
Volume	Low $\longleftarrow$			$\longrightarrow$	High
Volume	Variable $\longleftarrow$			$\longrightarrow$	Stable
Products	Special $\longleftarrow$			$\longrightarrow$	Standard
Variety	High $\longleftarrow$			$\longrightarrow$	Low
Orders	Make to order $\longleftarrow$			$\longrightarrow$	Out of stock
Scheduling	Each operation $\longleftarrow$			$\longrightarrow$	Once only

how workflow relates to layout. Traditional books refer to job, batch and line production systems. These terms refer to the pattern of workflow rather than layout, but are related to layout.

A job production system deals with one-off contracts. Job shops will normally be laid out as process layouts.

Line production systems process work along a line that is dedicated to the required type and sequence of work so a product layout is used.

A batch production system processes batches of work. A batch is processed for some time before the equipment is switched to another batch, which may be of a different component. A batch may be a pallet or boxful. Batch size is covered in Chapter 17, stock control. Batch production is not related to layout. Batches can be processed through process, cell or product layouts.

LAYOUT EVALUATION

Quantitative analysis

With a wide variety of products and process routes, work will need to flow between many different facilities. In a manufacturing system of some complexity it is necessary to determine these flows using a technique known as *production flow analysis*. This will translate product volume and process route data into the weekly volumes of flow between each pair of facilities, creating what is known as a volume matrix. Table 12.2 shows some product volume and process route data. Table 12.3 shows a volume matrix based on the volume and route data. This becomes the starting point for layout choice.

Table 12.2 Product volume and route data

Product	Volume	Route
1	10	A–B–C–D–E
2	15	A–B–D–E
3	6	A–C–B–D

Table 12.3 Volume matrix

From\to	A	B	C	D	E
A	–	25	6		
B		–	10	21	
C		6	–	10	
D				–	25
E					–

Because the volume matrix shows which facilities should be located near to each other it is sometimes called a 'proximity desirability matrix'.

A layout can be evaluated quantitatively in terms of the number of machines and operators required, the distance moved by material and operators, and the likely throughput time for the products. Comparison and choice of layout will be based on these quantitative measures plus management judgement of other factors and possible future developments.

The number of machines required can vary from one layout to another for the same output because of the different way in which machines are grouped. As shown in Chapter 11, the number of machines needed for a given capacity is likely to be a non-whole number. A given process may require 2.8 machines, so in a process layout three machines would be needed. However, in a product layout that has two lines, each line may require 1.4 machines, and so each line would need two machines, i.e. four in total. The efficiency of a layout can be measured as the utilization of the machines or the utilization of labour. These are calculated as follows:

$$\text{machine utilization} = \frac{\text{theoretical machines required}}{\text{actual no. of machines needed}} \tag{12.1}$$

$$\text{labour utilization} = \frac{\text{theoretical operators required}}{\text{actual no. of operators needed}} \tag{12.2}$$

It is possible to measure the distance between each pair of facilities in an existing or a proposed layout in a distance matrix as shown in Table 12.4.

Table 12.4 Distance matrix

From\to	A	B	C	D	E
A	–	10	15	20	12
B	10	–	10	15	18
C	15	10	–	10	12
D	20	15	10	–	10
E	12	18	12	10	–

The volume matrix referred to above is product data, but the distance matrix refers to a given layout. These data can be combined to provide a measure of the given layout processing the given products. This is obtained by weighting the distance between a pair of machines by the volume of work flowing between them. A volume–distance matrix is created by multiplying, cell by cell, the volume by the corresponding distance. Table 12.5 shows the volume–distance matrix for the example data.

Table 12.5 Volume–distance matrix

From\to	A	B	C	D	E
A	–	250	90		
B		–	100	315	
C		60	–	100	
D				–	250
E					–
Volume–distance total = 1165					

The sum of these volume–distances is another useful measure of the efficiency of the layout. For the example data the volume–distance total is 1165. This figure has little meaning on its own but can be compared for different layout proposals. A lower figure indicates a more efficient layout.

Clearly a layout will be more efficient if distances moved are minimized. Thus those sectors on which the flow is greatest should be located nearest to each other. By relocating facilities it should be possible to reduce the volume–distance total and generate a more efficient layout.

The flow diagram, introduced in Chapter 11, is used to show the movement of material or operators around a workshop and is useful in designing factory layouts. Unnecessarily long distances can be seen on such a chart.

Layout and productivity

Factory layout crucially affects performance. Layout decisions once made and implemented are not easy to change. Machines will have been screwed to the floor and have service ducting for electric power, computer links, water and perhaps fume extraction and automated handling equipment. Once set up it would be very expensive to rearrange them. Layout decisions are therefore medium to long term in their effect. An organization with an inefficient layout will suffer from inefficiency for a long time. It is important, therefore, to get the most efficient layout in the first place and to build-in some flexibility so that the layout can retain its efficiency with changing volumes and products.

Process layout is very flexible to changing products and often requires fewer machines than other forms of layout for a given set of products. It can, however, be inefficient in terms of distance and throughput time. Scheduling of work has to be done centrally and separately for each process.

Product layout may require more machines than process layout but distance moved and throughput time is much less. Work is scheduled centrally only once for the first operation. Work then flows automatically down the line. Throughput time should not greatly exceed the sum of the process times.

Cell layout combines the best features of both process and product layout. It has the speed of a product layout and the flexibility of a process layout. Throughput times are lower and so are both machine and labour requirements. Cell layout is able to cope with a changing mix of its products on the basis that, as some volumes go down it can accept higher volumes of other products or new products. The 'mix' of products is the proportion of volume of each product.

Fixed-position and flow layouts are rather specific in their application and limited to large products or flowing liquids or powders respectively. They are not widely used in general manufacturing but it is necessary to know they can be applied in appropriate situations.

Layout and organization

Organization must match the layout. Areas of supervisory responsibility relate to 'departments', which depend on the form of layout chosen. Scheduling information and control systems will be based on these departments. In a large manufacturing organization the separate parts of it may be laid out in different ways. Component manufacturing may be based on job layout, but assembly may use lines. Different products may occupy different parts of a factory and operate with different layout types.

With the physical split of manufacturing activities it is necessary to consider the decision structure, particularly the extent of centralized or delegated decisions. Information and decisions may be centralized. Certainly a centralized database is needed, but decisions can be delegated. Each 'department' created by layout planning will have a supervisor, but how much autonomy does the supervisor have? Modern practice would call for considerable autonomy, but not total independence since the different departments must work together effectively.

Each department can be treated as a separate division of the organization, virtually as separate organizations. Each department would then be accountable financially for its performance. Each department in this form of organization is described as a 'plant within a plant' or as a 'strategic business unit'. They are not legally separate organizations but have much of their autonomy and performance. This requires appropriate cost accounting systems, which are covered in Chapter 18.

LINE BALANCING

When manufacturing activities are laid out as a line in a product layout it is important to ensure, as far as is possible, that the capacity of each facility on the line is approximately the same; to do otherwise would be inefficient and wasteful. Facilities should therefore be planned to provide a similar capacity. This may involve buying larger or smaller machines to suit.

When designing assembly lines that involve many manual assembly tasks it is possible to organize where and when tasks are done to maximize efficiency. For example, the assembly of a product may require say twenty tasks, of differing standard times. Doing each in sequence with each being done by a single operator would be very inefficient. There may be some flexibility in the sequence in which the tasks are done, but much of the sequence will be fixed. By having, say, five workstations on the assembly line, it is possible to group tasks together at a workstation in a way that accommodates the given sequence constraints but brings the total standard time at each workstation to around the same value – a value that is feasible for the intended output volume. In this way each workstation, and hence each operator, will be equally loaded and thus have the same capacity. Work will therefore be able to flow smoothly along the line without 'bottlenecks' of lower capacity. Such a line would be balanced and would operate efficiently. In practice perfect balance is unlikely; the best solution is that in which the range of workstation standard times is a minimum. The efficiency and utilization of a line can be measured as follows:

$$\text{line efficiency} = \frac{\text{standard time for the product}}{\text{no. of workstations} \times \text{cycle time}} \qquad (12.3)$$

$$\text{line utilization} = \frac{\text{actual cycle time}}{\text{maximum cycle time}} \qquad (12.4)$$

In many cases there will not be a single feasible solution but several alternatives for consideration.

An example of line balancing

Table 12.6 Tasks required for assembly of a product

Task	1	2	3	4	5	6	7	8	9	Total
Time	1.2	0.7	0.9	0.9	1.5	1.8	0.4	0.4	1.5	9.3

Table 12.6 shows some data about a set of tasks needed to assemble a product. For each task a standard time is given. The dependence for each activity is an indication of which other activity or activities need to be done before the given activity can be done.

From the volume data it is possible to calculate the maximum cycle time. In this case the output per day is 240, so the time available to assemble each one is 1/240 of a day. If the working day is 8 hours or 480 minutes then the maximum cycle time is 2 minutes. This means that tasks can be combined together but the total time at a workstation must not exceed 2 minutes. The actual cycle time may be less than this. The actual cycle time for a line is the maximum of the cycle times at its workstations.

From the total work content, 9.3 minutes, it is possible to calculate the minimum number of workstations:

$$\text{minimum no. of workstations} = \frac{\text{total work content}}{\text{cycle time}} \qquad (12.5)$$

The value obtained is rounded up to the next highest whole number. In this case this is $9.3/2 = 4.65$, which is rounded up to 5. This does not means that the number of workstations has to be five, only that five is the minimum number needed to ensure that the cycle time is below that needed to achieve the required output. In some cases it may not be possible to achieve this minimum; in other cases a larger number could be preferred.

From the activity list and dependencies it is possible to draw a network showing the sequences of activities. This is shown in Figure 12.6.

By visual inspection it can be seen that some activities can be grouped together, up to the limit of the cycle time. Figure 12.7 shows a possible grouping

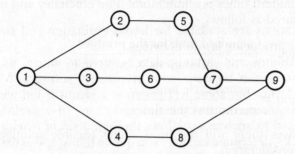

Figure 12.6 *Network diagram of assembly tasks.*

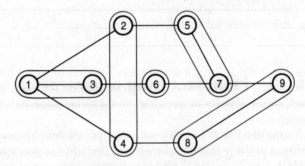

Figure 12.7 *Grouping of tasks in a line.*

of activities for this example, and Table 12.7 shows standard times for the five workstations.

Table 12.7 Workstations for assembly of a product

Station	1	2	3	4	5
Tasks	1	2	6	5	8
	3	4	–	7	9
Time	1.9	1.9	1.8	1.9	1.9

Measures of the efficiency of a line can then be measured and compared for alternative proposals.

line efficiency   = 9.3/(5 × 1.9) = 0.979   (12.6)

line utilization   = 1.9/2.0   = 0.95   (12.7)

line effectiveness = 0.979 × 0.95 = 0.93   (12.8)

INFORMATION TECHNOLOGY IN LAYOUT PLANNING

Computer applications are available for layout evaluation and simulation, line balancing and for producing layout drawings.

Using the volume and distance data referred to above, the quantitative evaluation of layouts can be done with proprietary software. Most will only evaluate a given layout but some will generate a layout if fed with data about machine space requirements and the size and shape of a workshop. Queuing models can be used to analyze workflows through a set of machines. This will identify bottlenecks, determine capacity and queue times, and assist in estimating throughput times. Queue analysis requires statistical analysis beyond the scope of this book.

Computer-aided design (CAD) can be used for drawing a proposed layout. Libraries of standard icons enable layout drawings to be produced quickly. Drawings of the electrical and other service requirements can be produced to form the working instructions for installation of a layout.

Task sheet 12   Manufacturing organization

Products X and Y require the processes and number of machines as shown in Table 12.8. Routing is in the sequence shown and the volume is output per week. Each machine requires an operator. Labour is flexible between machines within departments. Distances moved are 10 m within departments and 50 m between departments.

Table 12.8   Process route data for products X and Y

Process	1	2	3	4	Routing	Volume
Product X	1.3	0.4	0.3	0.6	1–2–3–4	100
Product Y	0.5	1.4	1.2	–	1–2–3	100

12.1 Analyze possible layouts of facilities if the manufacturing system is based on:
(a) process layout
(b) product layout
(c) cell layout

12.2 Measure the efficiency of each layout using the data given.

12.3 Consider the implications of layout for organization, control and throughput time.

12.4 A product is to be assembled on a line. The activities required and their standard times are shown in Table 12.9. Figure 12.8 shows the dependencies between the assembly activities. The output required is 100 units per 8 hour day. Determine the number of workstations and the activities that each should do. Calculate the efficiency and utilization of the line.

Table 12.9 Activities for assembling a product

Task	1	2	3	4	5	6	7	8	9	10
Time	1.2	2.1	2.3	2.5	4.0	3.2	1.2	1.1	1.6	3.6

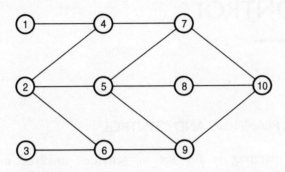

Figure 12.8 *Network of assembly activity dependencies.*

CHAPTER 13

MANUFACTURING PLANNING AND CONTROL

THE NEED FOR PLANNING AND CONTROL

Manufacturing planning is the set of activities involved in planning the machine, material and manpower resources to match the forecast level of future manufacturing activity. It plans the operation of a manufacturing system. This is distinct from process planning, which defines which processes are to be used.

Manufacturing control is the set of activities involved in monitoring the activity in the factory or operational area and taking corrective action when necessary. Reference to Figure 1.1 in Chapter 1 will enable the reader to visualize the information flows needed for this activity.

Although manufacturing planning and control is concerned with the operation of a manufacturing system, this cannot be done entirely within the manufacturing function. This function of management is the nerve-centre for planning and controlling the primary activity of the organization. It does so in a close relationship with both the purchasing and marketing functions of management covered in Chapters 6 and 16. Purchasing relates to suppliers of materials and marketing to customers who buy products. These functions of management work with manufacturing planning and control, exchanging data on product requirements and material availability in order to determine and achieve the required manufacturing output.

The objective is to manage the flow of materials through the manufacturing system to achieve output volumes of the required quality and quantity at the required times using minimum resources. Given the complexity of many manufacturing systems, this is not an easy task. Because complexity prevents efficiency, modern manufacturing practice tries to reduce complexity by splitting a large system into smaller systems, each with reduced variety. This concept is known as manufacturing focus. Cellular manufacturing, introduced in Chapter 12, is one way to split a large system. By delegating managerial authority and control down to cell level, each cell operates as an autonomous unit that can function effectively.

210

MANUFACTURING PLANNING

Levels of planning

Planning is done at various levels within an organization. The longer the time-span of planning, the higher the level of management involved. Table 13.1 gives a general indication of planning levels with typical periods and overall time-spans. The period of a plan is the unit of time for which data is stated and over which performance is measured. A plan will usually compromise several such periods, which together cover the span of the plan. Data for earlier periods is likely to be more certain, with estimates used with less confidence for later periods. Plans are updated each period, or more frequently, to incorporate more definite data about future activity.

Table 13.1 shows four levels at which planning might be done in an organization. Plans can be displayed visually using Gantt charts, introduced in Chapter 11.

Table 13.1 Levels of planning

Timescale	Content	Level	Period	Span
Long term	Market Finance	Board	1 year	5 years
Medium term	Capacity Manpower	Management	1 month	1 year
Short term	Materials Labour	Supervision	1 week	1 month
Day to day	Tasks	Operator	1 day	1 week

Capacity planning

This relates primarily to the provision of machines and other facilities needed. This task is simple if an organization produces a fixed volume of a single output, but is complex when a range of different outputs are to be provided and the levels of demand for them are variable and uncertain. A capacity model is needed which relates the capacity required for each resource to a variety of possible patterns of future demand.

In the long term capacity is variable – resources can be acquired or disposed of. Capacity can be planned to meet forecast needs. This stage of planning is concerned with the detail *design* of the manufacturing system. It is involved with determining how many machines are required to provide the capacity from the processes specified in the concept design of the manufacturing system. This form of planning was covered in Chapter 11.

In the short term capacity is fixed – it cannot be changed quickly. Work is scheduled within capacity limits. This stage of planning is concerned with managing the *operation* of the manufacturing system. It is involved with determining short-term capacity needs, and hence working hours and manpower requirements, to provide the capacity from the given manufacturing system. This form of planning is covered in this chapter.

Capacity planning for variable demand

Organizations must set the level of capacity to meet demand, but where demand is variable, e.g. seasonally, then a suitable level of capacity must be determined. The sales of ice-cream vary seasonally. If capacity is set to meet maximum demand, then the facilities will be idle for most of the year. If set at minimum demand the organization will be unable to meet maximum demand. By setting demand at some intermediate level, the annual capacity can be set to meet the annual demand. This involves making more than needed in a slack sales period and storing it so it can be sold in a peak sales period. The cost of storage must be taken into account. For foods, expensive refrigerated stores would be needed. A balance is needed between the higher cost of a larger facility incurring lower stocks, and the lower cost of a smaller facility but incurring higher stocks. The cost analysis associated with stocks is explained in Chapter 17. Figure 13.1 illustrates the decision to be made in setting capacity for seasonably variable demand. The capacity line is variable in its vertical position.

Tables 13.2 and 13.3 show some seasonally variable demand data, together with calculations of cumulative demand and cumulative output which enable decisions to be made of monthly output requirements to meet the variable demand.

In Table 13.2 the demand is set at 20 per month, the minimum needed to produce the annual volume of demand. The result of this approach is to save money on the machines and facilities providing the capacity, but to manufacture goods in advance of requirement and store them over a period of time. The average level of stock in this case is seen to be 14.4 units.

In Table 13.3 the capacity is revised to a higher level of 24 per month. This requires more machines but reduces the average level of stock to 2.7 units. Increasing the capacity to 28 would enable manufacturing to equal demand in

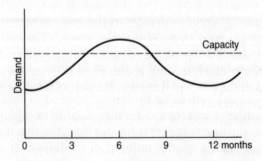

Figure 13.1   *Setting capacity for variable demand.*

Table 13.2 Setting capacity for variable demand

	Month												
	Jan	Feb	Mar	Apr	May	Jun	Jul	Aug	Sep	Oct	Nov	Dec	Total
Demand	12	15	19	22	25	27	28	25	21	17	14	15	240
Cumulative demand	12	27	46	68	93	120	148	173	194	211	225	240	
Capacity	20	20	20	20	20	20	20	20	20	20	20	20	240
Cumulative capacity	20	40	60	80	100	120	140	160	180	200	220	240	
Manufactured	20	20	20	20	20	20	20	20	20	20	20	20	240
Stock at end	22	27	28	26	21	14	6	1	0	3	9	14	

Assume the stock of 14 in December is available at the start of January. The average stock in this case is 14.4 units.

Table 13.3 Revised capacity for variable demand

	Month												
	Jan	Feb	Mar	Apr	May	Jun	Jul	Aug	Sep	Oct	Nov	Dec	Total
Demand	12	15	19	22	25	27	28	25	21	17	14	15	240
Cumulative demand	12	27	46	68	93	120	148	173	194	211	225	240	
Capacity	24	24	24	24	24	24	24	24	24	24	24	24	288
Cumulative capacity	24	48	72	96	120	144	168	192	216	240	264	288	
Manufactured	12	17	24	24	24	24	24	24	21	17	14	15	240
Stock at end	0	0	2	7	9	8	5	1	0	0	0	0	

Assume the stock of 0 in December is available at the start of January. The average stock in this case is 2.7 units.

each month with no carry-over stock from month to month. The financial analysis of such situations is covered in Chapter 21.

It may be possible to achieve a reduction in seasonal variation for some products by developing similar products for marketing in slack periods, e.g. introduce a range of ice-cream desserts which can be sold in winter. Alternatively, different products could be introduced. These may not use the same machines but would use the same labour and some of the same equipment. For example, some manufacturers of ice-cream also make sausages which have an inverse seasonal variation to ice-cream.

Capacity is not completely fixed in the short term. It can be varied, within limits, by changing the number of hours worked per week, by subcontracting some work during a peak period or by buying instead of making components. The capacity of existing resources can be increased by reorganizing work so the ancillary tasks done on-line (necessitating stopping the machine) are done off-line instead. In the longer term the number of facilities may need changing to meet changes in the level of demand.

Planning boards

In the pre-computer era, activity scheduling and control were done on large, wall-mounted planning boards. Work centres (section or machines) would be listed vertically and time horizontally. This type of chart is a Gantt chart, introduced in Chapter 11. Such charts can be used for short- or medium-term planning.

Activities would be represented by cards cut to length in scale to operation time and placed on the board at the relevant calendar time for the appropriate work centre. Tasks would be scheduled at a time to permit achievement of output due date and within the limit of the capacity of the work centre. Activity cards would be 'juggled around' to fit with the capacity limits and due date requirements. The sequential links between activities could not be seen; this was 'in the head' of the planner. Planning on this basis was a major and difficult task: it was not easy to achieve efficiency. Time delays in receipt of information back from the shopfloor led to long throughput times. To minimize this people known as 'progress chasers' were employed to find out what was going on the shopfloor, where work was being delayed and to get things moving faster. In some cases large numbers of people were employed in this way, adding considerably to the manufacturing overhead cost.

MATERIAL REQUIREMENTS PLANNING

The operation of a manufacturing system involved the processing of components and assemblies on facilities at particular times. This form of planning is referred to as scheduling. The objective of manufacturing scheduling is to operate the manufacturing system effectively and to achieve manufacturing outputs.

Scheduling involves planning the flow of materials and the activities in a manufacturing or operating system. Work is done on tasks and activities but outputs and forecasts are expressed in units of output. This form of planning is therefore one of 'explosion' of a product into its component parts, and of 'translation' of output requirements to the schedules of work required to achieve the output. This must take account of product structures, process routes, lead times and throughput times, and the capacity of the physical and manpower resources. As well as determining when work is to be done, this process also determines when raw materials need to be ordered and delivered to enable work to be done at the planned time.

This analysis needs to be done for all of an organization's products and schedules all manufacturing work to be done and determines the purchase schedule for manufacturing materials.

The process starts with the desired outputs and works backwards (back scheduling) to the required inputs. Given the throughput time, these will be scheduled for dates earlier than the product due dates.

The process of translation of a schedule of product demands into schedules of component-processing requirements and purchase determines the material requirements to achieve the objective outputs. This process was shown in Chapter 11, Tables 11.3–11.5, where it applied to long-term capacity planning using data three years apart. Exactly the same process is done at this stage but using forecasts one month apart. This is managing the flow of work through an existing manufacturing system.

The existing current stock, at various stages of processing are taken into account, together with any intended future stocks, in order to determine the *net* requirements for processing and purchase. This process is known as *materials requirements planning* (MRP). This is, however, only the start of the process since it initially ignores the capacity of the manufacturing system.

A second stage of planning, scheduling within capacity limits, and varying short-term capacity where necessary and possible, takes a more complete view of the whole of manufacturing scheduling and is known as *manufacturing resource planning*. Since this also has the initials MRP, this stage of planning is known as MRPII (MRP2). Figure 13.2 shows the information requirements for manufacturing resource planning.

Manpower planning

This involves assessing future manpower requirements to match the capacity

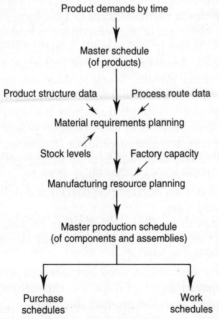

Figure 13.2 *Manufacturing resource planning.*

needed. It operates in parallel with the capacity planning of an organization's physical facilities. Although people are more flexible than machines, they are not universally interchangeable, but have differing skills, experience and competencies. Therefore each skill area and level must be analyzed separately in order to determine need by skill area and level.

A corporate manpower plan would cover all employees. It would take account of expected leavers and internal transfers and promotions, and enable the organization's training and recruitment needs to be determined. Manpower planning is developed further in Chapter 22.

MANUFACTURING CONTROL

Controlling the flow of work

When MRP calculations have determined the work and purchase schedules, these are implemented in the form of instructions. These instructions, normally from a planning and control office, trigger action in the factory and in the purchasing section. In product and cell layouts work is scheduled only once, in the form of a works order for the complete set of processes done on the line or in the cell. In a process layout each operation is scheduled separately. The first operation is scheduled. When it has been completed this is notified to the planning office, who then schedule the transportation to the next section and the next process. It is the delays in this information system which cause the throughput times in process layout to be much longer than in product or cell layouts. In the pre-computer era, planning was done only once per week, so the maximum rate of progress in a process layout was one process per week and the throughput time was one week per process. Modern computer systems have significantly reduced throughput times.

The instructions from the planning office operate on processes in sequence from first to final process. In doing so this system *pushes* work through the system, to achieve its output. In an ideal world the scheduling system would achieve the desired output of the manufacturing system. In practice delays may occur, machines may break down and components may be scrapped because of poor quality. A scheduling system needs to be able to cope with such disturbances and redirect the activity to recover from such problems and still achieve the desired outputs.

MRP, while valuable in planning the activity schedules, may be inefficient in controlling and replanning activities in response to these disturbances. Some additional techniques may be better at this short-term control stage.

Just-in-time (JIT) manufacturing

JIT is an alternative approach to short-term control of manufacturing activity, and is based on the philosophy that material should arrive at a factory and work be moved around it 'just-in-time' for its required processing. This should reduce levels of stocks of material and also the throughput time. This form of control *pulls* work through the system, work instructions being initiated by the customer order for the final product. This triggers demand for manufacture of the required volume backwards from the final to the first process.

The Japanese *kanban* system operates in this way. A *kanban* is literally a card – a document that initiates a demand on an earlier process for the manufacture of a given quantity of a component. The card may be attached to an empty box. The box is a request for it to be filled with components, the card identifying the part number and number required. The *kanban* quantity is the order quantity or batch size. This modern approach is based on JIT and minimum levels of stock. The question of stock levels and batch sizes is addressed in more detail in Chapter 17.

This approach cannot work on its own in a vacuum. It assumes that activities can be done when requested and that materials are available immediately when requested. It can work as a short-term control within a medium-term plan based on MRP. The actual volumes made are those determined by JIT but capacity is planned by MRP. Medium-term contracts with suppliers will be for planned volumes but actual volumes required will be 'called-off' on a daily basis. In this way the system responds to short-term changes and makes only the components required.

Line of balance (LOB)

This is a manufacturing activity control technique. (It must not be confused with line balancing, introduced in Chapter 12.) It uses product structure and process route data in the same way as MRP but sets in-process volume requirements for each stage of processing. These act as in-process due dates and enable output volumes to be monitored at each stage of processing. If delays or losses occur, then, by detecting them at an early stage the opportunity is created to correct the loss *before* they affect the final output (the completion of a given volume of a product by a given date).

The term 'line of balance' arises from a line drawn on a table or graph for a given date which identifies the expected volumes required to have been produced at each stage of processing if the actual volumes achieved match the expected volumes. The volumes measured are the cumulative volumes to that day from the beginning of a planning period. Cumulative volumes for each day are used, so that the amount to be produced on a given day is the difference between the cumulative planned volume for that day minus the cumulative actual volume up to the previous day. This measures the output requirement for a given day, including any deficiency in output in earlier days. This would restore the cumulative output to its planned level.

For a simple product (the same product as that introduced in Chapter 11) with a fixed daily output requirement a line of balance table would be as shown in Table 13.4. The right-hand column is the required output in a period (in this case a period is a day). To the left of the output is the cumulative output formed by adding the output from day 1.

Table 13.4 Calculations of a line of balance

Day	Components				Assembly		Product Y	
	C	D	E	F	3	2	cumulative output	
Reqd	3	6	2	3	1	3		
Start	1	2	3	3	4	5		6
1	30							
2	60	60						
3	90	120	20	30				
4	120	180	40	60	10			
5	150	240	60	90	20	30		
6	180	300	80	120	30	60	10	10
7	210	360	100	150	40	90	20	10
8	240	420	120	180	50	120	30	10
9		480	140	210	60	150	40	10
10			160	240	70	180	50	10
11					80	210	60	10
12						240	70	10
13							80	10

The line of balance for day 8 is shown in Table 13.4 as the set of requirements by day 8 for each stage of processing. This is the horizontal line of data for day 8.

Notice that a 'staircase' can be drawn across the data showing the lead times at each stage of processing. For example, the required cumulative output of 120 of item E by day 8 feeds forward as a requirement for 60 of assembly 3 by day 9, 180 of assembly 2 by day 10, and 60 of product 5 by day 11.

Actual outputs for day 8 for each item would be compared with these planned requirements. The differences would indicate problems requiring attention and rectification.

As an alternative to a digital representation, cumulative output requirements can be shown on a graph. This makes it easier to compare planned and actual outputs since they can both be drawn on the same graph.

Figure 13.3 illustrates this approach. Data is drawn here only for item E, assembly 3 and product 5 to avoid over-filling the graph. In this case, since the planned rate of manufacture is constant, 10 per day, the cumulative planned outputs are inclined straight lines. Actual output can be superimposed day by day and is shown up to day 8. The 'line of balance' for day 8 on the graph is then the vertical line at day 8 showing the planned and actual outputs for each part. This shows for item E the difference between planned and actual output.

By monitoring these differences, shortages at early stages of manufacture can be detected. If not corrected, these will cause shortages on later days or later

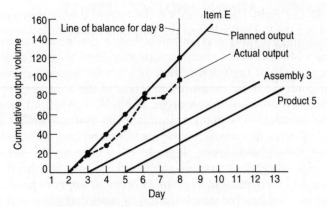

Figure 13.3 *Graph of output and line of balance.*

stages of manufacture. Knowing this in advance enables corrective action to be taken, such as extra working hours on specific processes, and output restored to objective levels before lateness to a customer occurs.

PROBLEMS IN MANUFACTURING SCHEDULING

The approaches outlined above all attempt to manage effectively the flow of work through a manufacturing system. No one system is always best: management must 'match' the approach to the complexity and configuration of the manufacturing system. To be effective, manufacturing planning and control systems must operate in real-time and be able to deal with changes and disturbances, i.e. they must be responsive.

A major potential problem with such systems is the possible lack of accuracy of data. This may arise from design modifications or other changes in the structural data or from errors or failures to take account of rejects or other losses. As with any system, if the input is wrong, then the output is wrong too. In management systems this can lead to the wrong decisions being taken and customers not getting their products on time. Clearly, to be effective, such problems must be identified and resolved.

In many manufacturing organizations several thousand parts may be involved. Product developments will cause parts lists to change. Keeping the data up to date is not a small problem. A change may be made in the design department on a given date but will only be effective for purchasing and for manufacturing at different later dates. These effectivity dates need to be known so that a change can be correctly phased in – otherwise chaos will occur, leading to the wrong parts being ordered or assembled.

MANUFACTURING PLANNING AND PRODUCTIVITY

The manufacturing planning and control activity is the 'nerve-centre' for control of the flow of work through a manufacturing system. Success in this activity is crucial to the success of the organization as a whole. Success in manufacturing planning and control can be measured in terms of the achievement of objective output volumes, with minimum resources, lateness, throughput times and levels of stock. The complexity of many manufacturing systems makes such control difficult. Inadequacies in manufacturing planning and control have been a major cause of long throughput times, high levels and costs of stocks, and low productivity. An integrated systems approach is needed to the design of manufacturing and management systems in order to improve productivity.

Some of the reasons for simplification of manufacturing and demerger of organizations is to reduce this complexity in order to achieve better control and greater effectiveness. Management systems are needed which provide for effective information and interaction between customers, manufacturing and suppliers. Changes in manufacturing systems and management systems are needed if Britain and Europe are to achieve world-class manufacturing standards of productivity.

INFORMATION TECHNOLOGY IN MANUFACTURING PLANNING AND CONTROL

Modern approaches to manufacturing planning and control use computers to perform these planning and scheduling tasks. These operate in 'real-time' (no delays in the computerized information system, the data is always up to date, being changed as changes occur) and significantly reduce throughput times and so contribute to manufacturing efficiency. Computer systems, however, operate on the same principles as the earlier manual systems and often produce outputs in the form of Gantt charts.

The volume of data involved in manufacturing planning and control means that computer databases are usually needed to handle the data. A multi-user system is required since several different management functions need access to the data and would be responsible for part of it. The manufacturing planning and control information system would have links not only with purchasing and marketing but also with stores, distribution, personnel and accounts. Electronic data or computer-generated data on paper will be used to order materials from suppliers, send delivery notes to customers and invoice them for the goods dispatched. Within the organization the information system can be used to generate work lists (lists of tasks do be done in a given period). When tasks are done, this can be fed back to the database. Where off-line inspection is done, the approval, or otherwise, of a batch of work can be recorded in the database. Information about hours worked and who has done which tasks may be needed by the personnel department for payment purposes.

It should be clear from this that a manufacturing database operates in real-time and monitors the progress and completion of all the tasks done in the factory. This provides information to other management functions and for effective management of the manufacturing function.

Modern electronic communication systems enable organizations to send and receive manufacturing data from other organizations at different sites.

Task sheet 13   Manufacturing planning

A product has a demand schedule as in Table 13.5 showing seasonal variation.

Table 13.5   Variable demand for capacity planning

Week	1	2	3	4	5	6	7
Demand	100	100	120	140	150	125	115

13.1   Plan production levels for each week, for different levels of capacity but assuming a fixed capacity each week.

13.2   Determine the average stock level for each level of capacity.

13.3   Reconsider the capacity if it is possible to work a maximum of 10 per cent overtime in any week.

13.4   Illustrate the relationship between demand and production graphically for each possibility you consider.

A product has the following total requirements of subassemblies and components as shown in Table 13.6 with given lead times for each activity (time before final assembly). The output schedule is also shown in Table 13.7.

Table 13.6   Data for line of balance calculations

Part no.	Product	Subassembly				Component							
	1	2	3	4	5	A	B	C	D	E	F	G	H
Total reqd	1	2	3	4	2	4	12	4	4	4	3	6	9
Lead time	0	1	1	2	2	4	5	3	3	1	2	3	4

Table 13.7   Output schedule for product 1

Week	1	2	3	4	5	6	7	8	9	10
Output	10	15	15	20	20	20	20	15	15	10

13.5 Calculate the cumulative output requirements for each week for the product, assemblies and components.

13.6 What are the output targets for completion by week 6?

13.7 Plot cumulative output graphically for the product and the assemblies.

CHAPTER 14

PROJECT MANAGEMENT

PROJECT MANAGEMENT

The need for project management

Apart from managing the continuing day-to-day activities of an organization, managers must also manage change. The nature of the activities and the style of management for managing change are quite different from managing routine activities. Change occurs as new products are introduced, new factories and offices established, new technology and new processes adopted, or when jobs or organization structures are redesigned. Change involves a set or related activities with a identifiable start and end. Such a set of activities is usually referred to as a project. The initial setting up of an organization also involves a set of one-off activities, so this too can be seen as a project.

Some management activities, such as design, are normally managed on the basis of managing a set of projects. Much of marketing involves sets of activities associated with promoting products. A sales promotion can be managed as a project. As continuing change becomes the 'norm', more organizations are managing all of their activities as projects.

Students too can organize their studies on the basis of projects. Each subject studied involves a set of activities which need to be planned, scheduled and done. Priorities need to be established when several activities compete for time. These competing demands can be managed more effectively using project scheduling techniques.

The construction industry is one in which work is done as projects. A project will involve the erection of a building, a bridge or a motorway. The biggest recent project in Europe has been the construction of the Channel Tunnel between England and France. This has involved many separate organizations, specialists in particular fields, being organized to build the tunnel. Big projects like this often overrun in time and cost more than originally planned. The Channel Tunnel was not too bad in this respect given its size, complexity and uncertainties. The bridge over the River Humber took several years longer than planned and cost several times as much as planned.

223

Project contracts

Some projects, such as those in construction referred to above, involve separate organizations being brought in to do a part of the project. Contracts are required – between the project-managing organization and contractors assigned specified tasks – to agree the work to be done, the timescale and the cost. For large projects taking a long time, stage payments may be negotiated so that some payment is made at specified times at stages of completion rather than payment at the end of the project. It may also be agreed that payment can be increased by the rate of inflation. This would only be done if the contract extended over a number of years. Contractors would not agree a fixed price over such a long period of time given the uncertainty over the future rate of inflation. Short-term contracts would normally be for a fixed price.

Conflicts can occur between project managers and contractors if contractors have to do tasks that were not part of their original contract. This may arise as problems occur as the project is done. Contracts should cover this aspect as far as possible and agree procedures for resolving any disputes. In the case of the Channel Tunnel, the amount of additional work disputed between the project managers and the contractors amounted to more than £1000 million.

Smaller projects within organizations may be managed in-house without the need for external contractors. Even so, the relationship between the project managers and the user departments is on a similar supplier–customer basis: the user departments are clients of the project manager. In this case the clients are internal rather than external.

Project teams

A project usually involves several different functions within an organization or at least several different people and resources. To co-ordinate their activities a project is usually managed by a project team comprising people from each relevant function. A project team – the term 'task force' is sometimes used – would be set up for the duration of the project. People in the team would be seconded from their normal job on either a part-time or full-time basis. In this way the different functions are involved at all stages of the project and the skills of each function are brought together and are available to the project. A project manager would be appointed, full-time except for the smallest of projects. Control is exercised by the board of directors or by a steering committee of the board. This board or committee would approve proposals from the project team and authorize the expenditure on the project at various stages. In a given organization there may be several projects going on at the same time.

Managing change

Only by introducing change can organizations improve their productivity and competitive advantage. Change is therefore vital to the continuing survival and development of an organization. However, change will be seen as a threat if it is feared that jobs will be lost. There is a natural tendency, therefore, for people to resist change. These fears must be recognized and understood if a change is to be successfully introduced. Part 7 of this book covers the human aspects of managing organizations. Managing change is an important part of this process.

To introduce change successfully the people to be affected by it must be involved in it. They have views about how they currently do their jobs and may have valuable ideas about how things could be improved. The project team involves some people directly; other people must be involved, perhaps through meetings with a project team member. If people are aware of the reasons for the change and the effect it will have on them, and have been involved in its introduction, then it is more likely to be successful. Since a change is likely to lead to an improvement in productivity, it should be possible to improve the pay, working conditions or job satisfaction of employees. It will be easier to introduce a change if employees can see some benefit for themselves from it.

One role of the project team, therefore, is to 'sell' the benefits of the change to employees. In this way historic and negative attitudes can be unfrozen so that the change can be introduced with the support of employees rather than conflict. In the past such conflict has prevented organizations from making change or getting the benefit from a change.

THE SCOPE OF PROJECT MANAGEMENT

Project management involves the following:

- *Project planning* The collection of data and determination of project duration and resources prior to the start of the project.
- *Resource scheduling* The procurement of resources or rescheduling within limits imposed by currently available resources (manpower, machines, materials and money).
- *Project budgeting* The initial planning of project expenditure by each department for each accounting period.
- *Project appraisal* The analysis of capital expenditure against future expected revenues to determine the financial viability of the project. This is not dealt with in this chapter but is deferred to Chapter 21.
- *Project control* The monitoring of activity progress and expenditure and any rescheduling that may become necessary as the result of delays.

The objectives of the project team are to complete a given project within time, cost and resource budgets. It may not always be possible to achieve all of these objectives simultaneously. Usually one of them will predominate and will determine the way the project is managed.

A project completion date is often externally imposed and may have cost penalties associated with lateness. The time available within which to complete a project may require additional resources (which will cost more). Decisions are needed as to whether to acquire additional resources or take longer to do the project using existing resources. This is the resource scheduling phase. Analytical techniques provide all of the data needed for such decisions.

The nature of projects – as one-offs and doing something new, is that despite good planning, delays can occur once a project has been planned and started. Techniques are required which enable management to monitor progress and to determine whether a project is going to be late sufficiently in advance of such potential lateness to enable the remainder of the project to be replanned with the objective of preventing lateness. This can be done by setting 'in-process due dates' using network analysis. Without such a technique an organization would not know whether it was going to be late finishing a project until it was late. It would then be too late to do anything about it.

PROJECT PLANNING USING NETWORK ANALYSIS

Network analysis is an analytical technique used for project planning. Given a list of activities and their durations it is *not* possible to determine how long it will take to complete the set of activities that form a project. The duration of the project depends on how the activities are related logically. Some activities can occur at the same time as others (in parallel), but some can only be done after others have been completed (in series or sequentially). The concurrent and consecutive relationships between activities enable a network diagram of the activities to be drawn. Only then can the project duration be determined.

A network diagram can be constructed when, for each activity, a set of preceding activities which must be completed before that activity can start is known, e.g. if activities A and B need to be completed before activity C can start. This can be written as follows:

C < A + B

meaning that C depends on A and B. C can only start when *both* A and B have been completed.

Figure 14.1 shows four examples of how a part of a network can be constructed from the logical dependencies. Notice in Figure 14.1 (d) that the dependencies are such that the logic cannot be represented in a single event node. Two event nodes are required, the upper event represents the completion of activity A only. This enables activity C to start. The lower event has inputs from activity B and a 'dummy' activity from the event at the end of activity A. In this way activities A and B are brought together logically to enable activity D to start.

On a network diagram activities are shown as arrows. The length of an arrow is *not* related to time; the diagram shows only the logical relationships between activities. These relationships indicate the series or parallel connections between

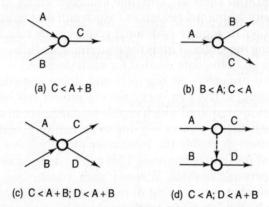

(a) C < A + B

(b) B < A; C < A

(c) C < A + B; D < A + B

(d) C < A; D < A + B

Figure 14.1 *Constructing a network from activity dependencies.*

activities. (An alternative type of diagram in which activities are shown as nodes is not used in this book.)

Activity arrows are joined by nodes, which are drawn as circles and represent 'events'. An event is the state of completion of a set of activities that must all be completed before another activity or set of activities can start.

A project will have a start and an end event representing the completion of none and all activities respectively. The start event is usually drawn at the left. The first activities that can be done are those that have no dependency. A network diagram can be built up by plotting those activities that can start when those on which they are dependent have already been plotted.

When two or more activities can start after the completion of a given activity, then two or more activity arrows will radiate rightwards from that activity. The geometric angle at which they are drawn has no meaning other than to separate the different activities. When one activity depends on two or more preceding activities being completed, then the arrows for these activities will converge rightwards to a single event. This event represents the completion of the two or more activities, not just one of them.

The final activities are those on which no other activities are dependent. These activities will converge rightwards to the single end event of the network. This represents the completion of the whole project.

In some cases a 'dummy activity' is needed to represent the logic in a network. A dummy is an activity of zero duration and zero resource. It is shown as a dashed arrow on the network. Dummies are needed where the logic relating a set of activities cannot be drawn using a single event node. An example is shown in Figure 14.1 (d).

An example of network analysis

The activities shown in Table 14.1 have been identified as being necessary for a project. Further information about their duration, resource type, cost and dependency are also collected and given in the table. The network diagram for this project is shown in Figure 14.2. Events are usually numbered – in this example from 1 to 8. The event circles are usually segmented in order to hold time data for the event.

Table 14.1   Activities, durations and dependencies

Activity	Duration	Resource	Cost	Dependency
A	4	X	2,400	–
B	12	Y	18,000	A
C	3	X	1,800	A
D	13	Y	–	A
E	4	Y	6,000	A
F	0	–	–	B
G	15	Y	15,000	C + F
H	1	X	600	D
I	11	X	8,250	D
J	5	X	1,500	E
K	15	Y	7,500	H + J

The following are notes on the activity data in Table 14.1.

1. A has no dependency and so is the starting activity.
2. No activities depend on G, I or K, so these are end activities.
3. D has no cost. It could be a 'lead time'. It is an activity for this analysis since a period of time is required for it to occur.
4. F has no duration. It is not a real activity. It is a 'dummy' activity introduced in order to resolve the logic relating a set of activities. (A set of overlapping dependencies cannot always be represented by a single event.)

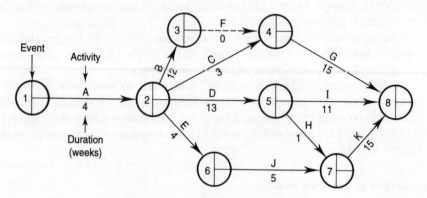

Figure 14.2 *Network diagram for a project.*

Project time analysis

Time analysis is based on analyzing the event times. Event times represent in-process due dates. For each event two times can be calculated:

- The earliest time by which the event can take place.
- The latest time by which it must take place if the overall project time is to be achieved.

A time of 0 is usually assigned to the start event, and by passing forward through the network the earliest time for each event is calculated. Where more than one activity leads to an event, the earliest event time is the latest of the set of earliest activity finish times. This is because all incoming activities must be completed for the event to occur. The earliest activity finish times are the earliest activity start times plus the activity duration. When the end event is reached the project duration can be determined as the time of the end event. Event times for this analysis are times from start, *not* dates. As a project is implemented it would be normal to state the event times as calendar dates.

By using the project duration and passing backwards through the network the latest time for each event is calculated. Where more than one activity starts from an event its latest time is the earliest of the set of latest activity start times. This is because the latest event time must be early enough to allow the activities on the longest path to be completed by the project completion time. The latest activity start times are the latest activity finish times minus the activity duration. When the start event is reached, its latest time should be 0, the same as its earliest time.

It will usually be the case that at some events there is *no* difference between its earliest and latest times, whereas for other events, there is a difference which is referred to as the 'slack time' at the event. This represents an amount of delay tolerance at the event. The time analysis for the example network is shown in Figure 14.3. This shows that the project duration is 33 weeks.

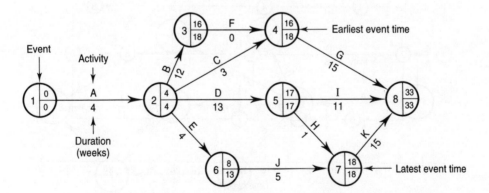

Figure 14.3 *Time analysis of the network.*

Critical path through a network

A network comprises several paths of activities from its start to its end, one (or more) of which is longer in duration than others. This path determines the project duration and is referred to as the 'critical path'. This is the path that defines the project duration. There is no delay tolerance on the critical path and no slack time at the events on it. For the example network the critical path is A–D–H–K.

Other paths through the network are not critical, and a degree of delay for these is tolerable without affecting the project time. A non-critical path comprises activities whose duration is less than the time available between the events where such a path joins the critical path. These activities can 'float' in time between these events. Float on an activity is calculated by referring to the times of the events at the start and end of the activity.

$$\text{float} = \text{latest time of end event} - \text{earliest time of start event} - \text{duration} \tag{14.1}$$

This float is only available for a given activity if some of it has not already been used by a delay of a preceding activity on the same path.

Network diagram on a time base

A network diagram is a logic diagram which does not itself show slack or float. These can be shown if the diagram is redrawn with the arrows to scale in time. This 'network on a time base' diagram combines the logic of a network diagram with the timescale of a Gantt chart. It is able therefore to show the float time on some of the activity paths. The network on a time base for the example network is shown in Figure 14.4. Notice that some event numbers appear more than once.

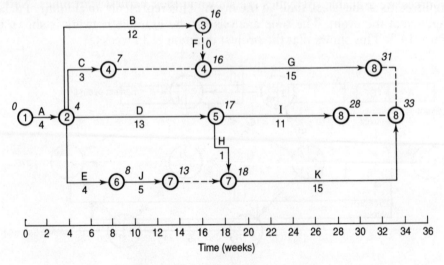

Figure 14.4 *Network on a time base.*

Logically these represent the same event but the separate nodes show the activity finish times separately. Dashed lines link the separate parts of an event together.

It can be seen from Figure 14.4 why A–D–H–K is the critical path. It is the path which causes the project duration to be 33. Other paths have float; there is no float on the path A–D–H–K.

On the time-based diagram, activities are shown at their earliest times. This maximizes the opportunity to recover from delays which would extend the activity duration and may affect the project duration. However, delaying an activity until its latest time postpones expenditure and so this may be beneficial. In this case the float time has been lost by the delayed start, so further delay would cause a problem. In practice, these factors must be considered, along with claims for alternative uses on scarce resources, in making a decision about when to start an activity.

RESOURCE SCHEDULING

From a time-based network diagram, resource schedules can be produced which show the activity schedules separately for each resource. A resource could be a type of machine, a type of labour or more usually a department that has financial responsibility for a set of activities.

A network diagram and resource schedules are usually calculated initially without regard to any limits imposed by current resource limits. It may be possible to achieve the project planned completion date with currently available resources. Figure 14.5 shows Gantt charts which show the resource schedules for each resource, X and Y, for the example network. These charts assume that the resources are available. Activities are shown at their earliest start times. Notice

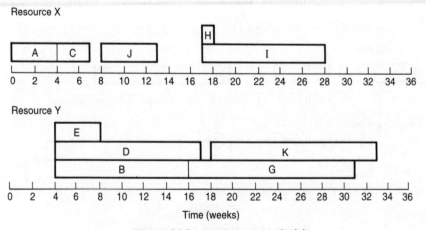

Figure 14.5 *Initial resource schedule.*

the peaks and gaps in resource usage. It may be possible to reschedule some activities to avoid peak resource use and to complete the project in the planned time, but by using less resources at a given time. This minimizes the maximum resource use.

In many cases an organization may not currently have the amount of resource required to complete the project according to its logical dependencies. If the resource requirement is greater than the current resource available then:

Either   more resources must be acquired (usually at a higher cost).

Or   some activities which logically can be done concurrently must instead be done consecutively since one must wait for the other to be completed for the resource to be available. In this case the schedule for the activity is determined by the resource and not just the logic. This will extend the duration of the project.

Figure 14.6 shows revised resource schedules for the example network for the case that the maximum availability of resource X is 1 and that for Y is 2. Notice that in this case the project duration is extended from 33 to 35 weeks. When two or more activities compete for the same resource at the same time, a decision is needed on which activity takes priority and which is delayed. Techniques are available for this but they are beyond the scope of this book. Figure 14.7 shows the network on a time base for this revised schedule. Notice that the critical path is now A–E–B–G, the link between E and B being a resource constraint rather than a logic constraint.

In some cases it may be acceptable for the project duration to be extended. When there is time to do so, the project could be started earlier. However, in other cases such an extension is not acceptable. In this case it is necessary to review the assumptions on which the activity durations were determined. It may be possible to reduce an activity by increasing the resource per unit time from that originally allocated, where this is possible within resource limits. For example

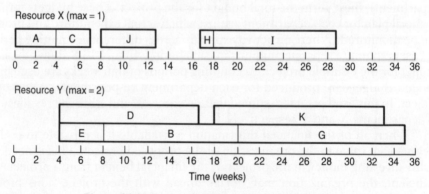

Figure 14.6 *Resource schedule for limited resources.*

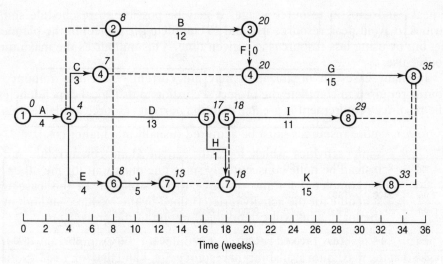

Figure 14.7 *Network on a time base for limited resources.*

an activity planned to take six weeks with one operator could be revised to take three weeks with two operators. In other cases it may be possible or necessary to subcontract some work to another organization.

Network methods give management all the information needed for such resource scheduling decisions.

PROJECT BUDGETING

When costs are attached to resource schedules, budgets can be established for each resource or department to show the planned expenditure by each department each month. When aggregated (added together for each resource or department), these form the total budget for the project. These budgets form the financial plan for each department against which actual expenditure and progress can be monitored. Where an activity spreads across different accounting periods, its cost is split into separate amounts in each period. Figure 14.8 shows the initial budget for the example project. Accounting periods of four weeks are used in this budget. Budgets are produced for each department responsible for a part of the project. In this example it is assumed that resources X and Y are the responsibility of departments X and Y respectively.

When all of the financial information is available it is possible to evaluate the financial value to the organization of the proposed project. This is needed to justify investing money in the project. If the financial benefit from a project is not sufficient, the organization may not go ahead with the project. This process, known as project appraisal, is covered in Chapter 21. The processes of budgeting

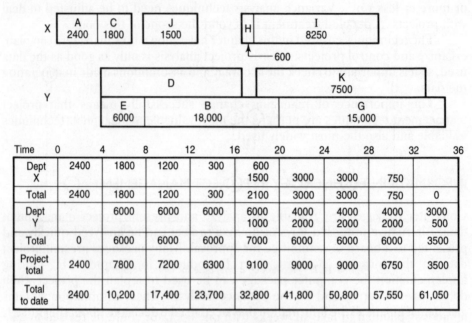

Time	0	4	8	12	16	20	24	28	32	36
Dept X	2400	1800	1200	300	600 / 1500	3000	3000	750		
Total	2400	1800	1200	300	2100	3000	3000	750	0	
Dept Y		6000	6000	6000	6000 / 1000	4000 / 2000	4000 / 2000	4000 / 2000	3000 / 500	
Total	0	6000	6000	6000	7000	6000	6000	6000	3500	
Project total	2400	7800	7200	6300	9100	9000	9000	6750	3500	
Total to date	2400	10,200	17,400	23,700	32,800	41,800	50,800	57,550	61,050	

Figure 14.8 *Project budget.*

and appraisal are covered in Part 6 of this book since they form part of the financial management of the organization as a whole. However, the techniques need to be adjusted in order to deal with a project rather than the day-to-day activities of the organization.

PROJECT CONTROL

Once a project plan has been completed and approved by the board or steering committee, it can start. The role of the project manager is then to manage the implementation of the project. This involves authorizing the start of activities and monitoring their progress.

At various times during the implementation of a project the actual amount of work done and money spent to date will be compared with that planned. When activities have been delayed, the network diagram needs to be redrawn using actual durations for the activities done. This may extend the project duration and require corrective action of the type referred to under resource scheduling above.

Monitoring and control of expenditure on activities is done using the methods of variance analysis covered in Chapter 20. It is important to monitor the amount of work done and not just the money spent. To know that £1000 was spent is not much use unless you know whether it was spent doing £1000 of work, or more or less work. Variance analysis techniques need to be adjusted to deal with projects. A detailed treatment is beyond the scope of this book.

The techniques covered in this chapter have shown how managers can plan, organize and control projects. Clearly project analysis is only as good as the data used, so it is important to check the accuracy and assumptions made in generating the data.

The importance of managing change successfully means that project management techniques are perhaps the most valuable management techniques available and also the most widely used.

INFORMATION TECHNOLOGY IN PROJECT MANAGEMENT

Computer packages are available to analyze and present project management data. Alternative proposed ways of doing a project can be analyzed quickly and presented to a board or committee for approval. Many packages present project data in the form of a Gantt chart (introduced in Chapter 11). During implementation the computer package can be used to replan the project when delays have occurred and do the variance analysis. These information systems give management all the information they need to take decisions about the project.

Task sheet 14   Project scheduling

14.1   Table 14.2 gives a list of logical dependencies between sets of activities. Draw a part of a network to represent the logic in each case.

Table 14.2   Dependencies between activities

1	B < A
	C < B
2	J < A
	K < A + B
3	J < A + B
	K < B
4	B < A
	C < A
	K < B + C
5	J < A
	K < A + B + C
	L < C

Table 14.2 *Continued*

6	J < A
	K < A + B
	L < A + B + C
7	J < A + B
	K < B
	L < B + C
8	J < A
	K < A + B
	L < B + C
9	J < A + B + C
	K < B + C
	L < C

A project comprises the activities shown in Table 14.3. For each activity its duration and dependencies are given.

Table 14.3 Data for network analysis

Activity	Duration	Dependency
A	2	–
B	4	–
C	5	A
D	1	B
E	3	B
F	2	C + D
G	4	E

14.2 Draw a network diagram for this project.

14.3 Calculate the project duration.

14.4 Determine the critical path.

14.5 Draw the network on a time base.

14.6 Determine the float on each activity.

PART 5

MATERIALS MANAGEMENT

LEARNING OBJECTIVES

5.1   Define the term 'logistics' in a manufacturing context.

5.2   Give the objectives of logistics management.

5.3   Understand the demand chain and the difference between demand-driven and supply-driven chains.

5.4   Understand the importance of plant location to the profitability of an organization.

5.5   Describe the various forms of transport and their suitability of different types of materials.

5.6   Explain the key strategy for transportation.

5.7   Understand the forms of internal transport and their suitability for different applications.

5.8   Understand the need to reduce handling.

5.9   Explain the need for storage facilities and their advantages and disadvantages.

5.10   Understand the concept of 'last-minute customizing' and its effect on stock holding.

5.11   Explain the use of EDI in purchasing and its advantages.

5.12   Explain the role and objectives of purchasing.

5.13   Explain the terms: quality, overspecification and underspecification.

5.14   Explain the importance of price in relation to quality, specification and purpose.

5.15   Understand the phrase 'comparing like with like'.

5.16   Explain the reasons for vendor assessment and some of the methods used.

5.17   Explain the terms: multisourcing, single sourcing, supplier development and group produce purchasing.

237

5.18 Explain the factors involved in make/buy decisions.

5.19 Identify the points at which material stocks may be held.

5.20 Understand the costs and benefits of holding material stocks.

5.21 Graph the holding and ordering costs against order quantity.

5.22 Define the economic order quantity and explain how it may be reduced.

5.23 Explain the concept of safety stock.

5.24 Define stock turnover and average stock time.

CHAPTER 15

LOGISTICS

LOGISTICS

Logistics is the management of the flow of materials and services into, within and out of an organization. Reference to Figure 1.1 in Chapter 1 will show how this fits into the organization as a whole. Logistics is not referred to by name in that figure but it shows the flow of materials from suppliers through the organization to customers as products. This includes the flow of information for management control. The function of logistics can be seen in the following rhyme:

> For want of a nail the shoe was lost.
> For want of a shoe the horse was lost.
> For want of a horse the man was lost.
> For want of a man the battle was lost.
> For want of a battle the war was lost.
> And all for the want of a nail.

This old children's rhyme shows the importance and military origins of the concept of logistics. More wars have been lost due to lack of supplies than the skill of generals and their soldiers. A major factor in the outcome of World War Two was the inability of Germany and Japan to supply their military vehicles and ships with fuel, the UK/US strategy being to cut off the supply of fuel.

Manufacturing industries also rely on logistics for survival, obtaining supplies or raw materials and bought-out items to enable them to manufacture their product and then get it to market. If supplies fail to arrive on time or of the right quality, expensive machinery and labour stand idle while costs escalate. If materials reach the factory on time but then get lost in inefficient stores and internal transport systems, or stand idle waiting to be progressed between operations, the result will be the same. If a manufacturer cannot get its product to the marketplace, the organization will lose customers to competitors who can.

Logistics takes a systems view of the organization and its relationships with the other organizations with which it relates. It covers the functions of purchasing of materials, other goods and services, capital purchases of equipment, storage

This chapter was co-written with Robert Mansfield.

of materials and finished goods, and the distribution of products to customers. Logistics is a part of materials management. Materials management interacts closely with manufacturing. Manufacturing decisions about production volumes and schedules determine materials requirements. Purchasing, the link with suppliers, is covered in Chapter 16. Effective materials management has a significant effect on the productivity and profitability of the organization. To achieve this an organization must develop a materials management strategy. Materials management is also vitally important within the organization, for managing internal material flows.

The demand chain

The material flow cycle can be seen as a chain linking the raw material supplier to the ultimate consumer. This is known either as the demand chain or as the supply chain. The basic links in this chain are: supply, manufacture, distribution, retailing and consumption. The concept of the demand chain also applies within the organization. The concept of the internal market, referred to in Chapter 5, treats departments within the organization as suppliers and customers to each other, creating an internal demand chain. Figure 15.1 shows the concept of a demand chain.

There may be several stages of manufacture, each forming a supplier–customer link. If the different stages of manufacture are undertaken at different sites or by separate organizations, then they are separate stages in this chain, so there can be many links in the chain. The demand chain for a steel sheet component could include the following:

Mining Extracting the ore-bearing rock, crushing and screening the rock to liberate the ore, separating the ore from the crushed rock (this would normally be done as close to the mine as possible to avoid transporting low-value rock).

Steel making Iron making, steel making, hot rolling, cold rolling, coiling.

Stock holding Storage, coil slitting, straightening, guillotining, packing.

Manufacturing Piercing, blanking, folding, bending, welding, painting, packaging.

Distribution Palletization, warehousing, transporting, load preparation.

Retailing Stocking, display, demonstration, sale, delivery.

Consumption Use by final customer, disposal.

Recycling Some scrap steel is recycled and introduced at the steel-making stage, in this way the chain becomes a loop.

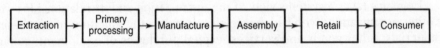

Figure 15.1 *The demand chain.*

Each of the above stages requires transport, both internal and external. The strategy, co-ordination and management of this is part of logistics. The marketing concept of 'place', detailing the distribution chain of agent, wholesaler, retailer, etc., to the eventual purchaser, is part of the same concept and was covered in Chapter 5.

The demand chain is normally driven by the ultimate consumer. However, each link in the chain usually deals only with the links on either side of itself – to the immediate supplier and the immediate customer. The demand chain is controlled by the flow of information between them. With perishable, seasonable items, such as fruit and vegetables, the chain is supply driven. The vegetables must be picked, transported and processed when they are ready, not when they are needed. With preservation technology – freezing, canning, bottling, etc. – it is now possible to store perishable items until demanded by the market.

With fresh produce, the task is to get it from grower to customer before it decays. Countries such as the former Soviet Union failed in this part of the chain and had food rotting in fields, barns and warehouses while people queued for food outside empty shops.

At each stage of the chain there are requirements, both internal and external, for identification, loading, transport, unloading and storing of materials. These contribute to delays and costs. Of the lead time quoted for supply of a product, most is taken up by the product waiting to move on to the next stage.

Transport systems

The location of an industry, and of the factories and warehouses within it, depends on a number of factors, including the availability of raw materials, energy, workers and the markets for the finished goods. The most significant factor in determining location is the cost of transport. Before the introduction of efficient transport systems, industry had to set up its factories where the raw materials were within easy reach. The same difficulty limited their market to the surrounding area.

Roads were crude and animal-drawn wagons were slow and limited in capacity. Goods in horse-drawn carts took more than a week to get from Birmingham to London. Passengers in horse-drawn coaches took four days. This was reduced in an express service to two days by using and changing horses in stages of 40 km, in what then became called stage coaches. Roads were important in the past for the development of markets. The Inca road in South America was 4000 km in length. The Incas had no wheels and no horses. They used runners running in stages to transmit information across the empire. The Silk Road from the Mediterranean to China was 6400 km in length and opened up markets at both ends and along its route. Silk came from China to Europe and woollen goods went the other way. The most significant effect of the Silk Road was that Aramaic Christianity spread along this route and with it the Aramaic alphabet, which became the basis of all the scripts in Asia except the Chinese and Japanese.

Rivers enabled easier transportation, and so towns and industry were built on their banks. Rivers also provided a supply of water and power. Where rivers were not available, canals were built to spread the transport network further, linking existing industries and enabling the development of new ones. Birmingham is the only major city of Britain not built on a river. Birmingham was an early site of industrialization in Britain and is the centre of a major canal network which was developed to transport industrial goods more effectively than the roads of the time. The building of the Bridgewater Canal to Manchester in 1761 reduced the cost of coal in Manchester to one-third of its previous value; this from a canal only 15 km in length! The Egyptians built a canal from the Red Sea to the River Nile as long ago as the 19th century BC. The Phoenicians rebuilt this canal a thousand years later and so too did the Persians. The current Suez Canal, built by the French in 1859–69 is the fourth canal in 4000 years to link the Mediterranean with the Red Sea.

The greatest advance in transportation was the introduction of the steam engine. This provided power to drive machinery, freeing industry from the need for water and wind power. It also led to the introduction in 1812 of Hedley's 'Puffing Billy' steam locomotive on the Middleton to Leeds railway, previously horse drawn. By the 1830s a rail network linked most towns and cities. The first 'intercity' routes were Manchester to Liverpool in 1830, and Birmingham to London in 1838. Factories could have direct access to the railway network by building a railway line right into the factory. Railways considerably reduced the time and cost of transportation. The Channel Tunnel from Britain to France now enables freight to cross Europe efficiently.

The introduction of tarmacadam roads speeded up horse-drawn road transport to compete with the railways. The invention of the internal combustion engine a century ago led to the modern motor vehicle. The consequent improvement in the road network has further reduced the costs of transport and hence the costs to industry.

Efficient transport systems enable an increase in the geographic size of the market that it is feasible to supply from a given place.

PLANT LOCATION

Just as with early industries, location is of vital importance to the profitability of modern industry. Introducing new industry to an area changes traffic flow patterns, so these must be considered. New or improved roads may be needed to cope with the change. Access to road and rail systems must be considered in selecting a site.

Decisions on location need to be taken separately for a manufacturing factory and for any warehouses. The number of facilities must also be considered. How many warehouses are needed for the efficient distribution of products to customers? Should a factory be located near to its source of supply of raw material or should it be located near to its markets? These aspects are shown in Figure 15.2.

At the primary stages of manufacture (upstream in the demand chain) it may be economic to have only one facility located near to supply, but further downstream the economic number is likely to increase and these facilities will be nearer to customers. Retail shops may have a facility in every town, post offices in every village.

Optimal location

The following is a simple example of a plant location calculation for a single facility:

> A mine has to extract 12 tonnes of rock to extract 1 tonne of ore. Rock and ore are separated at a separation plant before being smelted at a smelting plant. If the mine and the smelting plant are 52 km apart and a lorry can only carry 5 tonnes of material, where is the optimum location for the separating plant? The cost of transport is the same for carrying rock and ore.

The optimum location will be where the total distance travelled by both rock and ore lorries is the same. The lorry carrying rock from the mine to the separating plant will have to make 12 trips for every one trip by the lorry carrying ore from the separating plant to the smelting plant.

If the distance from the mine to the separating plant is d kilometres then:

$$12 \times 2 \times d = 1 \times 2 \times (52 - d) \tag{15.1}$$

so

$$26 \times d = 104 \quad \text{or } d = 4 \tag{15.2}$$

The optimum position for the separating plant is that which minimizes the total cost of transportation. It should therefore be located 4 km from the mine and 48 km from the smelting plant.

Site selection

Having decided on the theoretical optimum location, for a plant, there are other factors to consider under the heading of site selection: physical factors such as the nature of the available site, its shape, subsoil, drainage, etc.; legal factors such

Figure 15.2 *Plant location.*

as planning permission which restricts building in the 'Green Belt' and some urban areas; other planning regulations controlling noise, fumes, disposal facilities, etc., must be observed. The proposed site must be near enough to services such as water, electricity, gas, sewage, telephone, roads, etc., to enable economic connection. Human factors including worker availability and social facilities such as accommodation, shops, recreation, schooling, etc., must be considered. These must be available within reasonable travelling distance for the use of employees or must be provided as part of the project.

DISTRIBUTION

Selection of transport mode

The transport system used by an organization will depend on the nature of material, volume, value, urgency, frequency of demand, the physical separation of the links in the demand chain and the availability of suitable transport infrastructure. The key strategy is to reduce bulk (volume), increase value and minimize distances.

Bulk liquids and gases in large volumes can be transported by pipeline over long distances, examples are water, crude oil, refined petroleum products and natural gas. Medium quantities can be transported by train; small quantities by road or rail. Light goods can be transported by air freight.

With large volumes of bulk solids such as coal and ores, rail or waterway transport right into the processing plant are still viable methods. Smaller quantities over shorter distances can be carried by road. Some products, such as cars, use a combination of road and rail. They can be transported by a car transporter or individually to a rail depot, then by train over long distances to the next depot where they return to transporters or are driven individually to the dealers.

Until the Channel Tunnel between England and France came into operation, everything going or coming from outside Britain travelled by air or sea. With the opening of the Channel Tunnel more goods may take advantage of the integration of the British railway network with that of Europe to use rail for transport to Europe, reducing the amount carried by road.

Bulk cargoes with relatively low value can travel in the holds or tanks on board ship. High-value, low-bulk and perishable items can travel by air.

Transport strategy

Because of the costs of transportation involved in moving materials and products, it is important for an organization to develop a transport strategy that will minimize these transportation costs. This is not easy to achieve as different organizations are involved at different stages of the demand chain.

In order to reduce transportation costs, distances and trans-shipment from one form of transport to another should be kept to a minimum since these are the causes of cost. In order to reduce trans-shipment costs, transport systems should use standardized modular containers, designed to stack together and attach to flatbed lorries and rail trucks. These enable goods to be carried by sea, road and rail with easier trans-shipment between different forms of transport. Since government pays for part of the costs of road and rail transport, the costs to users depend on government transport policies.

The transport strategy adopted by a company can influence profitability. There are advantages and disadvantages to owning and running a transport fleet. Vehicles are expensive to buy and maintain. However, they are completely under the company's control and are a mobile advertisement for the company and its products. The use of haulage contractors may be a viable alternative. Contracts can ensure that either their vehicle, or a hired replacement will be available when needed. Most hauliers are prepared to paint their vehicles in the customer's livery if the contract is valuable enough. An alternative is vehicle leasing where the owner contracts to provide, maintain or replace the vehicle to ensure constant availability. Improvements in maintenance techniques are increasing vehicle availability and reducing costs. To make use of economies of scale, some manufacturers are grouping together to make up full loads on lorries or in containers.

INTERNAL MATERIALS MANAGEMENT

The principles of minimizing transportation costs explained above also apply at a lower level of the system to the organization of the movement of materials within a factory. Chapter 12 considered the relative location of machines as a means of reducing internal distances moved by materials. This chapter considers how materials should be handled, moved and stored within a factory. The internal transport system in manufacturing plants will depend on the type of material and processes involved. Assembly lines often use conveyor belts or feeder tracks to move part-assembled items through a set of processes. Materials can be fed to workstations by overhead conveyors which avoid congestion at ground level. Where manufacturing layout is on a job or process basis, work can be transported by hand truck, forklift truck or automatically by a robot truck following programmed instructions.

Materials handling

Where a material has to be loaded or unloaded there will be delay and cost. There is also the possibility of materials being contaminated or damaged by handling. With some cargoes there is a risk of loss due to spillage or pilfering. The emphasis must be on reducing the number of times an item is handled. Where it must be handled, the process should be as efficient a possible.

Goods can be stacked on standardized modular pallets that can be transported, loaded and unloaded without difficulty. These goods, depending on their nature and the conditions of the journey, can be shrink-wrapped in polythene or boxed and banded for protection. Pallets or separate items can be loaded into containers at their point of origin and not touched again until their destination. These containers can be locked and sealed to reduce the possibility of pilferage. Where the journey involves crossing national frontiers, a container sealed by the customs at the point of origin is only subject to paperwork checks and the occasional spot-check. The customs procedures still cause massive delays with reams of paperwork. However, the European Union 'Single Market' should eventually reduce this for transport within the EU.

A container is an example of a standard size 'unit load' for transportation. The same concept applies with a smaller unit for internal transport and for loading into lorries or into containers. Standard sized boxes or flat pallets are used for this purpose. This allows efficient stacking and storage.

Storage

Because of the economies of scale in transporting large loads, some storage will always be needed at the interfaces between transport systems and at processing plants. Even with the efficiency of containerization, large container depots are required at docks and rail terminals to cope with the different rates of arrival and departure.

Within factories and warehouses, storage racks can be designed to suit the standard size of the unit load. Given the cost of space, vertical storage, with units stored at several different levels, is becoming common. There are potential safety problems with vertical storage but modern equipment prevents such problems.

Some companies such as IBM are installing high-density computerized stores where there is only room for the automatic 'pick and place' machines that handle the movement of items within the store. These are rail-mounted robot forklift trucks which are able to store and retrieve items in the storage area.

INFORMATION FOR MATERIALS MANAGEMENT

The driving and controlling force in the demand chain is information. External documentation includes: *purchase orders* to request that certain quantities be dispatched to a specified location; *acknowledgement of order* to confirm contract terms, price and delivery date; *delivery notes* to acknowledge receipt of goods and trigger the invoice as a request for payment.

Internal documentation

Materials management involves processing data about materials at all stages from supply to distribution. This includes *goods received notes* to update stores and purchasing records and validate incoming invoices for payment. Where goods are not to standard, *goods inwards rejection notes* are used to inform the accounts department, who will stop payment, and purchasing, who will amend vendor quality ratings and ask for replacement items. Stores requisitions are needed to draw items out of stock for use in production, update stores records and warn purchasing of reducing stock levels.

Route cards tell the internal transport system where the items are required next. Operations cards tell the operator what to do with the item. Some systems use combined route and operations card with the previous operation, marked to identify its current state. Other systems allocate a new code for every change of state. Because of the paperwork involved, this works best with computerized systems.

Special documentation is required, including serial numbers and batch codes for items made for aircraft, defence and similar industries. These enable the item to be traced back through all stages of manufacture to the original batch of raw material. This concept of traceability was referred to in Chapter 9.

External documentation

Documentation for transport externally passes to and from several different organizations. Standard forms, acceptable to the different organizations, are beneficial.

When goods cross national frontiers, more documentation is required. Point of origin certificates, cargo manifests, VAT and Excise duty paperwork, health clearances, etc., must be shown at customs posts. Any irregularity can lead to long delays, seizure of goods and sometimes of drivers. The intention of the European Union is that goods can flow across Europe without these documents and delays.

INFORMATION TECHNOLOGY IN MATERIALS MANAGEMENT

All stages of transportation rely on accurate information on the contents of a load and its destination. Information is needed on where items are and how many there are of them.

Increasing use is being made of barcodes to capture data. This indicates an item type, location and perhaps quantity. When linked to real-time computer systems, scheduling and routing information can be collated and updated with speed and accuracy, reducing delays and expensive errors and providing management with information for the efficient management of the transportation of materials and products.

Some organizations interchange scheduling information by sending computer tapes or disks; however, there can be a problem with compatibility between computer systems. With modern data communications many large companies are introducing *electronic data interchange* (EDI). This transmits information electronically between supplier and customer. Orders are placed, acknowledged, billed and paid for without the need for paperwork. Internal documentation is being replaced by integrated computer systems that also provide analysis of stock levels and financial position as part of a management information system.

Routing and operation cards are replaced by workplace terminals and coding systems carried by the item or its pallet. Barcodes or magnetic stripcodes are allocated to materials by the goods inwards computer and attached to a component or a pallet or box of components. These codes are used by the automatic stores system to place the items in a rack until required. When requested by the manufacturing control computer, the stores computer remembers where it put the item, fetches it and puts it on the internal distribution conveyor. The code is read by sensors as the pallet travels along conveyors, and the item is routed automatically to where it is needed. When the next operation is completed the code is changed to confirm its change of state; the manufacturing control and stock control computers are then updated and the item either returned to stores or to the next operation.

Task sheet 15 Logistics

Manufacturer A supplies two customers. Customer X buys 600 units per week and is located 60 km away. Customer Y is 65 km away from A and buys 400 units per week. A lorry is used to carry goods from A to X or Y and can carry 200 items at a cost of £2/km. X and Y are 25 km apart and goods could be carried between them by a van which can carry 50 units at a cost of £1/km. The only routes available are AX, AY and XY.

15.1 Should a warehouse be located at A or at X or at Y or somewhere else?

15.2 What assumptions are made in your model?

A manufacturer produces 600 units per month at site A and 400 units per month at site B, which is 60 km from A. Customer X buys 300 units per month, is 50 km from A and 40 km from B. Customer Y buys 700 per month, is 70 km from A and 30 km from B. A lorry can carry 100 units.

15.3 How much should each factory supply to each customer?

15.4 Recalculate the data above for a lorry capable of carrying 500 units.

CHAPTER 16

PURCHASING

PURCHASING MANAGEMENT

The scope and functions of purchasing

Purchasing is the function of management which forms the interface between supplier and manufacturer. It is also called supply management, materials management or procurement. Figure 1.1 in Chapter 1 shows the relationships which purchasing has with the other functions of management. In a large organization there is likely to be a separate purchasing department; in a smaller organization it will form part of the manufacturing planning function. In its day-to-day operations a purchasing department buys not only the raw materials used directly in the manufacture of the organization's products, but also all of the indirect and consumable items. These include lubricants, cleaning fluids, cutting tools, office stationery, computers and all other goods and services purchased by the organization. In all of these cases an analysis is needed and choice made been alternative suppliers.

The purchasing department does not pay for the items ordered; that is the function of the finance department. Only for items for which there is no choice would invoices for goods and services be paid directly by the finance department without the involvement of the purchasing department. In the past, the telephone and electricity bills, for example, would be paid directly by the finance department. Nowadays there is choice of supply for these services, so the purchasing department would be involved in choosing the supplier.

In the case of investment in new machinery, the choice of equipment is made by manufacturing since that department has the technical knowledge required. However, the purchasing department may be involved in the search for possible suppliers and in managing the tendering process.

The role of the purchasing department is to have the knowledge about sources and prices of supply of the raw materials and components that an

---

This chapter was co-written with Robert Mansfield.

organization buys. It can then advise on suitable suppliers and process the information involved in purchasing these items.

Purchasing has close links with manufacturing, but it also interacts with finance, and with design. In design this interaction ensures that designers are aware about the availability and cost of standard parts which are likely to be bought. Without this link a designer may specify expensive items when cheaper items that achieve the required function are available.

The objectives of purchasing management

Purchasing management manages the purchase of material by an organization. Good knowledge and 'intelligence' about the supply market enables an organization to acquire its material inputs effectively and in doing so to contribute to the productivity and profitability of the organization. Briefly the objectives are as follows:

- to purchase the right quality of material
- delivered at the right time
- in the right quantity
- from the right source
- at the right price.

This is illustrated in Figure 16.1.

To do this the purchasing function must interact closely with the manufacturing planning function. Given that in an organization of any size the amount of money spent on the purchase of materials is large, probably measured in millions of pounds, then clearly it is important to manage this process carefully. A lot of money can be made or lost by good or bad purchasing. Profit is a small difference between large revenues and costs. A small difference in the cost of materials can have a large effect on profit. This makes profit very sensitive to the quality of management decisions. The sensitivity of profit is developed in Chapter 18.

Figure 16.1 *The objectives of purchasing.*

It is necessary, therefore, for an organization to develop a purchasing strategy.

Organization of purchasing management

The head of the purchasing function must have the skills and authority to act effectively if the organization is to operate effectively. This requires that the head of the purchasing function reports at a sufficiently high position in the organization structure. Because of the possibility that suppliers may offer perks or even bribes to get purchase orders, it is also important that the head of purchasing is a person of integrity and well paid.

The purchasing department must be seen as part of the industrial system, which includes, but goes beyond, the boundary of the manufacturing organization. It manages a part of the supply or demand chain. Because of its links with the world outside the organization it must be involved right from the start of innovation projects, searching for ideas, new materials, possible suppliers and prices. The purchasing department should be constantly monitoring prices and price trends and have a major input to the budget-making process. They are usually the first department to be notified by suppliers of proposed price changes, which must be used to revise budgets. Because of the lead time on most materials, the purchasing department must be notified as soon as possible of manufacturing schedules so that it can order and schedule the delivery of materials to match the requirements of production. The purchasing department is also involved in the stock control function, ensuring adequate but not excessive stock levels, geared to supplied lead times. This aspect is addressed in Chapter 17.

Purchasing strategy

The factors involved in purchasing, as outlined in this chapter, show that this activity is not simply a matter of buying materials. It requires an understanding of the organization's products, components and other goods and services required. It also requires an understanding of the supply markets in which the organization buys. It requires a systems view and a professional approach. This can only be ensured by the development of a purchasing strategy for the organization. This must be developed alongside the organization's marketing, design and manufacturing strategies so they work together to enable the organization to achieve its corporate objectives.

SUPPLY CONTRACT

The process and factors involved in forming a supply contract are shown in Figure 16.2.

Figure 16.2 *Supply contract.*

Price

When the purchasing department wishes to buy something, the buyer will usually send out a letter inviting potential suppliers to quote or tender for the business. There is an old saying that 'you get what you pay for'. This must be remembered when looking at the price quoted by a supplier for materials and goods. Although it may be tempting to accept the lowest price, the buyer must consider other factors including quality, reliability, experience, etc. The cheapest items may turn out to cost more in terms of reject work, consequential damage, guarantee or product liability claims and dissatisfied customers. Government purchasing procedures usually insist that the lowest price is accepted on the grounds that this saves money for the taxpayer. The problems mentioned above, however, often mean that poor quality or delay are the result of this purchasing strategy.

One of the most difficult things for the buyer to ensure is that 'like is being compared with like'. One example would be hi-fi systems. There must be hundreds of different models available, all with slightly different attributes and prices. The buyer must decide which are the important factors and compare price against specification, remembering the intended use. It is not worth paying extra for a sound frequency range that can only be heard by bats! Table 5.1 in Chapter 5 showed a preference matrix for a decision on choosing which television to buy. Industrial buyers need to use the same approach.

From time to time suppliers give notice of their intention to increase prices. The purchasing department can negotiate and, depending on the supplier's competition and the state of the economy in general, a final revised price will be agreed or an alternative supplier used. If the cost of the material is a large percentage of the total cost of a product, then the profit would be reduced significantly.

The buyer will often be offered a discount for quantity. The total costs and benefits must be analyzed before buying far more than needed to save 5 per cent off the purchase price. Another possibility is the chance to stockpile materials as an insurance against inflation and to save on possible price rises. The money spent to buy the stock must either be borrowed or could have been invested. Either way the total effect of these costs must be considered in order to assess the benefit of buying now or later.

With certain high-value items such as copper, tin and other metals, the price can fluctuate wildly as it is dependent on many factors including world events such as wars, changes of governments or just rumours of supply problems. Sellers and buyers need a stable, guaranteed price in order to plan production and set prices. A buyer can sign a contract for future delivery at a negotiated, stipulated price, regardless of what the market price does. If the price goes up the seller will lose while the buyer gains and vice versa. This is a form of insurance. What it does is to provide the advantage of certainty while only requiring a small percentage cash deposit. This risk can be transferred through the futures market to a futures broker who will buy and sell goods for future delivery at a price agreed.

SOURCE OF SUPPLY

A buyer must identify or develop suitable sources of supply. Before placing an order the financial background of the potential supplier is investigated by examining balance sheets, company reports and reports made by independent analysts. If the contract is to be long term it is worth finding out if the supplier is about to go bankrupt. In such an eventuality part-finished work and material not yet paid for is frozen by the liquidator. However, the liquidator is usually very happy for you to pay for the material.

The competence of suppliers may be assessed by their reputation in the market, by work that they have done for other companies or, by direct examination of their factory, machinery, staff and systems. Many large companies now insist on being able to send a team to assess and audit the vendor prior to the placing of an order. This is part of the vendor rating process used to select a suitable supplier. This information can be used in conjunction with the quoted price to compare sources.

In order to benefit themselves, manufacturing organizations sometimes assist suppliers. They are helping them in order to help themselves. Companies can provide the finance to help potential suppliers improve quality, develop new processes, install compatible systems to help with logistics, etc. Marks & Spencer plc are renowned for their work in supplier development. They will help to bring a company up to the point at which it is a good reliable supplier, even seconding their own managers to help. This requires long-term planning, and a good relationship between supplier and manufacturer. This type of relationship leads to a win–win situation since both sides win.

Location of supplier

The geographical location of a supplier is important in reducing costs, lead times and delays, and ensuring reliability of supply. Suppliers in the same town or district can usually deliver quickly and at a lower price because of reduced transport costs. Using overseas sources can introduce extra costs for transport and excise duties but for some items this can be cheaper than buying from a supplier in one's own country. Political changes overseas can cause unexpected price increases or a sudden withdrawal of supply. When the Suez Canal was closed in 1967 as the result of war, supplies previously using the canal had to go around South Africa instead. This not only cost a lot more but also took much longer, resulting in a gap in supply of several months.

When components come from outside the European Union, care must be taken to ensure that the percentage by value of components in a product does not exceed a certain level since that would define the product as being of foreign origin even if it is part-made and assembled within a country in the Union. Many overseas companies, especially the Japanese, are setting up in Britain as a means of their products being described as British for sale across Europe. Cars and electrical products made in East Asia for sale in Europe are subject to quotas and import duties. Cars and other products made in Britain by East Asian companies and using designs and some components from East Asia can be sold in the European Union as 'European made' if they contain a proportion of parts made in Europe. In this way import taxes are avoided, so East Asian companies can sell in Europe and support jobs and profit in their own country.

When overseas organizations manufacture in Britain and buy material from their own company in their own country, there is the danger that the price paid for such material, being an internal transfer price, will be carefully adjusted to reduce the profit in Britain as a means of avoiding corporation tax in Britain and increasing the profit in their own country.

Some countries that are short of 'hard currency' such as British pounds or American dollars demand reciprocal purchases whenever they buy products from European Union countries. They need this in order to acquire the hard currency with which to buy from Europe or America. This can be beneficial to the manufacturer if, as in the case of Russia, they are rich in raw materials and the manufacturer has advanced technology to sell which the Russians may want to buy.

Quality of supply

An organization buying materials will want to assure itself that the quality of goods it is buying is satisfactory. After the placing of an order the quality control department will monitor the supplier's quality. Random samples are taken of goods received. Defect rates are calculated and used to determine the supplier's status. This is used by the purchasing department, together with information on delivery reliability, etc., to determine the suitability for repeat orders. Excessive defect rates or poor delivery reliability can lead to the supplier losing the contract to supply. This process of monitoring the quality of a supplier is known as vendor rating.

Number of suppliers

The purchasing department have to decide how many sources they are going to use for the supply of material for a given component, and also the total number of suppliers with which they will be involved for the supply of all of the goods and services the organization needs. Purchasing is easier to manage if there are fewer suppliers to deal with. For a given item of raw material used directly in manufacture, the quantity required could be bought from a single supplier or from several suppliers. If several suppliers are used, the quantity for each would be small. Small quantity orders are usually done on a 'job' basis using basic machines. Many engineering jobs require money to be spent on tooling, jigs, fixtures and gauges. The cost of these will be included in the price paid. Using several suppliers means that these charges have to be paid several times. Only by increasing the quantity is it worth using 'line' processes, which are more efficient.

There are, therefore, good economic reasons for only using one supplier. This, however, carries a risk of creating a monopoly of supply and being open to pressure over price and quality, and being at risk of delays in delivery.

One possible strategy is to have two suppliers for each item. This is known as 'dual sourcing'. It improves security of supply, reduces vulnerability to delay and results in lower prices because of the competition between the two suppliers. Suppliers know that if there are problems of quality or delays in delivery, then the amount ordered from them will decrease or cease as the manufacturer has an alternative source of supply.

In looking at the total range of goods and services bought it is sensible, where possible, to buy from suppliers who are able to supply a range of products. This reduces the total number of suppliers and the number of invoices. It also means that a supplier can deliver several items at the same time, reducing transportation costs. This should be reflected in cheaper prices.

SUPPLY OF MATERIALS

The flows of information involved in the supply of materials to a manufacturer is shown in Figure 16.3. This shows only the flows of information immediately and directly involved in a purchase transaction. The determination of price and the decision to buy will depend on subjective decisions on quality and also on the relative power balance between the supplier and manufacturer.

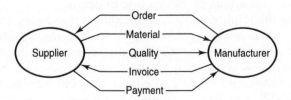

Figure 16.3 *Supply of materials.*

Quality of goods supplied

Quality is fundamental to the success of an organization. Quality has already been covered in Chapter 9 but it is relevant here to consider particularly the quality of materials coming into the organization. The quality of raw materials must be specified and based on what is needed for the purpose and market for the product. Overspecification costs money and will not help to sell the product if it inflates the selling price above what the market is prepared to pay. Underspecification may cause problems with the manufacturing processes, resulting in expensive delays and excessive scrap rates. Poor quality will result in dissatisfied customers. Since raw materials are one of the inputs to the manufacturing system, it is important to ensure that these materials are of the requisite quality before they are used within the organization. Once they have been used and found to be of poor quality, then not only is the cost of the material lost but so too is all of the work done on the material within the factory – this is costed as the value added to the material and includes labour, processing and overhead costs.

Having specified a quality standard, the purchasing department must ensure that the supplier is capable of meeting it. Many suppliers now work to quality assurance procedures, such as BS 5750 or ISO 9000, which should guarantee that the supplied material is within the required specification. In many cases, this enables the manufacturing organization, as purchaser, to reduce the amount and cost of goods-inwards inspection to a minimum.

Delivery

Delivery schedules must be based on the needs of manufacturing and driven by the demand from the market. The ideal schedule would provide the material 'just in time' for manufacturing to use it. Materials that are purchased too soon must be stored until required. This can be expensive and there is always the possibility of damage, loss or deterioration. Goods must usually be paid for within thirty days of delivery. If they are not going to be used for some time, they will not generate the revenue to cover the purchase cost until much later. This ties up more capital in material stocks and can cause an adverse cash flow situation. Materials ordered too late can cause manufacturing delays, dissatisfied customers and lost orders. A price premium (higher price) can be paid to speed up delivery but this adds to the cost and may reduce the profit. This course of action may be justified, however, if it avoids profit loss due to delays.

Forecasting the requirements and scheduling purchases are a vital part of the purchasing function and require inputs from manufacturing planning to ensure accurate estimates of materials requirements in terms of volumes and dates. In the case of new products, purchasing should be involved with marketing, design, manufacturing and finance to ensure that full information is available about the availability and cost of components proposed for the new product. Design in isolation may cause expensive components to be specified where cheaper, but equally functional components could be specified. The knowledge that the purchasing function has about the supply market needs to incorporated into the design and specification of components.

The purchasing department can negotiate with the supplier, on the basis of these specifications and schedules, to agree the volume and timing of deliveries to minimize costs for both supplier and manufacturer. Late delivery of raw materials causes difficulties and loss of output by the manufacturer. Uncertainty about delivery causes manufacturers to carry more stocks 'just in case' deliveries are late. This increases the manufacturer's costs. This aspect is explored further in Chapter 17.

In modern, large manufacturing organizations delivery intervals are reduced in order to reduce the level and cost of stock and may be measured in hours. In smaller organizations it would be days or even weeks.

Quantity and schedule of purchases

The ideal quantity to order at one time from a supplier would be just the right amount needed at that instant of time. Unfortunately, most materials are made and delivered in batches, the size of which depends on the economics of the process and of transportation. The main element determining batch size is the time required to change from one process to another and the cost of transport. This aspect is developed further in Chapter 17.

When an item is required repetitively it is possible to negotiate a contract for a large quantity over a period of time and to negotiate within this a 'call off' or schedule of delivery of small quantities more frequently. This will enable the manufacturer to obtain a lower price and deliveries when they are needed. The amounts ordered and delivered will depend on variations in the amount manufactured. These will depend on customer demand which varies in the short term and will not be known at the time of negotiation of the contracts with suppliers. Because the purchase order in total is for a large quantity, the suppliers can better organize their own manufacturing and so benefit themselves. This sort of arrangement is one that benefits both manufacturer and supplier. It is an example of manufacturer and supplier working together for mutual benefit – an example of a win–win relationship.

With some materials, such as fabrics, the colour will vary very slightly between batches. The only way to ensure a colour match is to purchase the total quantity required in one order. This can be very expensive and damaging to the cash flow of an organization. Where more material is required than can be made in a single batch, measures must be taken to ensure that different colour batches are not used close together where a colour difference would be detected.

Excessive stock holding can cause problems in a volatile market such as clothing. If the fashion changes, the manufacturer can be left with large quantities of unsaleable items. In an industrial context, if modifications are required, they can either be brought in immediately, making large stocks obsolete, or delayed until existing stocks are used up.

All of the above factors must be considered when deciding what quantities to order. Accurate estimates of usage rates and forecasts of future demand will aid decision making.

LAW IN PURCHASING

The law in purchasing is the same as that in selling, but with the manufacturer on the opposite side. Purchases are governed by the law of contract and specifically by the Fair Trading Act of 1977. Purchasers have rights to a refund if the goods are not of merchantable quality. A purchase takes place when the seller accepts the order, not when it is delivered or paid for.

INFORMATION TECHNOLOGY IN PURCHASING

The basic activity of purchasing involves the placing of orders for materials and other items, the receipt of goods ordered and the payment for these goods, all at different dates. These transactions need to be recorded, processed and progressed (checked to ensure that intended actions have occurred). This data involves purchasing, stores, manufacturing and finance. Databases are suitable for processing such data, although specific purchase order processing packages are available. Such systems are modular and can be integrated with systems in manufacturing and finance to form an integrated management information system.

Databases and CD-ROMs can contain details of suppliers of components and their prices. These enable purchase managers to quickly find available suppliers.

SUPPLIER RELATIONSHIP

The old style of purchasing management was adversarial: the buyer used the purchasing power of the manufacturer to coerce suppliers and force down prices. This was done on the basis that lower prices meant higher profits. This is counter-productive since suppliers will look for better customers, suppliers will buy cheaper materials themselves and use cheaper processes which could adversely affect quality. Co-operation will be grudging and liaison poor. Deliveries may be late. In extreme cases suppliers can close down if profits are squeezed too much. This style of supplier relationship was based on the idea that for a given sale price of a product the wealth available was fixed, so profits could only be raised by reducing material costs. This is a fixed-sum, win–lose view: the manufacturer wins; the supplier loses. However, the result often is that due to these pressures on the supplier, quality and delivery are threatened, so the manufacturer suffers quality problems and delays in delivery which shut down manufacturing. The effect therefore is a lose–lose result: both sides lose. Suppliers cannot invest in modern equipment, so productivity is low. Profit for both organizations is reduced. Eventually such a purchasing strategy could lead to the bankruptcy of the supplier. In the event of bankruptcy of a supplier the manufacturing organization is left without a supplier, and if manufacturing stops, the manufacturer may also go bankrupt.

The modern approach to purchasing is to create a common objective which enables a climate of mutual co-operation and assistance in which both parties win. Techniques such as just-in-time material delivery will only work if there is a good supplier–manufacturer relationship.

Some manufacturers are negotiating single-source agreements which last for one or two years. This gives the supplier the confidence to invest in equipment and labour for a large order and stability of its output over the medium term. This leads to efficiency for the supplier and hence to lower prices. This is a win–win result. The incentive for the supplier to perform effectively also comes from the existence of competitors who are working to get the next contract. In such cases single sourcing does not lead to supplier monopoly.

Some companies are part of larger organizations and have a policy of purchasing from within the group whenever possible. This can put artificial, and sometimes damaging, constraints on the design of the product, forcing it to use components manufactured in the group and restrict the buyer's choice, but has the effect of increasing group added value and profits. At the level of the operating organization it may cause higher prices and lower profit.

MAKE–BUY DECISIONS

One decision that all organizations have to make when the bill of materials for a product is prepared is whether a component will be made or bought. This is not merely a simple cost decision but a strategic decision as it determines the range of processes, the space and number of employees needed. It therefore significantly affects the profitability of the organization. Figure 16.4 shows the alternatives of making or buying. Figure 16.5 shows the effects of make–buy decisions.

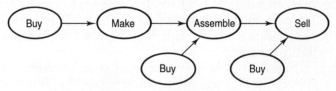

Figure 16.4 *Make–buy decision.*

Make or buy

Manufacturing costs

Lead time

Range of processes

Technology skills

Confidentiality

Number of employees

Added value

Profit

Figure 16.5   *Effects of a make–buy decision.*

All organizations buy raw materials and/or services upstream from their primary activities. They may, or may not, buy other materials or services further downstream, prior to assembly for example. Some organizations buy everything and make nothing. They are simply traders, buyers and sellers.

If a company has the machinery and the skills required, they could make the components themselves. However, it is usually not economic to make such standard items as nuts, bolts, washers and many electronic components. These are made by the million by specialists for prices that the organization could not match because of the efficiency of large-scale manufacture by the supplier. The supplier may also pay much lower wages in some East Asian countries. A company must decide on the basis not only of costs but also the strategic implications of having, not having or acquiring the necessary technology and skills. A strategy often used is to concentrate on a narrow range of technologies and to buy components which require other technologies. For example, a manufacturer may make all of the metal components in a product but buy all the plastic components, or vice versa.

Some parts, such as bearings, motors, pumps, gears and printed circuit boards, require special machinery and processes in manufacture. Where a special item could be replaced by a cheaper standard item, the buyer should liaise with the design department to review the design to incorporate the standard part. If for genuine reasons a non-standard part is required, it is often still worth getting the item made by the specialists. An alternative to this is to buy a standard item and then modify it to suit the special requirements.

Other items, such as electronic components, require specialized technology that most engineering companies do not have. It is possible to acquire technology by recruiting skilled personnel and setting up facilities. This is usually very expensive and the organization must decide that there is sufficient need over a period of time to justify the cost.

One strategy is to take over the specialist supplier so it becomes part of the manufacturing organization. This was often done in the past by car manufacturers to ensure continuity of supply, better control of the supplier and to deny the product to competitors. It was also done to achieve better control of supply and better prices. Organizations found, however, that they did not have the technical or managerial skills to manage such complex organizations.

The problems of managing such diversified organizations operating differing technologies on different sites has led more recently to demergers. These create smaller organizations focused on their technical specialisms. These smaller concerns can be more effective and more profitable. Manufacturers then create win–win relationships with such independent supplier organizations rather than controlling them as part of the organization.

Subcontracting

An intermediate stage between making a part 'in-house' or 'buying out' is the use of subcontractors. This would be used when a component required several processing operations using different technologies. For example, a metal component requiring machining may also require an operation such as heat treatment or chrome plating. It would probably not be economic to cater for such processes internally. These processes require expensive equipment and consideration of safety and pollution. Such processes are only economic when dealing with very large volumes which can cover the capital costs involved. In such cases the component would be part processed internally, then sent out to a subcontractor for the specialized processes, before being returned to the manufacturer for further processing.

Subcontracting can also be used to cope with variable demand even where a manufacturer has the requisite processes in-house. At peak periods work will be subcontracted. In times of recession subcontract work can be brought back 'in-house' to provide work for the organization's own employees. In this way an organization can stabilize its levels of capacity and employment and deal effectively with variable levels of demand for its products.

Task sheet 16   Purchasing

16.1   List the relative advantages and disadvantages of making or buying components.

16.2   List the advantages and disadvantages of single or dual sourcing components.

16.3   Consider ways of developing a win–win supplier relationship.

16.4   Identify the information flows for the supply of materials between the supplier and the purchasing, manufacturing and financial functions in an integrated management information system.

CHAPTER 17

STOCK CONTROL

STOCKS OF MATERIAL

Industrial organizations must decide how much stock of material (called inventory in American textbooks) they should keep. Keeping stocks costs money, so there is a tendency to reduce stocks in order to save money. Accountants only see the cost of holding stocks. Decisions based on this narrow approach will not be beneficial to the organization. This is suboptimal since *not* keeping stocks also costs money – but not in a form that can easily be seen in the accounts. Taking a systems view, it is necessary to optimize the total costs of stock holding in order to minimize costs for the organization as a whole.

Figure 1.1 in Chapter 1 shows stores of raw material prior to processing and of products. This chapter considers whether stocks of materials and products are needed.

First it is necessary to identify where and why organizations may want to keep stocks of material and analyze all the costs involved. The *economic order quantity* (EOQ) technique is used to design cost-optimal stock-holding systems.

Where may stocks be required?

Stocks may be beneficial at boundary points between different stages of processing, particularly at input to and output from a work area. In addition to stocks of raw materials and finished goods, it may be beneficial to hold stocks of part-finished goods between various stages of processing – especially between manufacturing and assembly. Components purchased (bought out) directly for use in assembly would also be stored at this location. This enables the collection, 'kitting' or 'marshalling' of manufactured and bought-out components ready for assembly.

There are also stocks of material within the processing areas – being processed, waiting for processing or for movement between processes. This is known as 'work in progress' or WIP. Work in progress may be in 'buffer stocks' between processes or just lying around waiting for something to be done. Buffer stocks decouple one process from another and reduce interdependence. If there

262

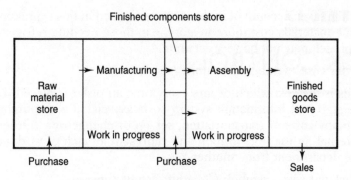

Figure 17.1 *Location of stocks in a factory.*

is a stoppage of work at a process, the next process may run out of material to work on. Buffer stocks provide a buffer or cushion against such knock-on effects. With long throughput times, work in progress can be very high and very costly.

Costs of holding stocks

Holding and handling stocks of material costs money. In seeking to save money it is tempting to reduce stock levels. This, however, may *not* reduce total stock-holding costs since other costs may rise. It is necessary to identify and quantify all the costs involved in stock holding.

- Purchase cost   symbol: C_p unit: £/unit

  This is the cost of purchasing one unit of material. This is paid when material is bought and not repaid until a product in which the material is used is sold. This may be a long time later. The money spent purchasing stock represents capital which is tied-up in stock and thus not available for other, more profitable purposes.

- Carrying cost   symbol: C_i unit: £/unit/year or % C_p/year

  This is cost of having one unit of the stock in a stores. This cost is calculated for each type of material since its value depends on the value of the material. It includes the cost of interest on, or lost profit from an alternative investment of, the capital used to purchase it (but not the capital itself since that is included in the unit purchase cost), the cost of the storage facility (space, power, equipment) and the cost of storekeeping (people, information systems, insurance and obsolescence). This may be expressed as a cost per unit per year. In this form a separate value is need for each material.

    It can instead be expressed as a percentage of the unit purchase cost per year. This usually requires only a single value since the cost of capital is likely to be the same for all materials. The variation in material value is then taken account of in the value of C_p used in this expression. Where some materials cost more to store, e.g. those needing refrigeration, this approach may not be very accurate.

- Order cost   symbol: C_o unit: £/order

  This is the cost which occurs every time an order is placed. It includes the staff and information system costs associated with purchasing, but also any costs of transportation, set-up or changeover. (Orders may be 'internal' to the organization and represent a batch of work ordered by one department from another.)

- Stock out cost   symbol: C_s unit: £/unit/time

  This is the cost of going out of stock by one unit of stock. This occurs when an order from a customer cannot be fulfilled because there is no stock. It includes the cost of customer dissatisfaction, loss of immediate and future sales revenue and/or the extra costs of change incurred in avoiding a stock out.

Notice that these costs are unit costs, the cost of one unit. The actual cost is the unit cost multiplied by the number of units. Note also that these costs are measured in different units. Some are costs per time, others are costs per order. Costs can only be added when they are all measured in the same units.

Benefits of holding stocks

Raw material stocks

These enable an organization to respond more quickly to customer demand since delivery time does not have to include supply lead time. Stocks will be higher if materials are bought in larger quantities, but this cost may be offset by discounts for bulk purchases. Having stocks of raw material enables a factory to continue manufacturing when a supplier has failed to deliver on time. 'Safety stock' is an amount of stock carried to cover for such contingencies.

Work in progress

Since each process can be seen as the 'supplier' of the next process, then material within the factory represents raw material to the next process and similar benefits accrue. Work in progress 'decouples' one process from another and provides a tolerance to breakdowns and delays. Stocks of material held between processes are known as 'buffer stocks'. Batch sizes are the equivalent of order sizes. (In America they are called 'lot sizes'.) Reducing the number of batch changeovers not only reduces changeover costs, but, by reducing changeover time, increases utilization and output. However, larger batches means higher stock levels and stock-carrying costs, so a balance is needed.

Finished goods

Customers can be supplied 'off-the-shelf' if stocks of finished goods are held. Without them, delivery times to customers may be too long to win the order. Short delivery times may provide an organization with competitive advantage. Where demand is seasonally variable it may be necessary to manufacture in advance of demand and store until required as a means of stabilizing output and satisfying the variable demand at a lower cost.

DECISIONS ON ORDERING MATERIALS

The benefits of stock holding do *not* imply that it is necessarily beneficial to hold high levels of stocks. A cost balance must be made between the costs and benefits of stock holding. This can only be done when all of the costs and benefits have been identified and quantified.

Ordering covers not only the ordering of raw materials from suppliers, but also the dispatch of orders for products to customer and also 'works orders' on manufacturing departments. An organization is its customers' supplier and its suppliers' customer. All are part of the 'supply chain'. Similar decisions are needed in both relationships.

An organization must decide *when* to place an order. A variety of ordering procedures are possible. Orders may be placed at prescribed intervals of time – known as the reorder interval; or when the level of stocks have fallen to a prescribed level – known as the *reorder level* (ROL). It is easier to organize a fixed reorder interval, but a fixed reorder level responds better to variations in the rate of use of a material.

Another decision is *how much* to order at a time. A *reorder quantity* (ROQ) may be prescribed or a maximum stock determined so that the amount ordered brings the stock to that level, the amount ordered depending on the stock level prior to ordering. (In America the reorder quantity is called the 'lot size'.)

Variation of stock over time

Figure 17.2 is drawn for the case of a raw material. A different picture arises for work in progress and for finished goods. Steady-state conditions are assumed, so this variation repeats over time. The slope of the line is the rate of use of the material. In these models these usage rates are assumed to be continuous and constant. In practice they will vary and occur in a series of steps as material is moved out of a stores.

This diagram is called a 'saw-tooth' diagram because its shape is similar to that of the teeth on a carpenter's saw. The reorder level is above the base of the saw-tooth due to the lead time between placing and receiving the order. The amount of stock at the base of the saw-tooth, just prior to replenishment is the

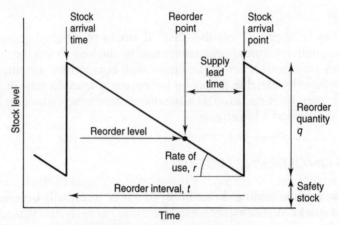

Figure 17.2 *Variation of stock over time.*

safety stock. In theory this is never used; it is there to provide 'insurance' for a lead time being longer than expected or a rate of use during the lead time being greater than expected.

The saw-tooth pattern of material stock may not be familiar to readers. It is, however, a pattern commonly experienced personally with the stock of money at a bank. Employees receive a stock of money (salary) monthly, which increases the level of money stock at the bank. Spending during the month gradually reduces the stock until it is replenished again at the end of the month. Spending is not at a fixed daily rate but in small chunks, each being a payment for something. For students the termly grant replaces the monthly salary, but the concept is the same.

In industry too the reduction of stock is in small chunks, possibly of uneven size.

ANALYSIS OF STOCK-HOLDING COSTS

A simple stock-holding model

This model shows the relationship between the costs of holding and of ordering stock. An assumption is made that the rate of use of an item of material is constant. In practice it is randomly variable, but the complex models for dealing with this are beyond the scope of this book. It is also assumed that the saw-tooth sits on the baseline, i.e. at the base of the saw-tooth the level of stock is zero. The following symbols are used in Figures 17.2 and 17.3:

q = reorder quantity
t = period of reorder cycle
r = rate of use

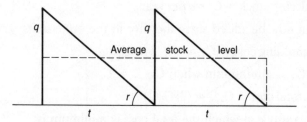

Figure 17.3 *Average stock over a period of time.*

In the triangle in Figure 17.3, q is the height in units of stock, t is the length of the base in units of time and r is the slope of the triangle in units of stock per time. These are therefore related as follows:

$$r = q/t \tag{17.1}$$
$$t = q/r \tag{17.2}$$
$$q = r \times t \tag{17.3}$$

It is important that the units of time are checked. Some may be in weeks, others will be per year. Errors will occur if the unit of time is not the same in each part of the calculation. t may be expressed in days or weeks, but costs are normally analyzed over a period of a year. Therefore it is better to express all times in years.

The actual level of stock falls over the period of the reorder cycle. The average stock over the period of the cycle is the average of the starting and finishing stocks. If the base of the saw-tooth is at the zero stock level, this is half the height of the triangle, so:

$$\text{average stock over the cycle} = q/2 \tag{17.4}$$

The period of one cycle is t. However, it is necessary to find the number of cycles in a longer time period, such as a year. If the cycle time is measured in the same units as the overall time period, then:

$$\text{number of cycles per year} = 1/t = r/q \tag{17.5}$$

If not, an adjustment is needed to express it in the same units. It is better to express all times in years since costs per year are usually calculated.
Then:

$$\text{cost of carrying stock} = C_i\, q/2 \text{ per year} \tag{17.6}$$
$$\text{cost of ordering stock} = C_o \text{ per order} \tag{17.7}$$

The total cost is the sum of the holding and ordering costs. However, the costs above are measured in different units. One is a cost per year and the other is a cost per order cycle. Therefore they cannot be added in this form. The cost of ordering must be converted to a cost per year.

Since the number of cycles per year has been calculated as r/q, then:

$$\text{cost of ordering stock} = C_o\, r/q \text{ per year} \tag{17.8}$$

These costs can now be added since they are in the same units, £/year:

$$\text{total stockholding cost} = C_T = C_i q/2 + C_o\, r/q \tag{17.9}$$

The total cost C_T is a minimum when $C_i q/2 = C_o\, r/q$:

$$\text{minimum total cost} = C_{\min} = (2rC_oC_i)^{1/2} \tag{17.10}$$

The reorder quantity q at which the total cost is minimum is:

$$q_{\text{opt}} = (2rC_o/C_i)^{1/2} \tag{17.11}$$

q_{opt} is known as the *economic order quantity* (EOQ) or, in the case of placing orders on manufacturing, as the *economic batch quantity* (EBQ).

These calculations have determined the optimal size of the saw-tooth.

Variation of cost with order quantity

Figure 17.4 shows how the carrying cost, ordering cost and the total stock-holding cost vary with order quantity. The total stock-holding cost increases on both sides of the EOQ. It is, however, a shallow-bottomed curve, so a reorder quantity, ROQ, does not have to be exactly the EOQ, since any convenient value near to EOQ would be acceptable. It would be silly to use an ROQ of 94 or a reorder cycle of 4.3 days when values of 100 or 1 week are far more convenient to organize.

It must be stressed that EOQ calculations are only as good as the data that goes into them. The concept does show, however, that reducing order quantities, while it will reduce stock levels and stock-carrying costs, does not always reduce total stock-holding costs. A systems approach is needed to identify all of the costs and benefits. This model does not include all costs but it does include two costs and is better than models and decisions based on only one cost, usually the carrying cost.

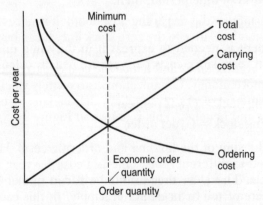

Figure 17.4 *Variation of costs with order quantity.*

An example of EOQ analysis

An organization uses 1000 items of a material per year at a purchase cost of £4 each. The unit carrying cost per year is 25 per cent of the purchase cost, and the cost of placing an order is £3.20. Assume a 50-week year.

The material is currently ordered once a year. Calculate the EOQ and the money saved by using the EOQ.

$$\text{EOQ} = (2 \times 1000 \times 3.2/0.25 \times 4)^{1/2} = 80 \text{ units} \tag{17.12}$$

This means placing an order every t weeks where:

$$t = q/r = 80/1000 = 0.08 \text{ years} = 4 \text{ weeks} \tag{17.13}$$
$$C_{min} = (2 \times 1000 \times 3.2 \times 0.25 \times 4)^{1/2} = £80 \tag{17.14}$$

The current order quantity is given by:

$$q = r \times t$$
$$= 1000 \times 50/50 = 1000 \text{ units} \tag{17.15}$$

(Dividing by 50 converts 50 weeks to 50/50 of a year.)

The current cost uses equation (17.9) and is:

$$C_T = (0.25 \times 4 \times 1000/2) + (3.2 \times 1000/1000)$$
$$= 500 + 3.2 = £503.2 \tag{17.16}$$

Notice that the two parts are not now the same value.

$$\text{cost saving} = £503.20 - £80.00 = £423.20 \tag{17.17}$$

By changing the order quantity from 1000 to 80 the organization saves £423.2 or 84 per cent of its total stock-holding cost.

A simple safety stock model

In addition to determining the *size* of the 'saw-tooth' it is necessary to decide the *position* of the saw-tooth relative to the zero stock line. It can be seen from Figure 17.2 that as safety stock is increased, the saw-tooth pattern is higher on the graph. This increases the average stock and hence the cost of carrying it. Holding more safety stock costs more money. Organizations carry safety stock as an insurance, 'just in case' demand increases or suppliers deliver late. A balance is needed between these cost and benefits. This is shown in Figure 17.5.

Service level

As safety stock is reduced, an organization will run out of stock due to the variability in the rate of use or in delays in supply. In this case the base of the saw-tooth falls below the zero stock line, implying negative stock. Physical stock

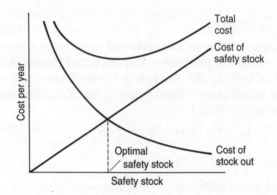

Figure 17.5 *Variation of costs with safety stock.*

cannot be negative but demand can exceed supply. This 'negative' stock can be quantified. This is the same as having negative cash at the bank, i.e. an overdraft.

A higher level of safety stock will prevent stock-out but will incur a higher cost for carrying the safety stock. Organizations may prefer to allow for some time when they have no stock rather than pay the cost of preventing a stock-out. This leads to the concept of a 'service level', which is expressed as the percentage of order cycle time that stocks are available. An organization must therefore determine what is an acceptable service level and what safety stock is needed to achieve this. Because of the random variability involved, the calculations required in a safety stock model require statistical analysis or a stock level simulation model, which are not dealt with in this book.

Limitations of stock control models

Stock control models include quantifiable costs and derive a result which is optimal for the data and model used. The result is only as good as the data and the model used. The quantitative analysis of stock holding is based on the following assumptions:

- that the rate of usage is constant;
- that the rate of future usage is known;
- that all the costs have been included and measured.

In reality there is uncertainty. Future rates of usage can only be estimated. Supply lead times can only be estimated and may extend if suppliers fail to deliver on time.

It should not be difficult to measure the purchase and carrying costs. Order costs are not so easy to measure since they should include transportation costs. These may not be paid directly by an organization but would be reflected in a supplier charging a higher price to cover their transportation costs. Very few organizations would be able to measure their stock-out costs because this includes subjective judgements of the attitude and action of dissatisfied customers. Nevertheless, organizations need to take decisions on stock holding and need to determine these costs accurately in order to do so.

The costs used are those incurred by an organization, but do not include the costs borne by the supplier. To this extent the analysis is not of the whole material supply system, but as indicated above, some of these costs are included where they affect the purchase cost.

The models used above look separately at the size and position of the saw-tooth. Simulation models can include random variables and the effects of all costs together.

The EOQ model looks at costs for a single item. This could lead to ROQs of various sizes and many different reorder intervals. If several items come from the same supplier, it is preferable to arrange sensible intervals for the whole set of items.

The EOQ calculated by a manufacturer is likely to be different, and in most cases smaller, than the EOQ of the supplier for the same item. It is for this reason that suppliers offer quantity discounts to a manufacturer for buying larger quantities at a time. This reduces the supplier's transportation costs, so the discount is an incentive designed to achieve this.

The mathematical results from these models should be used as starting points for a managerial judgement which is able to include other intangible factors. There are additional benefits from reducing throughput time which are not included in these models.

MANAGEMENT OF MATERIAL STOCKS

Traditional approaches to stock control

Because of the uncertainties inherent in material supply the traditional approach has been to keep a lot of safety stock 'just in case' things go wrong.

As stock can be physically seen and its cost appears in the accounts, approaches by accountants have attempted to save money by reducing stocks. This is a suboptimal approach which may not reduce total system costs. Merely reducing stocks may not save money if other costs rise. Multi-cost models are needed to provide a full analysis of material stock systems and decisions which are optimal at the system level.

The result of traditional approaches is high levels of stock and of money tied up in stock.

The traditional approach takes a view that sees the costs of carrying and ordering stock, safety stock and stock-out as fixed, and finds a balance between them. By focusing on balancing the existing costs, no effort is applied to the ways of reducing these costs.

'Just-in-time' approaches seek to reduce stock costs by arranging for suppliers to deliver materials just in time for use in the factory. The same idea applies within the factory to transfers between processes. This reduces buffer stocks so that if problems arise, production stops. To prevent this, supply contracts force suppliers to carry, say, two or three days of stock. This may appear to save money for the manufacturer but the problem is not being solved, it is merely being pushed onto someone else (the supplier). This is an example of an adversarial relationship with suppliers. Manufacturers use their power to force additional costs on to the supplier. The result is likely to be 'lose–lose'.

Modern approaches to stock control

A modern approach is based on the concept of *kaizen*, continuous improvement, and seeks a win–win result in which carrying and ordering costs *and* supplier uncertainty are all reduced simultaneously.

This systems-based approach recognizes the complexity of the stock-holding system. Attempts are made to reduce uncertainty, and improve supplier relationships so that the risk of late delivery is minimized. 'Just-in-time' (JIT) approaches seek to create optimal systems in which economic batch sizes are as low as one and stock levels are reduced to virtually nil as supplies arrive 'just in time' to be loaded into the factory without putting them into stores. Modern manufacturing systems need to be 'flexible'. This necessitates small batch sizes and order quantities. JIT should not be used to push problems onto suppliers, it should be used when these problems have been solved. A 'win–win' strategy is needed in which both manufacturer and supplier benefit.

Kanban systems of workflow control are based on small batches of work; a *kanban* box contains a given number of components, which is in effect the reorder quantity. The factors used to determine *kanban* order size relate to shopfloor workflow patterns rather than EOQs. The costs of ordering stock should be reduced so that the EOQ is the *kanban* quantity.

Merely reducing stock may not be beneficial since it can lead to stock-outs, lower utilization, idle labour and machines, and 'crisis management' as the problems are resolved in a non-optimal way. This is a suboptimal approach which fails to examine total system costs. Means must be found to reduce safety stock and the risk of stock-out together. This can be seen to necessitate an improvement in supplier relationship as since this can reduce uncertainty.

Merely reducing batch sizes may not be cost-effective. Means must be found to *enable* small batches to be cost-effective, by reducing the economic order quantity.

Analysis of the EOQ equation and graph shows that this can be achieved by reducing the order cost, C_o. In purchasing materials, the order cost is that associated with office processes and transportation, but when applied internally to manufacture the order cost is the changeover cost, which derives from the changeover time, the time taken to change a machine from making one product to making another. This cost includes not only the direct costs of the changeover but also the cost of lost production while the machine is stopped.

The Japanese have used EOQ concepts to analyze stock-holding and ordering costs. They showed that reducing the order cost was the way to reduce the EOQ to the small value needed in modern manufacturing. Increased productivity is the result of this approach.

Reduction of changeover time

One part of the order cost, when applied within the organization, is the time, hence cost, of changing from the production of one product to another.

The Japanese engineer Shingo has shown that the changeover time can be reduced significantly by prepreparing dies for the next job so that changeover merely involves lifting out one die and putting in another. Previously the change of tools or dies would be done *on* the machine, which would stop producing for the time it took to make the change. Short changeover times *enable* small yet economic batch sizes. In this way small batch sizes can be used which are the EOQ, or at least close to it. Shingo reduced die changeover time from 8 hours to a single minute. This concept is known as SMED, single minute exchange of dies. This approach was seen to be necessary to achieve the economic 'batch of one' which gives the ultimate flexibility and efficiency in a modern manufacturing system. The Japanese have used the concept of EOQ to show how to achieve small batches cost-effectively.

In some specific cases the concept of SMED can be taken to ZMED, zero changeover time. In a steel rolling mill, steel bars are put through a pair of rolls into which a profile is cut which creates the shape of the steel bar at output. If several different profiles are cut on the rolls, then the changeover from one shape of product to another can be done anytime without changing the rolls, so no changeover time is necessary. Such rolls would be more expensive initially but would save money in use since the machine would not have to stop to change product. The financial analysis of such investments is covered in Chapter 21.

Modern approaches are therefore based on reducing the order cost by the following methods:

- Simplifying clerical and information systems.
- Changing 'on-line' set-ups to 'off-line' set-ups.
- Rationalizing distribution to reduce transportation costs.

Supplier relationships

Improving supplier relationships improves information on material supply, reduces uncertainty and *enables* safety stock to be reduced without incurring stock-outs. This is a win–win outcome. Again it is the Japanese who have shown that improved supplier relationships can lead to lower stock levels and lower costs with benefits to both manufacturers and suppliers. This aspect of supplier relationship reinforces the comments made in Chapter 16 on this topic.

Other methods of reducing stock levels

Stock levels can be reduced by reducing the variety of items stocked. For example, a typical computer keyboard contains 81 keys, 74 of them the same size. Every language requires slight differences, some use totally different scripts. If every different key had to be kept in stock, the stock level would be enormous, each key would need its own stock reference and the opportunity for error on assembly would be astronomic. IBM avoids these problems by assembling keyboards with blank keys. These are then etched at the last moment when the computer tells the system to make, for example, a Greek language keyboard for immediate dispatch. This method of last-minute customizing enables increased variety for the customer but with reduced stock levels of finished goods for the manufacturer.

Measurement of stock-holding efficiency

The average value of stock at any one time can be compared with the annual value passing through the stores. The value of the stock is the volume of the stock multiplied by the unit value – the unit value being purchase cost for materials and cost of sales (*not* sales value) for finished goods (products). The value of work in progress is often taken to be the average of the material and finished goods values. For example, if average finished goods stock/annual cost of sales = 0.1, then there is 0.1 of a year's sales in stock (or 1.2 months). This ratio can be inverted to give a value of 10, which is called the 'stock turnover', meaning that on average the stock in the stores is turned over, or replaced, 10 times per year. This use of the word turnover differs from its use in 'sales turnover', which means the value of sales in a year.

It was not uncommon in the past for several weeks' worth of stock to be held in stores. In Britain it is down to a week or so, and less in some cases. In Japan average levels of stock are about two days, but less than a day in some cases.

Japanese approaches to stock control are reducing average stock levels by reducing changeover times and supplier uncertainty. These approaches are based on the concept of EOQ analysis in which stock costs are not just accepted, but are driven down so that although the organization operates at EOQs, the EOQs are smaller than before.

Task sheet 17   Stock control

17.1   An organization uses an item at the rate of 40 units/week. The carrying cost is £5/unit/year and the order cost is £50. Assume a 50-week year. What is the economic order quantity and the reorder cycle time?

17.2   An organization places orders for 1000 items every two weeks. Its carrying cost is £10/unit/year and the order cost is £50. Assume a 50-week year. What is the economic order quantity and how much would the organization save if it used this order quantity?

17.3   An organization uses an item at the rate of 25,000 per year and orders them every two days. The purchase cost is £40/unit, the carrying cost is 25 per cent per year of the purchase cost and the order cost is £50. Assume a five-day week and 50-week year. Is this the economic order quantity?

17.4   The cost of purchasing an item appears in the accounts as £1050/month. Its carrying cost is £7/unit/year and its order cost is £49. What is the economic order quantity? What is the economic order frequency?

17.5   An item is used at the rate of 4000 per week. The carrying cost is £12/unit/year and the order cost is £80. Assume a 48-week year. What is the economic order quantity? To what level must the order cost reduce if the economic order quantity is to be reduced to one day's requirement?

PART 6

MANAGING FINANCIAL RESOURCES

LEARNING OBJECTIVES

6.1 Classify costs according to sources, elements, and their capital or recurrent nature.

6.2 Lay out a set of accounts for recording transactions using the double-entry bookkeeping system.

6.3 Explain the difference between a cost centre and a profit centre.

6.4 Describe fixed and variable costs, and classify costs according to their relationship to volume.

6.5 Distinguish between direct costs, indirect costs and overheads.

6.6 Explain and perform the processes of allocation, apportionment and absorption of overheads.

6.7 Explain job costing and process costing.

6.8 Explain the term standard cost and relate it to standard time.

6.9 Define marginal cost and contribution.

6.10 Perform break-even analysis.

6.11 Construct a cash flow forecast and explain its use.

6.12 Define depreciation and explain its purpose.

6.13 Construct a profit and loss account and explain its derivation from the manufacturing and trading accounts.

6.14 Explain the difference between cash and profit.

6.15 Construct a balance sheet.

6.16 Calculate the standard ratios from the accounts.

6.17 Describe the process of budget preparation and the value to management of budgets.

6.18 Calculate variances, explain the process of variance analysis and construct a variance hierarchy.

6.19 Carry out project appraisal and measure the accounting rate of return and payback period.

6.20 Understand and explain the concept of the time value of money.

6.21 Use discounted cash flow analysis to calculate net present value and internal rate of return.

CHAPTER 18

COST ACCOUNTING

THE NEED FOR ACCOUNTING

Accounting and accountability (Chapter 4) have a common origin. Both refer to the need to report and provide information to a higher authority. Accounting involves the collection and systematic recording of cost and financial data within an organization. Given the importance of ensuring that money is used effectively, accounting techniques are used not only to analyze how money is spent but also to plan and control expenditure.

Figure 1.1 in Chapter 1 shows the main activities of a manufacturing organization. Part 4 of this book showed how manufacturing activities are controlled and performance measured using physical measures. This part of the book explains the cost and financial analysis of manufacturing activities and measures performance using financial measures and ratios. This is not restricted to manufacturing but embraces all of the money flows within the organization and also to and from other organizations. The principles and concepts apply to non-manufacturing organizations and, with some variations, to non-profit organizations.

Cost accounting is covered in this chapter. It provides for the recording of income and expenditure, and enables the classification and analysis of costs. Some form of cost analysis is needed in all functions of management, wherever money is spent. The accounts function within an organization will collect this data, receive money from customers and pay the bills for materials, wages and all other expenditure.

Financial accounting, Chapter 19, prepares the financial accounts required by law for presentation, in the case of companies, to the shareholders, auditors and the tax authorities. Similar procedures are practised in non-profit organizations and the public sector, the information being required by trustees or the government. Financial accounting presents financial information outside the organization. It also enables organizations to compare their performance with other organizations which publish their accounts in the same form.

This book provides only an introduction to cost and financial accounting as a foundation for further study, but develops further the area of management accounting.

Management accounting uses cost data within the organization for management decisions. Managers need to plan and control their activities, monitor their expenditure and measure their performance. Budgetary control, in Chapter 20, provides the analysis for this. Financial models can be constructed which allow managers to simulate the effects of proposals for price and volume changes.

Managers also need to be able to appraise a proposal for investment of money in a project involving change, i.e. to assess the likely future income for comparison with the intended initial expenditure. Project appraisal, Chapter 21, provides the analytical techniques for assessing investment in making major change within the organization.

COST RECORDING

In order to record costs – and income – usefully a set of accounts is established according to the type of cost or income. Accounts are sometimes called ledgers or simply 'books', giving rise to the term bookkeeper and the procedure of bookkeeping. An account is simply a list of items of income or expenditure – noting the date, the amount, to whom it went or from whom it came and what it was for. Income and expenditure will usually be in separate columns. Formal accounts will be totalled and presented annually, but for management control purposes within the organization accounts are normally presented and analyzed monthly.

Separate accounts are kept for each type of cost or income. These enable costs to be totalled by type. Income will be received from customers, and there will be an account for each customer so that the customer can be invoiced. Similarly there will be an account for each supplier. There will be accounts for the material in stores and in 'work in progress' and for office expenditure. Expenditure of capital is separated from day-to-day expenditure. Every income, expenditure and transfer of material will be recorded. A set of accounts must be constructed which is appropriate to the organization and useful to managers for analysis of performance.

Table 18.1 shows a list of the different accounts that a typical manufacturing organization would keep. Their use is indicated by their title.

Double-entry bookkeeping

Every transaction is recorded. Each transaction involves someone giving and someone receiving. Therefore each transaction involves two accounts, the giver and the receiver, so two entries are made for each transaction, one in each of two accounts. This is the basis of the double-entry bookkeeping system. Transactions include not only the flows of money but also of materials, which are expressed in terms of their financial value. Transactions therefore include the sale of goods as well as the later payment for these goods.

Table 18.1 List of accounts for a manufacturing organization

Capital accounts	Trading accounts	Expenditure accounts
Share capital	Sales	Salaries
Loan	Purchasing	Wages
Reserves	Creditors	Computer
Dividend	Debtors	Office expenses
Depreciation	Cash	Sales expenses
Buildings	Customer	Telephone
Machinery	Supplier	Electricity
Vehicle	Stores	Furniture

This list is illustrative but not a complete list of accounts.

Every account will have entries for receiving (credit) and transferring out (debit), so each account will have two columns. Table 18.2 shows how transactions are recorded in the two columns of an account. An example of the recording of a set of transactions involved in the purchase of materials is shown in Table 18.3.

Table 18.2 Recording transactions in the accounts

Transaction	Left side (debit)	Right side (credit)
Purchase of material	Purchase account	Supplier account
Payment for material	Supplier account	Cash account
Sale of goods	Customer account	Sales account
Payment for goods	Cash account	Customer account
Payment of wages	Wages account	Cash account

Table 18.3 Example of a purchase transaction in the accounts

Purchase account

Date	Debit	£	Date	Credit	£
15 Jan	Nuts and bolts from Murphy Co.	75.45			

Supplier account (J. Murphy Co. Ltd)

Date	Debit	£	Date	Credit	£
			15 Jan	Nuts and bolts supplied	75.45
12 Feb	Nuts and bolts paid for	75.45			

Cash account

Date	Debit	£	Date	Credit	£
			12 Feb	To J. Murphy for nuts and bolts	75.45

Transactions are recorded during a financial period and totalled at the end of the period. At the end of a financial period the accounts will be balanced, i.e. the total in each column will be made the same, by a closing entry that represents the value in the account at the end of the period.

For example, in a stores account the total receipts and issues during a financial period are likely to be different. Assuming the stores were empty at the beginning of the financial period, then the difference (receipts − issues) is the value of stock at the end of the period. This is the 'closing stock', which is also the 'opening stock' for the next financial period.

The closing value of an account will be transferred to a higher-level account and the subsidiary account closed. Eventually all items will be transferred to the profit and loss account or balance sheet, as shown in the next chapter.

Cost centres

Within the organization a set of cost centres will be established. A cost centre is a department or section for which costs are separately ascertained. Each cost centre will have its own accounts, and each type of expenditure will be recorded in separate accounts. In this way, for example, not only is the total wage bill known but also how much of it is attributed to each cost centre. This enables the measurement and responsibility for costs to be delegated to different sections of the organization.

Profit centres

The concept of delegated responsibility can be taken to the extent that, for internal purposes only, each section is an autonomous unit. The plant-within-a-plant concept was introduced in Chapter 12. This requires each cost centre to have income as well as expenditure so that its profit can be calculated. If this is done, a cost centre becomes a profit centre. This approach extends the concept of the demand chain to relationships between departments within an organization. Each department sees others as suppliers and customers.

The way in which manufacturing sections have income is to sell their products to the next section and buy their components from the previous section. This is the concept of the internal market introduced in Chapter 5. The price of such internal sales, known as the transfer price, is not easy to determine. Prices are normally determined by the external market and include competitive pressures. There is no such pressure internally; prices have to be negotiated or imposed from above. Competition could be introduced if managers were free to buy their inputs from external or internal sellers, taking make–buy decisions at profit centre level. However, the strategic implications of make–buy decisions often cause them to be taken at a higher level.

Internal services such as accounting, marketing and personnel rarely sell their services, but manufacturing sections carry the cost of these services through accounting procedures that allocate these costs to the income-generating sections.

The profit centre concept is beginning to be used in the public sector where competitive tendering has been introduced.

COST CLASSIFICATION

Before costs can be controlled it is necessary to understand what they are, where they come from, how they behave and who is responsible for them.

Cost elements

There is a fundamental distinction between money invested in an organization which is used to buy the buildings and machines, and the money needed for the day-to-day activities. Money invested and used long term is called capital. Money received by an organization, for sales, is called revenue, and money spent by the organization on its activities is called its revenue costs. This term is confusing since revenue means income and cost means expenditure. The term means money spent out of revenue. To avoid confusion, the term 'recurrent costs' is used in this book.

Capital costs

Capital is money provided by shareholders, banks or government for the purpose of buying the long-term assets of the organization such as buildings and machines. The money spent on such assets is the capital cost of such assets. Chapter 21 introduces techniques for appraising whether it is financially worthwhile to invest capital in a machine or a factory.

Working capital

This is the amount of capital needed for running the organization. Some money will be tied up in stocks of material. Money is needed to pay recurrent costs before revenue is received. Working capital is measured in Chapter 19. Working capital plus the capital invested in machines forms the total capital of an organization.

Recurrent costs

It is possible to identify three separate elements of recurrent cost, namely materials, labour and expenses. These in turn can be subdivided into direct and indirect costs, fixed costs and variable costs.

Material costs

In manufacturing, a raw material or part-processed material such as a casting is worked on to convert it eventually into the finished product. This material is known as direct material. Any such materials such as lubricants, even though they are an essential part of the process are known as indirect material – they are not an inherent part of the finished product.

Labour costs

The machine operator who operates the machine that makes the item is direct labour, while the storekeeper, supervisor and inspector are indirect labour.

Expenses

An expense or service that is part of the manufacturing process, such as the cost of a subcontracted heat treatment process, would be a direct expense because the process is inherent to manufacturing the product. The skip lorry which takes away the scrap material would be a indirect service. The cost of stationery and the telephone bill are administrative expenses.

Direct costs

Costs which can be identified with and directly related to the particular products that the organization produces and sells to its customers are known as direct costs. The labour and material used directly in the product come into this category. The sum of the direct costs is called the prime cost.

Indirect costs

Indirect costs are those incurred in carrying out activities but which cannot be directly related to products. Some indirect costs occur within manufacturing but most non-manufacturing costs are indirect. Such costs are described as overheads – costs that are above or add onto the direct costs of manufacturing. These overheads can be classified by type, e.g. power, indirect labour, vehicles, insurance, and rent; and by function, e.g. production, research and development, accounting, marketing, selling or distribution. Procedures for apportioning overheads to products are needed in order to determine the full cost of a product and are explained later in this chapter. It will be seen in Chapter 19 that selling and administrative costs are treated separately from the direct and indirect manufacturing costs in the financial accounts.

Fixed costs

The term 'fixed' in this context does not mean fixed over time but rather fixed relative to changing volume of production. Most indirect costs are fixed. It so happens that most fixed costs are also fixed over time, if inflation is excluded.

The changes in volume referred to here are short-term changes, i.e. variations in volume from week to week caused by changing demand. Long-term changes in volume, such as doubling production over a two-year period, will cause some indirect costs to increase, so in the long term fixed costs are variable.

Variable costs

In contrast to the above, variable costs are those that vary with short-term changes in volume. These are mainly the direct costs. The amount of direct labour and material used varies proportionally with the volume of production. In order to increase volume in the short term, extra hours may need to be worked or additional operators taken on temporarily. Accounting procedures treat operators as casually employed and disposable, so that if output falls, operators are made redundant. While direct labour is a variable cost, sick pay and holiday pay are fixed costs. This practice dates from the time when manual employees did not receive pay for holidays. This also assumes that if no output is produced, employees are not paid. These practices have long since passed but the accounting convention remains. Accounting procedures treat wages differently from salaries, as will be seen when examining financial accounts in Chapter 19.

However, modern management practice treats employees as assets and would not dispose of them to suit short-term variations. In some organizations, but not yet many, all employees are salaried and treated similarly. Accounting practice does not yet reflect this.

The result of this approach is that in the short term variable costs are fixed.

The interaction of fixed and variable costs is examined later in this chapter.

OVERHEAD ABSORPTION

In order to determine the full cost of a product it is necessary to include indirect costs as well as the direct costs. This requires the splitting and spreading of centrally incurred indirect costs to the manufacturing cost centres and to products. Overheads may be allocated or apportioned to a cost centre. These processes operate historically, i.e. they are done at the end of an accounting period.

On the basis of historic expenditure it is possible to relate overhead to direct costs and to set what are known as overhead absorption rates. These are used to predict future overhead costs and are used in budgetary control (see Chapter 20).

Overhead allocation

Where an overhead such as machine maintenance can be directly traced back to a cost centre, the full cost of the service can be allocated to that cost centre. As another example, a supervisor would be an indirect cost or overhead allocated to a manufacturing cost centre.

Overhead apportionment

Where an overhead such as a rent, rates or office salaries cannot be directly traced to any one cost centre, the cost must be shared or apportioned between them on some fair and equitable basis. In the case of rent and rates, it could be apportioned on the basis of the area occupied by a cost centre. This means that the proportion of the rent and rates bill charged to a department would be the same as the proportion of the total space which the department occupies. Heating could be apportioned by the volume or the number and output of heaters; lighting by the number and wattage of light fittings, etc. Insurance costs can be apportioned according to the value of assets in a cost centre.

Overhead absorption

Once indirect cost allocation or apportionment to the various cost centres has been analyzed, it is possible to calculate an overhead absorption rate for future use. In this way overhead costs are absorbed (i.e. paid for) by manufacturing cost centres as they occur. To do this a cost per unit of output is added on top of the direct cost per unit to cover the overheads. The ratio of this additional cost per unit to the direct cost of the base factor is known as the overhead absorption rate.

As an example, if an overhead cost to be absorbed is £10,000 per year and the cost centre expects to make 5000 components per year, then the absorption rate would be £10,000/5000 = £2.00 per component. Alternatively, the overhead could be related to direct labour cost. If the direct labour in the above example were £4000, the absorption rate would be £2.50 per £1.00 of direct labour.

The basis of absorption must be chosen according to the type of overhead and the type of cost centre in which it is to be absorbed.

At the end of a financial year it is likely that the amount of overhead charged to and absorbed by the operating departments is not the amount actually spent on the overhead. This situation is known as under- or overabsorption. This can be analyzed using budgetary control and variance analysis as explained in Chapter 20, but this is not covered in this book, being left to a more advanced study of the subject.

COSTING

Job costing

Where an organization undertakes subcontract work where each order is unique and requires different amounts of labour, material and overhead, the cost of each order must be calculated separately. Such a cost accumulation system is known as a job costing system. This system gives accurate production costs because costs are accumulated for each specific order; however, this is expensive and time consuming in terms of the clerical work involved in attaching costs to a specific job.

Process costing

Where an organization makes large quantities of the same product, an alternative system of cost accumulation known as process costing can be used. It is not necessary to allocate costs to a specific order, instead the cost per unit can be obtained by dividing the costs of production over the period by the number of units made in that period.

Product costing

Where organizations have a range of standard products, they will wish to ascertain costs by product, and so a product costing system will be used. Costs are determined by accounting period for each product.

Project costing

Where a set of activities is organized as a project, as explained in Chapter 14, the costs will be attached to the project as a whole and to smaller cost centres within it.

Activity-based costing

Activity-based costing (ABC) is a modern approach to costing which pushes costs down in more detail to cost centres and products. Overheads are allocated to a cost centre on the basis of factors relating to the overhead itself rather than, as above, being apportioned to the department absorbing it on the basis of a factor related to that department. A more detailed analysis of the electricity bill, for example, will determine the causes of its cost so that they may be allocated to a cost centre. In this way overheads are more accurately absorbed. This leads to more accurate costing of products, processes and departments, and to more accurate measures of performance.

Standard costing

This costing system is based on the generation of standard costs. These are the expected costs based on expected volumes of output at expected efficiencies. Standard costs are costs per unit of output, not the total cost of an item in a period. Standard costs are used predictively to create budgets and to compare standard with actual costs after they have occurred.

Unit costs

Unit costs are costs per unit of output. Output can be at the component or product level. Unit costs can be calculated by dividing costs by the volume produced. It is more informative to know the unit costs and not just total costs. Total costs will change as volume changes, but it is not clear how much of the cost change is simply due to volume and how much is due to changes in the price or cost per unit. Unit costs enable this analysis to be done. The price of a product is its income per unit, so this is comparable with a unit cost.

This approach is developed in variance analysis in Chapter 20.

Standard costs

A standard cost is a budgeted unit cost. Standard costs can be derived for direct materials, labour, expenses and overheads. These can be added to find the total standard cost for a component. Subtracting the standard cost from the price gives the standard profit per unit. Standard costs are used predictively in budgetary control (Chapter 20). The standard cost of labour for a task can be derived from the standard time for the task, derived in Chapter 11, multiplied by the pay rate.

COST ANALYSIS

Variation of cost with volume

A basic analysis of costs covered here examines how costs vary with volume of output. The variations in volume referred to are the short-term variations, not the changes that may occur due to planned growth over a longer period of time. Figure 18.1 shows the four different ways in which costs can vary, or not, with volume.

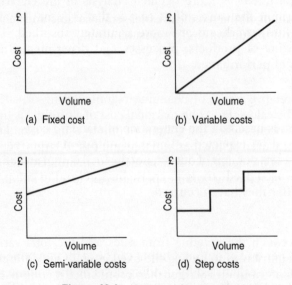

Figure 18.1 *Variation of costs with volume.*

Fixed costs

As explained above, fixed costs remain fixed at a constant level independent of the volume of production. Rent, rates, loan interest and most administrative costs are in this category.

Variable costs

These increase as production volume increases. For this purpose, over the volume range under consideration, this is considered to be a linear relationship. Direct costs are in this category.

Semi-variable costs

Some costs have both a fixed and variable element and therefore do not have a direct relationship to production volume. Electricity costs, for example, have a fixed standing charge on top of which the variable cost per unit is charged.

Step costs

Some fixed costs only remain fixed up to certain levels of production. If, for example, the capacity of the machine is exceeded, a second machine would be needed. In this way the cost behaves as a step when related to volume.

Total costs

Total cost is the sum of all the fixed, variable, semi-variable and step costs at the particular volume of production under consideration.

Marginal costs

To calculate the relative effect of changing manufacturing or sales volume, a marginal cost can be calculated. The marginal cost of a product is defined as the change in cost that occurs when the volume of output is increased or reduced by one unit. This is measured by the total variable unit cost. Marginal costing is used to assess whether it is financially feasible to increase manufacturing volume or to calculate the effect of reducing volume, perhaps due to a decline in the market.

Contribution

The difference between the revenue from sales and the total variable cost of making that item contributes to paying the fixed costs. This difference is called the contribution. Contribution is larger than profit since it comprises profit plus fixed costs. It is possible for the contribution to be positive even though the profit is negative. With a negative profit an organization may be tempted to stop producing a product. If this is done, the contribution to fixed costs is also lost. Only if the product is replaced with another is some contribution retained. If it is not, then the fixed cost has to be borne by other products, raising their cost.

BREAK-EVEN ANALYSIS

The total costs of a product can be plotted against volume and compared with the income or revenue from sales of the product. Figure 18.2 shows how revenue and costs vary with volume of production. This is used to determine the minimum volume of production needed to make a profit. This is known as the break-even volume.

The sales price should be higher than the variable cost so will have a steeper slope when plotted against volume. The sales revenue line will cross the total cost line when they are equal. At this volume the profit is zero; this is the break-even volume. Below this volume there is a loss or negative profit, above it there is a profit. It is important for an organization to know what its break-even volumes are, for comparison with the actual or operating volumes.

The break-even volume can be calculated as follows. If s = sales price per unit, F = total fixed cost and v = variable cost per unit, then at the break-even volume, B:

$$s \times B = \text{revenue} = \text{cost} = F + v \times B \tag{18.1}$$

$$B \times (s - v) = F \text{ or } B = F/(s - v) \tag{18.2}$$

Since $(s - v)$ is the contribution per unit then:

$$\text{break-even volume} = \text{fixed cost/unit contribution} \tag{18.3}$$

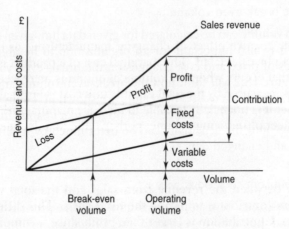

Figure 18.2 *Break-even analysis.*

Example of break-even analysis

The fixed costs for a product are £60,000. The product sells for £8 and its total unit variable cost is £5.

$$\text{break-even volume} = B = 60,000/(8 - 5) = 20,000 \qquad (18.4)$$

The sale of 20,000 units will just cover the fixed costs. If the operating volume, V, is 25,000, then the revenue and costs can be calculated and the profit determined for this manufacturing volume. Results for the example product at different volumes are shown in Table 18.4.

Table 18.4 Variation of revenue and costs with volume

Volume	Revenue	Fixed cost	Variable cost	Contribution	Profit
10	80	60	50	30	(30)
15	120	60	75	45	(15)
20	160	60	100	60	0
25	200	60	125	75	15
30	240	60	150	90	30
35	280	60	175	105	45

Figures in £000s, parentheses indicate negative values.

Given the uncertainty of future volumes, it is important for an organization to know whether its operating volume is near to its break-even volume. If it is, then a future loss of sales could push the organization below its break-even volume. The ratio of operating volume to break-even volume is known as the margin of safety. The larger this ratio, the safer the organization is from a loss-making situation.

$$\text{margin of safety} = M = B/V \qquad (18.5)$$

Sensitivity of the break-even volume

The break-even volume can be calculated for given data but, given the uncertainty of future data, it is also important for an organization to be able to assess the effect of changes of price and cost changes on the break-even volume. A visual inspection of Figure 18.2 will show how the break-even volume and profit change with changes in price and costs. The results are shown in Table 18.5.

Sensitivity is the ratio of change in an output to a change in an input. The amount of change and the sensitivity can be determined by making calculations for the original and changed data.

The sensitivity of profit to productivity

Break-even analysis can be used to show how sensitive profit is to productivity. Managers should be able to estimate the effect on profit of changes in productivity.

Table 18.5 Sensitivity of break-even volume

Price variation			Cost variation		
Price	Contribution	Break-even	Cost	Contribution	Break-even
5	0	infinite	2	6	10
6	1	60	3	5	12
7	2	30	4	4	15
8	3	20	5	3	20
9	4	15	6	2	30
10	5	12	7	1	60
11	6	10	8	0	infinite

Price, variable cost and contribution are unit values. Break-even volume is measured in 000s.

If productivity improvement is defined as an increase in output using the same resources, then fixed costs remain fixed and a percentage productivity increase is measured by the percentage volume increase.

Table 18.4 shows for the example product that an increase in volume from 25,000 to 30,000 has resulted in an increase in profit from £15,000 to £30,000. The increase in productivity is 20 per cent and the resulting increase in profit is 100 per cent, so the sensitivity of profit to productivity is:

$$\text{sensitivity} = 100\%/20\% = 5 \tag{18.6}$$

Sensitivity increases if the operating volume, O, is close to the break-even volume, B. Sensitivity can be related to the margin of safety M. It can be shown that:

$$\text{sensitivity} = M/(M - 1) \tag{18.7}$$

In the example the margin of safety at the operating volume of 25,000 = 25,000/20,000 = 1.25. So:

$$\text{sensitivity} = 1.25/0.25 = 5 \tag{18.8}$$

If an organization operates at 10 per cent above its break-even volume, then its sensitivity is 1.1/0.1 or 11. This means that an increase of productivity of X per cent will lead to an increase in profit of $11X$ per cent. The significance of increasing productivity as a means of increasing profit can be seen from this analysis.

Task sheet 18 Cost accounting

18.1 Transfer the following set of transactions into the appropriate accounts.

03 Feb   order £100 of materials from Joe Bloggs.
07 Feb   receive £100 of materials into raw material stores.
10 Feb   issue £100 of materials to factory.
(10–15 Feb   make product worth £300 from this material.)
16 Feb   receive £300 of product into finished goods stores.
21 Feb   sell £300 of product to Jack Jones.
24 Feb   deliver £300 of product to Jack Jones.
27 Feb   pay Bill Smith £75 for wages.
07 Mar   pay Joe Bloggs £100 for material.
24 Mar   receive £300 from Jack Jones for products.

An organization sells 15,000 units per year of a product at a price of £20. The unit variable cost is £10 and the fixed costs are £100,000 per year.

18.2 Calculate the break-even volume and the margin of safety.

18.3 Calculate the effect of changing price and variable cost on break-even volume.

18.4 Determine the price at which profit is reduced to zero at the operating volume (assume variable and fixed costs are unchanged).

18.5 Determine the variable cost at which profit is reduced to zero at the operating volume (assume that price and fixed costs are unchanged).

CHAPTER 19

FINANCIAL ACCOUNTING

FINANCIAL ACCOUNTING

History of financial accounting

All societies have found it necessary to regulate and control the financial affairs of companies and other organizations. This is done in the public sector to ensure that taxpayers' money is properly spent and not misappropriated for unauthorized purposes. In the private sector it is done to protect shareholders from misdemeanour or fraud by directors.

Many organizations are required to publish their accounts and to have them audited by independent accountants. This creates work in Britain for 250,000 accountants.

The earliest writings on stone or clay tablets of the Sumerians and Egyptians are the accounts of the state. They date back to 3000 BC and many of these still exist. Figure 19.1 shows an account from an Egyptian tomb of the 13th dynasty, about 1750 BC. The layout of the account is very similar to that used today. Accounting rules and conventions today seem to be based on these very ancient accounts. It must be realized that money was not invented until the fourth century BC, but accounts were found to be necessary long before that.

Law in financial accounting

Financial accounting is concerned with the external requirements of shareholders, prospective investors, financial analysts, creditors, the Registrar of Companies, the Inland Revenue (taxation authority) and the government, as well as with the internal requirements of the management of a company. There is a legal obligation in Britain, under the various Companies Acts, for companies with limited liability. They are required to keep books of accounts and to produce annual accounts. These must be audited, presented to the Registrar of Companies and made available to shareholders. For public limited companies (plc), these accounts must be made public. For companies formed as groups, as mentioned in Chapter 4, only the holding company is required to publish its

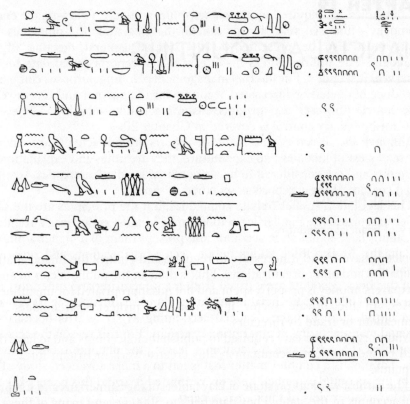

Figure 19.1 *Account from an Egyptian tomb. Source: A. H. Gardiner (1927), Egyptian Grammar, Clarendon Press, Oxford. Reproduced with permission.*

accounts. Accounts are therefore not legally presented for Ltds which form part of a plc.

The law also requires the use of standard formats for balance sheets and profit and loss accounts, although there is a choice of formats for some accounts. The main principle is that the accounts should present a true and fair view of the state of affairs and the profit and loss of a company. The accounting principles are collected as *statements of standard accounting practice* (SSAP).

Financial accounts

The accounts that a company is required to present to the Registrar of Companies are as follows:

- The balance sheet.
- The profit and loss account.
- The directors' report.
- The auditors' report.

Companies are required to present accounts annually, following the end of the company's financial year. In addition to the Registrar of Companies, the accounts must also be presented to the annual general meeting of the shareholders. The information is historic in that it reports actual data for the financial year just ended. For internal management planning purposes companies may produce budgeted or forecasted accounts for the current and future years, and for control purposes may produce historic accounts monthly. The use of budgets for budgetary control is covered in Chapter 20.

Although the directors' and auditors' reports are not actually accounts, in the sense of a list of incomes and expenditure, they are notes and explanations of the accounts and are considered to be accounts under the Companies Act. The required content of these reports is specified in the Act.

This legislation applies only to companies, but the principles are the same for other organizations. Similar legislation exists in most countries. Virtually all organizations have to maintain accounts and present them to a higher authority. In Britain the Audit Office audits public sector organizations and reports to government.

In this book accounts are derived from manufacturing and other cost data as a means of collecting financial information about manufacturing. It provides only an introduction to basic financial accounting as far as it is necessary to understand manufacturing of other primary activity. For this reason the accounts are simplified. They focus on the concepts, leaving the detail to more advanced studies. Taxation is a complex matter that is left to a more advanced study of the topic. In the accounts explained here tax is omitted to avoid this complexity and to provide a clearer view of the basic concepts.

The directors' report

The directors of a company are required to report any significant changes in the nature of the business and in its assets. The names and financial interests of directors must be reported together with the recommendations for appropriation of the profit (i.e. the distribution between the dividends paid to shareholders and the profit retained in the company and transferred to reserves).

This report is also required to disclose considerable information that does not seem to be directly related to the accounts, e.g. the number and total pay of employees. This is a means by which the government collects statistical information about companies for analysis of the state of the economy.

The auditors' report

Auditing is a process required by law in which the accounts of a company are examined by external and independent professional accountants. SSAPs are binding on accountants and auditors who must mention any infringements of standards in their audit report. The purpose of auditing is to ensure that a company is trading legally and that the accounts presented are the 'true and fair view' of the financial health of the company. Auditors are required to sign a statement to that effect and to present it to the shareholders.

Audited accounts form the basis on which companies pay Corporation Tax, which is in effect a tax on profits. Companies arrange their accounts to minimize tax. Auditors must ensure that what is done is legal. Because of their expertise in auditing, accountants offer advice to companies on how to minimize tax. Auditing is done to protect shareholders from fraud by company directors and to give them sufficient information to judge whether to invest in the company.

THE CASH FLOW STATEMENT

The cash flow statement is an account of the cash flows into and out of the company. It can be based on a monthly or weekly accounting period. It is a financial account but it is not legally required as part of the annual accounts. It is, however, useful as a starting point for the preparation of the annual accounts as well as being a valuable document for monitoring the flow of money and managing the organization. The cash flow statement is a record of actual cash flows on an historic or real-time basis. It is, however, also used predictively to forecast expenditure and plan the availability of future cash to meet future needs. In this form the account is known as a cash flow forecast.

The personal equivalent of the cash flow statement is the record that some individuals keep of their income, expenditure and bank balance. This is done to ensure that there is enough money in the bank before using a cheque or debit card to pay for something. Authorized lending can be obtained through a loan, but if a bank balance goes negative, high charges for unauthorized overdrafts are incurred. Individuals who wish to avoid overdraft charges need to plan their expenditure and monitor their bank balance.

Companies have the same need as individuals to monitor their bank balance. Money must be available when needed to pay wages and other expenditures. Because goods are bought and sold on credit there is a difference in time between sales and cash inflows and between purchases and cash outflows. The usual credit period is thirty days, so material received one month is paid for in the next month after delivery. Every source of income and expenditure must be included in the cash flow statement. Any shortfall must be made up by either withdrawing cash from deposit accounts or by the use of an overdraft facility. Any surplus cash can be put on short-term deposit until needed.

Having a cash flow forecast enables a manager to determine future cash requirements and decide whether to put cash into short-term savings or to spend it in machines or other assets. With large bills such as rates or insurance that become due once or twice a year, or fuel bills that come every three months, sufficient funds must be retained to cover them when they occur.

A cash flow forecast starts with a forecast of sales volume, which is expressed as sales value. From this the various types of expenditure can be forecast, or budgeted. Sales and costs are usually calculated for monthly periods.

The current European calender is based on the Roman calender and has months of unequal lengths and month starts on different days of the week. (The Roman Emperor Augustus insisted that his month had as many days as that of Julius Caesar!) This is not convenient for comparing data for different months, so organizations will define their own 'months'. Usually each month will start on a Monday. Some organizations use thirteen months of four weeks, others use twelve months or five, four and four weeks for each quarter of the year. Quarterly accounts are then of equal length, aiding comparability.

As a financial year progresses, the forecast figures in a cash flow forecast can be replaced by actual figures to produce a cash flow statement at the end of the year. The totals for the year can be used to produce the profit and loss account since it comprises many of the figures needed. It cannot, however, be used to calculate the profit as other factors such as depreciation are not included in the cash flow statement. Depreciation is not a cash flow, so it is not shown in the cash flow forecast; it is, however, included in the profit and loss account.

Other financial accounts comprise either capital items (the balance sheet) or recurrent items (the profit and loss account). The cash flow statement is an exception, as it includes both capital and recurrent items. Table 19.1 presents some manufacturing and cost data which will be used to construct a set of accounts. A cash flow statement is shown in Table 19.2. The cash flow statement shown here is for a 'straightforward' manufacturing business without any of the complexities that can occur in practice.

Table 19.1 Example of manufacturing and cost data

Admin. salaries/year	50,000	Discount rate	0.12
Admin. expenses/year	18,000	Interest %/month	0.01
Sales salaries/year	30,000	Dividend/profit	0.33
Sales expenses/year	12,000	FG Stock %price	0.80
		Depreciation/year	0.20
Direct material/unit	20		
Direct labour/hour	5	Direct operators	8
Direct expenses/year	4,000	Indirects	2
		Office staff	5
Indirect material/year	5,000	Total employees	15
Indirect labour/hour	4.50	Wages + salaries	181,920
Factory expenses/year	10,000	Std. min/unit	22.4
Buildings	80,000	Sales price	45.00
Machinery	50,000	Prod. output/year	20,000
		Prod. hours/week	40
Share capital	80,000	No. weeks/year	50
Loan capital	180,000	Production/week	400

The top four rows in Table 19.2 are not strictly a part of the cash flow statement; they record production volumes and the volume of materials ordered

Table 19.2 Cash flow statement

	Month												
	Apr	May	June	July	Aug	Sept	Oct	Nov	Dec	Jan	Feb	Mar	Total
No. of weeks	5	4	4	5	2	4	5	4	4	5	4	4	50
Volume													
Material ordered	1,600	1,600	2,000	800	1,600	2,000	1,600	2,000	1,600	1,600	1,600	2,000	20,000
Production/month	0	1,600	1,600	2,000	800	1,600	2,000	1,600	1,600	2,000	1,600	1,600	18,000
Sales/month	0	0	1,600	1,600	2,000	800	1,600	2,000	1,600	1,600	2,000	1,600	16,400
Cash inflows													
Loan capital	180,000												180,000
Share capital	80,000												80,000
Sales revenue	0	0	0	72,000	72,000	90,000	36,000	72,000	90,000	72,000	72,000	90,000	666,000
Total cash inflow	260,000	0	0	72,000	72,000	90,000	36,000	72,000	90,000	72,000	72,000	90,000	926,000
Cash outflows													
Capital													
Buildings	80,000	0	0	0	0	0	0	0	0	0	0	0	80,000
Machinery	50,000	0	0	0	0	0	0	0	0	0	0	0	50,000
Direct costs													
Direct materials	32,000	32,000	32,000	40,000	16,000	32,000	40,000	32,000	32,000	40,000	32,000	32,000	360,000
Direct labour	8,000	6,400	6,400	8,000	6,400	6,400	8,000	6,400	6,400	8,000	6,400	6,400	83,200
Direct expenses	400	320	320	400	160	320	400	320	320	400	320	320	4,000
Manufacturing overheads													
Indirect materials	500	400	400	500	200	400	500	400	400	500	400	400	5,000
Indirect labour	1,800	1,440	1,440	1,800	1,440	1,440	1,800	1,440	1,440	1,800	1,440	1,440	18,720
Factory expenses	1,000	800	800	1,000	400	800	1,000	800	800	1,000	800	800	10,000
Selling and distribution costs													
Sales salaries	2,500	2,500	2,500	2,500	2,500	2,500	2,500	2,500	2,500	2,500	2,500	2,500	30,000
Sales expenses	1,000	1,000	1,000	1,000	1,000	1,000	1,000	1,000	1,000	1,000	1,000	1,000	12,000
Administration expenses													
Admin. salaries	4,167	4,167	4,167	4,167	4,167	4,167	4,167	4,167	4,167	4,167	4,167	4,167	50,000
Admin. expenses	1,500	1,500	1,500	1,500	1,500	1,500	1,500	1,500	1,500	1,500	1,500	1,500	18,000
Loan interest	1,800	1,800	1,800	1,800	1,800	1,600	1,400	1,200	1,000	800	600	400	16,000
Loan repayment	0	0	0	0	20,000	20,000	20,000	20,000	20,000	20,000	20,000	20,000	160,000
Total cash outflow	152,667	52,327	52,327	62,667	55,567	72,127	82,267	71,727	71,527	81,667	71,127	70,927	896,920
Net cash inflow	107,333	−52,327	−52,327	9,333	16,433	17,873	−46,267	273	18,473	−9,667	873	19,073	29,080
Cumulative net cash inflow	107,333	55,007	2,680	12,013	28,447	46,320	53	327	18,800	9,233	10,007	29,080	

and received. However, it is useful to show these since the material expenditure and sales revenues will relate to the volume of materials ordered and products manufactured.

The main body of the statement contains the incomes and expenditures of each type for each time period. The total cash inflows and outflows are shown for each period, and the cumulative net inflow calculated for each period of the year.

The 'bottom line' of the cash flow forecast is the cumulative net cash inflow. This is calculated in order to determine whether there is sufficient net inflow to date to cover expenditure to date, and how much capital is needed to avoid an overdraft (negative cashflow). Figure 19.2 shows a graph of cumulative net cash inflow. Its vertical position can be adjusted by varying the amount of initial capital, from shares or loans. This graph shows a typical 'cash trough', the lowest value on the curve. For a new business this will be a few months after start up as costs occur before revenue starts to flow in. This data is needed to determine how much initial capital is required.

This cash flow statement is for a new business. The capital income and expenditure is therefore shown in the first month. All values are shown as positive, but some are inflows and others are outflows. Care must be taken when adding values. In a spreadsheet inflows would be shown as positive and outflows negative. Accountants sometimes use brackets instead of a negative sign.

As can be seen from the data in Table 19.2 the net cash inflows for May, June, October and January are negative. This is common at a business start up and when large bills arrive occasionally. This is not a problem in the example since the amount of loan has be calculated to ensure that the cumulative net cash inflow remains positive at all times. Had this not been the case it would have been necessary to increase the loan in order to avoid an overdraft, or accept the costs of an overdraft.

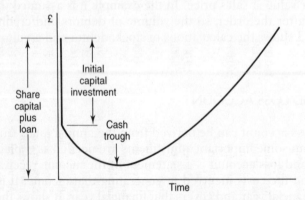

Figure 19.2 *Cumulative net cash inflow.*

Stocks, debtors, creditors, sales and materials

The values of stocks held at the end of the year will appear in the balance sheet since they are assets, but they are also needed to make adjustments in the profit and loss account. They are calculated here because they can be derived from the cash flow statement. Values are needed for raw material, work in progress and finished goods stocks. Raw materials will be valued at purchase cost per unit, the volume being that in stock on the last day of the financial year. The volume of stock can be estimated as the average time that materials remain in stock, multiplied by the rate of use. Finished goods, i.e. products, are not valued at sales price since that includes profit, which is not made until the goods are sold. They are usually valued at total manufacturing cost derived from the manufacturing account. In the example it is assumed that the volume of raw material and finished goods stocks are two weeks' usage.

Work in progress is the material lying around the factory, being processed or waiting to be processed or moved. Its value will change as it moves through the factory and value is added to it. It may be appropriate, therefore, to value work in progress at the average of the material and total manufacturing cost. In the example it is assumed that there is four weeks of work in progress.

Money owed by debtors is included as an asset even though it has not yet been received since in law it is owed to and owned by the company. The main form of debtors are customers who have not yet paid for the products sent to them. The point of measurement of a sale is when the order is received. Delivery and payment will occur later. The value of debtors will depend on the volume and value of the products, valued at total production cost.

Similarly money owed to creditors is included as a liability. The point of measurement of a purchase is when the order is sent. Delivery and payment will take place later. The main creditors are material and other suppliers who have not yet been paid for the materials they have delivered or services they have supplied. The value of creditors will depend on the volume of materials and services and their value at sales price. In the example it is assumed that bills are paid four weeks after the order, so the volume of debtors and creditors is four weeks. Table 19.3 shows the calculations of stock, debtor and creditor values for the example data.

THE PROFIT AND LOSS ACCOUNT

The profit and loss account can be derived from the annual totals from the cash flow statement, but some important adjustments are needed as explained below.

The profit and loss account is a statement of the income receivable during a given period and the costs incurred in generating that income. It is produced at the end of a financial year and covers that financial year. It shows the recurrent items of income and expenditure but not the capital items, which are shown in

Table 19.3 Stocks, debtors and creditors

Stocks, debtors and creditors

	Opening value	No. weeks	Value /unit	Closing value
Debtors	0	4	36	5,760
Creditors	0	4	20	3,200
RM stock	0	2	20	1,600
WIP stock	0	4	28	4,480
FG stock	0	2	36	2,880

Sales and materials values

Value in	Cash flow	Adjustment	Value in profit and loss
Materials	360,000	+ creditors 3,200	363,200
Sales	666,000	+ debtors 5,760	671,760

the balance sheet. Some recurrent items, however, are related to capital. The interest paid on capital loans and depreciation of capital assets are included.

The prime purpose of the profit and loss account is to calculate the profit, which is the prime measure of a company's performance. However, what matters from a management point of view is that a profit and loss account should give an accurate picture which enables the various component elements to be tracked and the appropriate decisions taken.

The profit and loss account can, of course, show either a profit or a loss but not both. The name of this account reflects the traditional accountant's view of the organization in which profit for shareholders is the sole objective and costs, including wages and salaries, are considered as losses.

The profit and loss account is made up of four parts:

1. The *manufacturing account*, which includes all of the costs incurred in actually manufacturing the products.
2. The *trading account*, which includes the sales revenue and the manufacturing cost of all the goods actually sold.
3. The *profit and loss account* itself, which includes all of the non-manufacturing costs together with the tax paid.
4. The *profit and loss appropriation account*, which shows the distribution or retention of the profit.

As can be seen, the depreciation is subtracted before profit is calculated to represent that part of the fixed assets consumed in generating the profit. The profit and loss account shows cash and depreciation, not only cash. Cash and profit are not the same thing and must be kept separate.

Calculation of profit

The stages by which profit is calculated follow the ancient traditions of accountants, which have existed for many centuries. In the cash flow statement the values of materials and sales are the cash flows, i.e. the payments made. In constructing the manufacturing account the value of material is the material ordered in the year, so the value in the cash flow statement has to be adjusted by adding the materials ordered but not yet paid, i.e. the increase in the value of creditors over the year. Similarly the value of sales in the trading account is the sales made in the year. To obtain this the value in the cash flow statement is increased by the increase in the value of debtors over the year to measure the sales made in the year. These adjustments are made before entry into the profit and loss accounts.

In the manufacturing account the direct costs and manufacturing overheads are totalled for the year. These show the direct expenditure in the year.

However, what is needed to calculate the profit in the year is not the direct costs in the year but the direct costs in the sales in the year. This is done to avoid distortions of profit that would otherwise occur if levels of stocks changed during the year. Therefore the direct costs of materials are adjusted by subtracting the increase in stock during the year from the purchases in the year. This adjustment is shown within the account, i.e. the original value is shown, then the adjustment. The result is the value of materials used in manufacturing in the year. This adjustment is made for raw materials and work in progress in the manufacturing account and for finished goods in the trading account. This gives the cost of material used in the products sold during the year.

What these adjustments are doing is moving the time and level of measurement from the level of a set of transactions to the level of the organization as a financial system.

It should also be noted that the sales in the year are the goods delivered to customers, but these goods may not yet have been paid for. They count as sales in the accounts, the outstanding money due from customers being shown as debtors in the balance sheet. Similarly materials purchased will include those delivered but not yet paid for. The money owed to suppliers is shown as creditors in the balance sheet.

Gross profit is calculated in the trading account by subtracting manufacturing costs from sales revenue. This is not real profit since sales and administrative costs have not yet been deducted. This is done in the profit and loss account giving the net pre-tax profit. Corporation Tax is then deducted to calculate the net profit – the profit available to the shareholders. This profit may be either distributed to the shareholders as dividend or retained in the business as reserves. As mentioned above, tax is ignored in these accounts.

The manufacturing account

The manufacturing account, or similar account in non-manufacturing organizations, shows no income but lists the direct costs of manufacture plus the manufacturing overheads which can be directly allocated to manufacturing. The manufacturing, and other annual accounts are produced from the various accounts in which transactions during the year were recorded. This was covered in Chapter 18.

Accounting traditions include only direct labour as a direct cost. Holiday and sick pay for direct operators is not a part of the direct labour cost, so has to be accounted for separately and would be included in factory overheads. The cost of labourers, crane drivers, storekeepers and floor sweepers is indirect labour and is included with the salary of the supervisor as factory overheads. Depreciation of the machines is included here as a cost. Factory buildings in most cases appreciate rather than depreciate. The 'bottom line' of the manufacturing account is the cost of finished goods transferred to the warehouse. This value transfers to the trading account. A manufacturing account is shown in Table 19.4.

Table 19.4   Manufacturing account

Cost of finished goods		488,040
Direct costs		
Direct materials	363,200	
+ Opening raw matl. stock	0	
− Closing raw matl. stock	−1,600	
Direct labour	83,200	
Direct expenses	4,000	
Prime cost of production		448,800
Manufacturing overheads		
Indirect material	5,000	
Indirect labour	18,720	
Factory expenses	10,000	
Total mfg overheads		32,720
Machine depreciation		10,000
Total cost of production		492,520
+ Opening work-in-progress		0
− Closing work-in-progress		−4,480
Cost of finished goods		488,040	488,040

The trading account

This account includes the revenue from sales less the cost of finished goods, which was transferred from the manufacturing account. It calculates and shows on its bottom line the gross or trading profit. Notice that the term 'profit' is used here before the sales and administrative costs have been accounted for. A trading account is shown in Table 19.5.

The profit and loss account

This account takes the gross profit from the trading account and subtracts the sales and administrative costs, loan interest and Corporation Tax. Depreciation

Table 19.5 Trading account

Sales		671,760
Cost of finished goods	488,040	
+ Opening finished goods stock	0	
− Closing finished goods stock	−2,880	
Cost of sales	485,160	
Gross profit	186,600	
Sales	671,760	671,760

of non-factory assets is included at this stage. The profit and loss account calculates the net post-tax profit and shows this on the bottom line. This is available to the shareholders. A profit and loss account is shown in Table 19.6. Figure 19.3 shows the same data in the form of a bar chart. It is useful at this stage to see how the value of a product builds up during manufacture. This is shown in Figure 19.4. The concept of added value is developed further later in this chapter.

Table 19.6 Profit and loss account

Gross profit			186,600
Selling and distribution costs			
Sales salaries	30,000		
Sales expenses	12,000		
Total selling and distribution costs		42,000	
Administration expenses			
Admin. salaries	50,000		
Admin. expenses	18,000		
Loan interest	16,000		
Total administration expenses		84,000	
Total selling and admin. costs		126,000	
Net profit		60,600	
Gross profit		186,600	186,600

The profit and loss appropriation account

This account takes the net profit and appropriates it, i.e. shows how much is distributed to the shareholders as dividend for the financial year and how much is retained within the company. Money retained within the company is used for growth and increases the assets of the company, which are available to the shareholders if the company were to be closed down. Table 19.7 shows a profit and loss appropriation account. In this example the proportion of profit distributed as dividend is given in the original data to allow the accounts to be constructed. In practice, however, the split would only be made after the accounts have been produced. The annual general meeting of shareholders has to approve the split of profit, so technically the dividend is a proposed dividend until it is approved by the AGM. Accounts are, however, produced on the assumption that the dividend has been paid.

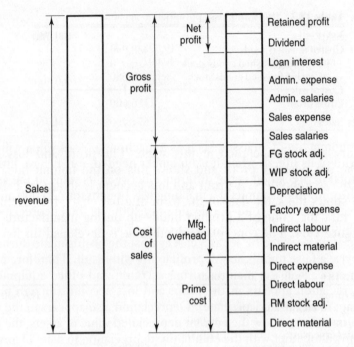

Figure 19.3 *Profit and loss account as a chart.*

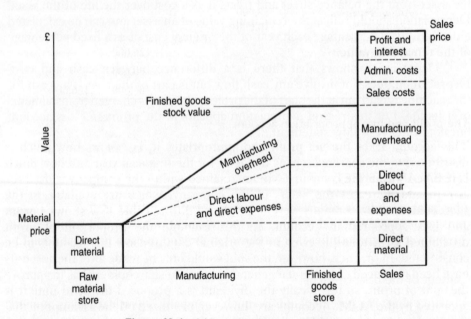

Figure 19.4 *Adding value during manufacture.*

Table 19.7 Profit and loss appropriation account

Net profit		60,600
Dividend	19,998	
Transfer to reserves	40,602	
Net profit	60,600	60,600

Depreciation

Over the economic life of an asset, such as a machine tool, its value will reduce until at the end of its life, or when it is no longer needed, it may have little or no value. The total depreciation is the difference between its value at the start and its value at the end. For example, if a machine was bought for £20,000 and at the end of ten years it had a second-hand value of £2000, its total depreciation over ten years would be £18,000.

Depreciation needs to be set as a cost against the profits made during those years, otherwise a false impression of profitability will result. Therefore machine depreciation appears in the manufacturing account, and office equipment and building depreciation appears in the profit and loss account (though buildings may appreciate rather than depreciate). Depreciation also appears in the balance sheet where, in order to show the current depreciated value of assets, the original values are shown together with the cumulative depreciation to date. Depreciation is an internal transfer from the balance sheet to the profit and loss account to account for the decrease in value of an asset. It gradually removes the value of the asset from the balance sheet and treats it as a cost over the life of the asset. Depreciation should reflect the remaining value of an asset and can be calculated either as a fixed percentage each year of the original cost or as a fixed percentage of the remaining value.

Depreciation shows that there is a difference between cash and cost. Depreciation does not involve any cash flow but is part of the cost. The cash for the machine is paid out at the start of the project; the cost is charged by instalments over its life. For this reason depreciation appears in the profit and loss account but not in the cash flow statement.

THE BALANCE SHEET

The balance sheet is different from other accounts in that it does not record transactions over a period of time. It is a statement of assets and liabilities of a company at a point in time. It is produced at the end of each financial period.

Content of the balance sheet

A balance sheet is a statement of how much a company is worth at a given moment in time. It is called a balance sheet because it is a measure of the assets that a company owns and the liabilities it owes. These two values must be the same, i.e. they must balance. The balance sheet therefore shows the capital items and not the recurrent items, which are shown in the profit and loss account. The balance sheet will be produced for the final day of the financial year and will represent the value of the company at that date. A financial year can be seen to have balance sheets for its start and end points and a profit and loss account for the period between them.

It is convenient to see the balance sheet in terms of who owns and owes what if the company were to be closed down that day. Were this to happen the receiver would collect in debts and pay creditors. What is left is given to the shareholders. For this reason debtors and creditors are included in the balance sheet.

The balance sheets for the previous and current years contain the values of opening and closing stocks, which are needed in the profit and loss account for the period between them.

Layout of the balance sheet

The balance sheet consists of two lists – one of assets and one of liabilities. As indicated above, these must balance. Sometimes these two lists are presented side by side but they can be presented one above the other, as in this book.

Assets are separated into long-term or fixed assets, and short-term or current assets. Liabilities are similarly split into long-term liabilities to shareholders and banks, and short-term liabilities.

Two alternative layouts are possible within the balance sheet. In the traditional format all of the assets are in one list and all of the liabilities are in the other. This is the simplest form.

In the alternative layout current liabilities are listed with and subtracted from current assets on the asset side of the balance sheet. This gives a better view of the real value of the company. This layout has two advantages. The 'bottom line' value of the balance sheet in this form is the net value of the company and the net capital invested in the company (also called 'capital employed'). It is the amount the shareholders could expect to get if the company closed down that day. It also shows the value of working capital, which is the amount of capital needed from shareholders to finance the short-term 'working' assets used to run the company. This finances stocks and short-term cash needs. This is in addition to the long-term capital need to finance the long-term assets such as buildings and machines.

This is the form used in this book. Table 19.8 shows a balance sheet laid out in this form. Figure 19.5 shows a balance sheet in bar chart form.

Fixed Assets

Fixed assets are purchased and owned by the company. They represent the means by which the company earns its profits. They are called fixed because they are

Table 19.8   Balance sheet

Balance sheet: end of year 1
Assets
  Fixed
    Buildings   80,000
    Machines
      Original cost   50,000
      Depreciation   −10,000
      Current value   40,000

    Total fixed assets   120,000

  Current
    Cash   9,082
    Debtors   5,760
    Raw material stocks   1,600
    Work in progress   4,480
    Finished goods stocks   2,880

    Total current assets   23,802

  Current liabilities
    Creditors   −3,200
    Net current assets = Working capital   20,602

  Total assets (− current liabilities)   140,602

Liabilities (source of funds)
    Loan outstanding   20,000
    Share capital   80,000
    Reserves   40,602

    Shareholders' funds   120,602   120,602

  Total liabilities (− current liabilities)   140,602

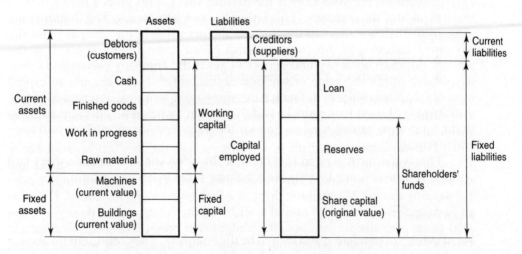

Figure 19.5   *Balance sheet as a chart.*

not for sale in the normal course of business activity. Fixed assets include freehold land, buildings, plant, machinery, vehicles, etc. These are known as tangible assets because they can be seen and touched. The value of fixed assets is usually taken to be the original purchase value less depreciation.

Another type of fixed assets are known as intangible assets. The main intangible asset is called goodwill; this is the difference between the 'book value' of the company and the value placed upon it by investors when a company is sold. For example, a newsagents with tangible assets worth £200,000 might be sold for £250,000 because it has a customer base of regular clients and newspaper deliveries. The value of the business as a whole is estimated by reference to the expected future flows of income from the business. The £50,000 difference is accounted for on the balance sheet as goodwill.

Working capital

Working capital is the term used for the net current assets available to the company:

$$\text{working capital} = \text{current assets} - \text{current liabilities} \qquad (19.1)$$

These are the short-term current assets that provide funds for the day-to-day running of the company in terms of raw materials, components, services and labour. At any one time the working capital will be invested in the following:

- Stocks of raw materials and components
- Stocks of work in progress
- Stocks of finished goods
- Trade debtors (customers who have not yet paid)
- Cash in hand and bank deposits

From this must be subtracted the current liabilities which are usually the following:

- Trade creditors (suppliers owed but not yet paid)
- Other creditors (tax and dividends not yet paid)

The value of cash in the cash flow statement is before any dividend is paid. Any dividend would be paid from cash, so cash is reduced by the amount of the dividend to show the cash remaining after payment of the dividend.

The more money tied up in working capital, the less will be available to invest in growth of fixed assets. It is important, therefore, to keep stocks and debtors as low as possible. Current liabilities represent a reduction in the need for working capital. Maintaining these at a high level can be beneficial. Delaying paying suppliers is a form of capital available to the company, but if excessive can lead to supplier dissatisfaction and withdrawal of credit facilities. In estimating the needs of a company for capital, provision must be made for working capital, as well as for the fixed assets.

Share capital

The issued share capital is the money invested by shareholders in shares. Shares will be issued for a nominal value, usually £1. The nominal value of one million £1 shares is £1,000,000. The balance sheet shows the original, nominal, value of shares issued.

The authorized capital is the amount of money that the company is authorized by its memorandum of association to raise by the sale of shares. In practice the company may not require the amount immediately, but it can issue shares up to the limit specified at any time when it wishes to do so. The issued share capital is that part of the authorized capital actually issued.

Over a period of years retained profit from trading will be converted into capital assets. These will appear as reserves on the asset side of the balance sheet. This increased asset value is owed to the shareholders, so is shown as a liability in the balance sheet. These reserves, together with the issued share capital, represent the shareholders' funds. These would be paid to the shareholders if the company were to be closed down.

The value of shares

The increase in the asset value of a company over a period of years increases the likely dividend payable to shareholders. This makes the shares more attractive to potential investors and increases the value of the shares. The actual value of shares may therefore increase when bought and sold on the stock exchange.

The actual value of shares is determined in the money market but will be based on an investor's assessment of the current value and future prospects of the company. This is the share value quoted daily in the newspapers for public companies. The share value multiplied by the number of shares is the value of the company as seen in the market – known as its 'market capitalization'. This value is not directly related to the value of shareholders funds and is not shown in the balance sheet, or in any other account.

Reserves

Once the trading profit is produced and any tax and interest paid, the remaining net profit belongs to the shareholders. This profit may be paid out to shareholders as dividends, or retained in the business to fund future growth. Usually, some is paid out and some retained. The amount retained is known as the revenue reserve for the year and forms part of the shareholders' funds in the balance sheet. Usually the reserves are reinvested in the business in the form of fixed assets and therefore do not exist as cash. Shareholders benefit from both dividend and retained profit. Dividends provide current income. Retained profit will, indirectly, increase the value of the shares, so the shareholder gains an increase in capital value, which is received when the shares are sold.

Calculation of reserves

The reserves shown in the balance sheet are cumulative since the formation of the company. They complement the nominal value of the shares and together show the current value of the shareholders funds.

Reserves are derived from the retained profit, the increase in reserves from the previous year being the retained profit for the year.

THE ADDED VALUE STATEMENT

The concept of added value was introduced in Chapter 2. Now that the accounts have been examined it is possible to calculate added value more precisely. This can be done in the added value statement. Companies are not required by law to produce an added value statement and few of them do. It has, however, been recommended by the accounting profession as being of some value.

It can be deduced from Chapter 2 that added value is a different way of categorizing the incomes and expenditures of an organization. An added value statement contains all of the items in the profit and loss account but arranged in a different way. This is done to focus on the common objectives of the participants in the organization. Traditional accounting procedures treat employees' wages and salaries as a loss, reducing the shareholders' profit. This approach is based on an adversarial style of management which causes conflict and holds down productivity and profit. Chapter 2 showed that a win–win strategy of common objectives can be developed in which improving productivity enables both profit and wages to rise simultaneously. The purpose of the added value statement is to calculate the wealth of the organization and to show how it is distributed.

It is relevant to re-read Chapter 2 and to refer to Figure 1.1 in Chapter 1 in considering the meaning of added value.

The added value statement subtracts from sales revenue the cost of materials and services bought from other organizations. For this purpose, rates, Value Added Tax and auditors' fees are included as services. Depreciation is also considered to be cost. Since stock adjustments are made in the profit and loss account, they must be made here too.

The added value statement defines the investors, the employees and government as participants in the organization. Notice that for this purpose loan interest is treated similarly to profit as a return for use of money. The government is included here as the recipient of Corporation Tax, which can be seen as the contribution that the company makes to society as a whole or as a contribution to the infrastructure – education, health, roads and defence – provided by government. In the example Corporation Tax is ignored, and so does not appear. These participants share the wealth created by the organization.

Added value can be calculated either as:

sales − (materials + services + depreciation)   (19.2) or as:

$$\text{wages} + \text{salaries} + \text{tax} + \text{net profit} + \text{loan interest} \qquad (19.3)$$

It would be prudent to calculate it both ways to make sure they are the same. An added value statement is shown in Table 19.9. Figure 19.6 shows the same statement as a bar chart.

Table 19.9 Added value statement

Sales revenue	671,760
+ Increase finished goods stock	2,880
+ Increase work in progress	4,480
Direct materials	−363,200
− Increase raw material stock	1,600
Indirect materials	−5,000
Direct expenses	−4,000
Factory expenses	−10,000
Sales expenses	−12,000
Admin. expenses	−18,000
Depreciation	−10,000
Wealth created	**258,520**
Net profit	60,600
Loan interest	16,000
Direct labour	83,200
Indirect labour	18,720
Sales salaries	30,000
Admin. salaries	50,000
Wealth distributed	**258,520**

Clearly added value is very different from profit. The bottom line of the added value statement is the wealth created by the company during the year by the activities of its participants – and the wealth distributed by the company to its participants. Its categorization of costs is different from the profit and loss account because it takes a different view of the company. It is not seen as a property of the shareholders but as a wealth-creating system involving jobholders, shareholders and government making inputs and deriving outputs from the company. This 'stakeholder' concept was introduced in Chapter 2 and can now be seen in the accounts.

Added value is a different philosophy from profit, and its use requires a different psychological perception of a company as a system. A change of attitude and culture is needed to use it, but its use can generate changes of attitude and motivation, release the full potential of employees and achieve the higher productivity on which both profit and wages depend. The non-use of added value in Britain shows that neanderthal management is not yet dead.

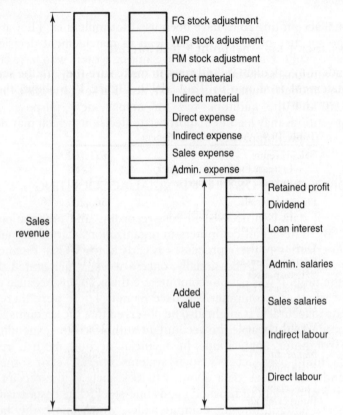

Figure 19.6 *Added value statement as a chart.*

USE OF FINANCIAL ACCOUNTS BY MANAGEMENT

The financial accounts are produced primarily for the shareholders and the tax authorities. They are, however, also of value to management. It has already been mentioned that companies find it useful to prepare forecast accounts for planning purposes and monthly accounts for control purposes. With delegated responsibility, managers need to know the financial details for the section they manage. Their objectives are likely to be set in terms of financial performance. Where employees are paid performance-related pay, pay may be related to some form of financial performance. Chapter 25 develops this concept.

A company's accounts can be compared with others in the same industry as a benchmark against which to compare company performance.

From the financial accounts, various ratios can be calculated which indicate the performance of the company and its financial wellbeing. Some of these ratios are developed in Chapter 20.

Traditional accounting rules have been used for millennia. They are based on a narrow perspective of profit but may lead to management decisions that prevent or reduce profit. Drucker and Hill are among writers who have criticized accounting conventions as being inadequate and inappropriate in a modern society operating modern manufacturing systems. These conventions may suit the government and the auditors, but for management purposes current accounting conventions may lead to inappropriate decisions which may not be in the best interests of the company.

INFORMATION TECHNOLOGY IN FINANCIAL ACCOUNTING

A large volume of data is processed in cost recording and preparation of the financial accounts. The use of computers in organizations started in most cases with the computerization of these processes as early as the 1950s. Financial data only needed to be processed occasionally, once a week, which suited the slow computers of the time. It was easy to computerize the accounts because the data had to be right whether a computer was used or not. There were therefore no problems of data inaccuracy. It might also be observed that the accounts were the first to be computerized because the accountant authorized the expenditure!

Modern computer applications in financial accounting use relational databases with multi-user access. Such systems operate across the whole organization and interchange data about activities such as the receipt of raw materials and the movement of finished goods into stock. The costs arising from these activities are entered into the computer files, which mimic the 'books' of old. From these cost records the monthly accounts can be produced for internal management use as well as the annual accounts.

With electronic funds transfer (EFT) this information system can be used to pay material suppliers, employees' wages and to receive payments from customers.

Given the confidentiality of some financial information, security of data is an important factor and access to data will be limited to those who need to know. Fraud prevention measures must also be taken.

The accounting application will communicate with similar information systems in marketing, manufacturing and personnel. Together they form an integrated information system for the whole organization. Such a system is not only used historically to report events that have already happened but can also be used for planning purposes.

Managers will wish to identify the sensitivity and risk attached to managerial decisions. Accounting systems, by their nature, record definite data and do not normally handle uncertainties. Nevertheless managers can run models of the financial results of particular decisions or particular scenarios in order to determine what the result would be. In this way the effects of decisions and possible external events can be simulated in advance. This is a major aid to managerial decision making.

Task sheet 19 Financial accounting

Table 19.10 Manufacturing and cost data for task

Admin. salaries/year	70,000	Discount rate	0.10
Admin. expenses/year	24,000	Interest %/month	0.01
Sales salaries/year	36,000	Dividend/profit	0.40
Sales expenses/year	18,000	FG stock %price	0.80
		Depreciation/year	0.20
Direct material/unit	30		
Direct labour/hour	5	Direct operators	14
Direct expenses/year	10,000	Indirects	6
		Office staff	10
Indirect material/year	12,000	Total employees	30
Indirect labour/hour	4.00	Wages + salaries	301,520
Factory expenses/year	8,000	Std. min/unit	32.7
Buildings	100,000	Sales price	60.00
Machinery	70,000	Prod. output/year	30,000
		Prod. hours/week	40
Share capital	150,000	No. weeks/year	50
Loan capital	275,000	Production/week	600

Table 19.10 shows some manufacturing and cost data in the same form as that presented in Table 19.1. From this data construct the following:

19.1 A cash flow forecast for the year.

19.2 A table of values of stocks, debtors and creditors at the end of the year.

19.3 A manufacturing account for the year.

19.4 A trading account for the year.

19.5 A profit and loss account for the year.

19.6 A profit and loss appropriation account.

19.7 A balance sheet for the end of the year.

19.8 An added value statement for the year.

CHAPTER 20

MANAGEMENT ACCOUNTING

MANAGEMENT ACCOUNTING

Objectives

From the Companies Act of 1844 until comparatively recently, the main emphasis of accounting was on the statutory requirement for submission of an audited balance sheet and profit and loss account annually to the Registrar of Companies. This is financial accounting and was explained in Chapter 19. Financial accounting, while legally necessary, is of little value to management. The use of archaic accounting conventions produces data that may reduce taxation but can produce cost data that is inappropriate and can lead to inaccurate decisions. With the increasing size and complexity of business organizations there is a need for more and better financial information to assist in the day-to-day management of a business. Financial accounting is not designed or intended to provide this information. Management needs accounting information for decision making and for assessing the performance of the system they are managing. This branch of accounting is now known as 'management accounting'. Its purpose can best be defined as the application of accounting techniques to the provision of information designed to assist all levels of management in planning, in making decisions, and in controlling the activities of an organization. This form of accounting is of value to managers in all functions of the organization and forms the basis of their interaction and of measuring performance.

This chapter covers the analysis of financial ratios, and the formation and use of budgets in comparing actual with planned activity levels and costs. The application of accounting to financial appraisal of projects is covered in Chapter 21.

RATIO ANALYSIS

Having constructed the financial accounts for legal purposes, management can then utilize them to evaluate whether they are satisfactory or not.

Because financial ratios derive from the financial accounts there use could be seen as a part of financial accounting. They are included in this chapter, however, because they are used by management to assess performance. This is the major role of management accounting.

Ratios can be used to compare the actual accounts with those budgeted and to compare the company with others in the same industry. Such data is compiled by organizations such as the Centre for Inter-Firm Comparison.

Overall ratios

The first ratios to consider are those that provide an overview of the budgeted and actual performance, highlighting profitability, solvency, capital structure and use, and cash flow. An understanding of these ratios is valuable to all levels of management. These are, however, the sort of ratios that would be of particular interest to directors. Other ratios, derived from internal accounts at sectional level, are more appropriate for management.

Return on capital employed (ROCE)

The best measure of profitability and budgeted monetary performance is probably pre-tax profit before interest charges have been deducted. Taking the pre-tax profit removes the uncertainty of tax rate fluctuations and differences in capital allowances. Return on capital is the best measure to compare budgeted return with the return from investing the capital elsewhere; current and past performance; and the company's performance with the performance of other similar companies in the same industry.

$$\text{ROCE} = (\text{pre-tax profit} + \text{interest})/\text{capital employed} \qquad (20.1)$$

Capital employed is the total investment in the company as shown by the balance sheet, i.e. total assets minus current liabilities. An overdraft would only be included if it is regarded as a part of the long-term funding of the company. Notice that in this form the return includes both profit and interest, i.e. the return to both the shareholder and the bank. As an alternative, profit only could be ratioed to shareholders' capital only. This ratio would be different but would be valid.

Return on total assets (ROTA)

This is very similar to ROCE but the figure for asset value comes from the opposite, capital, side of the balance sheet. It is a useful measure of real profitability. Companies with very low assets, such as a sole market trader, will make relatively higher returns on assets than a chemical company such as ICI with vast assets in relation to their profit.

$$\text{ROTA} = (\text{pre-tax profit} + \text{interest})/\text{assets} \qquad (20.2)$$

Current ratio

For a company to be successful it must not only make a satisfactory level of profit, but must also remain solvent, i.e. have enough cash to pay its bills. If a company is unable to pay its debts it may be made bankrupt and the receivers called in to take over the company. The company may have to sell its assets or even the whole company as a going concern to cover the debts. The current ratio gives a measure of liquidity – the ability of a company to pay its debts.

$$\text{Current ratio} = \text{current assets/current liabilities} \qquad (20.3)$$

Current assets includes stock and work in progress as well as cash and debtors.

For a company to pay its bills, its current ratio should be greater than one. However, some current assets are not immediately available, so a higher ratio is needed. Expert financial opinion argues that the current ratio should be around 2:1 (i.e. current assets should not be less than twice current liabilities).

Quick ratio (also called the acid test ratio)

Because stock and work in progress cannot always be quickly turned into cash, the quick ratio excludes them from the current assets. This is the real 'acid test' for liquidity – how much cash can you get your hands on quickly?

$$\text{quick ratio} = \text{liquid assets/current liabilities} \qquad (20.4)$$

Liquid assets includes cash and debtors but excludes stocks. This ratio is also sometimes called liquidity.

Once again, expert opinion says that the quick ratio should be around 1:1 (i.e. liquid assets not less than current liabilities). A supermarket with high stock levels that can be readily sold for cash can afford a lower quick ratio than an engineering company with a long working capital cycle and extended credit periods. The safe level therefore depends on the nature of the business.

Solvency

This is the ability of an organization to pay all of its liabilities if it were to be closed on that day. It shows how much of the total assets are funded by the shareholders.

$$\text{solvency} = \text{shareholders funds/total assets} \qquad (20.5)$$

In normal trading the solvency ratio should be less than one.

Gearing

This is a measure of the extent to which long-term funding of the company is owned by the shareholders as opposed to outsiders, i.e. other investors such as banks. It indicates the extent to which outsiders have potential (if not actual) say in how the company is run. For example, a banker who had provided more than 50 per cent of the long-term funding could expect some say in how the company was being run by appointing a director. (In America gearing is called 'leverage'.) Gearing can be measured in different ways. The following is used in Britain:

$$\text{gearing} = \text{total investment/shareholders' funds} \tag{20.6}$$

A similar ratio of loans/shareholders' funds is called the 'debt:equity' ratio. The higher the gearing, the greater the proportion of funding from loans as opposed to shares. Since the cost of interest on loans is usually less than the dividend on shares, a higher-geared company should have a lower cost of capital. However, in a poor year dividends can be reduced and may be less than the cost of capital. Interest on loans must be paid, so a higher-geared company has a greater outgoing in this situation.

Subsidiary ratios

The previous ratios, and especially the return of capital, only highlight overall problems. To provide detailed information that enables performance to be improved, more detailed ratios are needed. The following ratios give some insight into why the overall ratios might be showing an unsatisfactory picture. They are of particular concern to management.

The profit/assets ratio can be split into two parts, profit/sales and sales/assets, which when multiplied together form the profit/assets ratio. Both of these ratios can be split into their component parts – sales, manufacturing costs and overhead costs – which can be further split into more and more layers of detail, forming a hierarchy of financial ratios as shown in Figure 20.1.

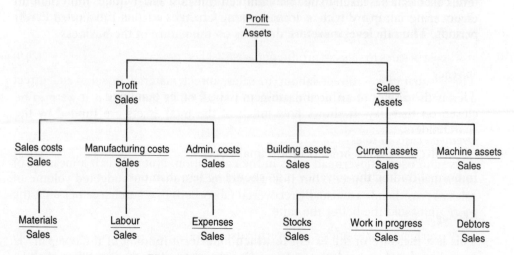

Figure 20.1 *Hierarchy of financial ratios.*

Return on sales

In any particular industry it is common for there to be norms for what is an acceptable return on sales. Comparison with other companies in the sector, and with the company's own previous performance, will give guidance as to whether the budgeted performance ratio is satisfactory. In accounting, the term 'sales' refers to the financial value of sales (in contrast to manufacturing where sales usually means sales volume). Return on sales gives an indication of the profit margin obtained in the market by expressing it as a proportion of the sales value. It measures the profit on £1 of sales.

$$\text{return on sales} = (\text{pre-tax profit} + \text{interest})/\text{sales} \qquad (20.7)$$

The profit figure can be broken down into its component parts – sales revenue and various costs – for further analysis, which is useful for identifying the causes of a low profit/sales ratio. This is expanded later in this chapter and shown in Figure 20.5

Sales over assets

This gives an indication of how effectively the assets are used to make goods which are sold. It measures the sales from £1 of assets. This ratio is often inverted so that all ratios have a common base of sales.

$$\text{sales/net total assets} \qquad (20.8)$$

Cost of sales to sales

The cost of sales is taken to be the manufacturing cost and includes direct labour, direct material and direct overhead costs. Changes in this ratio can identify wasteful or inefficient processes.

$$\text{cost of sales to sales} = \text{production costs/sales} \qquad (20.9)$$

Three sub-ratios – direct labour to sales, direct material to sales and direct overheads to sales – enable the problem to be further examined.

Overhead costs to sales

This ratio uses all overhead costs in the calculation. Not only is it a measure of efficient use of resources, but it is a good measure of the budgeted volume of sales. As overheads are usually recovered (absorbed) by a surcharge per unit, the more units sold the better the ratio.

$$\text{overhead cost to sales} = \text{all overheads/sales} \qquad (20.10)$$

In assessing the acceptability of the overhead cost of sales ratio, it will be valuable, as with the previous ratio, to subdivide it into its component ratios, by taking each overhead in turn and relating it to sales.

Sales per employee

A measure of efficiency is the amount of sales generated per employee. When comparing this ratio it must be remembered that different types of business have different norms. The sales per employee for an exclusive shop where personal service is given will be lower than for a self-service hypermarket.

$$\text{sales per employee} = \text{sales/no. of employees} \tag{20.11}$$

Added value per employee

Because some of the value of sales is used to pay for materials and services, the added value is a more relevant measure of output per employee than sales.

$$\text{added value/employee} = \text{added value/no. of employees} \tag{20.12}$$

Debtor turnover

It is useful to compare the value of debtors with that of sales. This can be done in two ways.

$$\text{debtor turnover} = \text{sales/debtors} \tag{20.13}$$

This measures the number of times per year the debt is 'turned over'. A ratio of 8 would indicate that in a year the amount of sales is 8 times the debtors. An alternative measure is made by inverting the ratio to debtors/sales. A value of 1/8 would indicate that the amount of debt on average was 1/8 of a year's sales, or that the average period between selling and being paid is 1/8 of a year. Often this value would be multiplied by 365 and expressed as 'debtor days'.

Stock turnover

Similarly, stock values can be ratioed to sales.

$$\text{stock turnover} = \text{sales/stock} \tag{20.14}$$

or

$$\text{average days of stock} = (\text{stock} \times 365)/\text{sales} \tag{20.15}$$

Each type of stock, raw materials, work in progress and finished goods would be measured separately.

A hierarchy of financial ratios

The ratios above can be related to each other hierarchically as shown in Figure 20.1. This approach has the following advantages:

- It prevents the possibility of individual managers budgeting and working towards incompatible or conflicting goals.
- It focuses the attention of all managers upon the organization as a whole and ensures that they see their contribution in that context and in relation to those or their colleagues.
- It provides a means whereby unsatisfactory performance can be investigated systematically.
- It makes possible direct comparisons with similar organizations if standardized definitions of ratios are used.

Trend analysis

Apart from measuring these ratios and comparing them externally with other organizations, there is value in using them internally for comparison of the values now with those at an earlier point in time.

The comparison with the previous year can be made to see if there is a trend. Are overheads increasing year on year with no real improvement in sales? This can be a warning that the organization is becoming complacent and top heavy. Trends in profit and sales may be related to trends in the economic environment; such information is a useful aid to forecasting. However, any ratio is only an aid to decision making. The key point for managers is to draw the correct interpretations from the ratios and to use them as the basis of understanding and improving performance.

BUDGETING

Managers need to be forward looking. They must attempt not just to predict the future, but to shape it. The proper function of budgeting is to plan, measure and control future activities. A budget is a financial plan, based on the organization's objectives for a future period. A budget is a co-ordinated estimate of the organization's costs, revenues, working capital movements, capital requirements and cash requirements. Budgets will normally be produced for each month and totalled for the financial year.

Budget preparation

The budget should not be the work of one person, alone in an 'ivory tower'. Budget preparation should be an iterative process that secures the involvement of everyone with responsibility in setting realistic but challenging targets and co-ordinating different plans of action for every function and level of the company. In this way, advantage is taken of each individual manager's knowledge, initiative and experience. The process also generates involvement, awareness and commitment, with the result that the targets implied by the budget are more likely to be accepted by all those whose task it will be to achieve them.

Stages in the budgeting process

Communicating details of budget policy and guidelines

Decisions taken as part of the long-term corporate planning process will act as the starting point for the preparation of the annual budget. Therefore directors must communicate the policy effects of the long-term plan to those responsible for preparing the current year's budgets.

Determining the factor that restricts output

In every organization there is some factor that restricts performance for a given period. This factor could be sales volume; however, it is possible instead for production capacity to restrict output. This restricting factor will determine the point at which the annual budgeting process should begin.

Preparation of the sales budget

The volume of sales and the sales mix determine the level of a company's operations, when sales demand is the factor that restricts output. For this reason the sales budget is the starting point. Because total sales revenue depends on the actions of customers, the state of the economy and the actions of competitors, it is the most difficult to produce. This emphasizes the importance of good market research.

Initial preparation of individual budgets

The managers responsible for meeting the budgeted performance should prepare the budget for those areas for which they are responsible. The budget should originate at the lowest levels of management and be refined and co-ordinated at higher levels. The guidelines provided by directors, together with past data, expected changes in the prices of purchases of materials and services, and standard costs may be used as the basis of costing the activity volumes planned in the budget.

Negotiation of budgets with superiors

The 'bottom-up' approach of budgeting requires each level of budget to be approved by the superior manager before being combined with other budgets and being passed up the hierarchy. At each of these stages the budgets will be negotiated between the budgetees and their superiors, and eventually will be agreed by both parties. It is important that the budgetees should participate in arriving at the final budget, and that the superior does not revise the budget without consultation – otherwise real participation will not be taking place and it is unlikely that the subordinates will be motivated to achieve a budget that they did not accept.

Co-ordination and review of budgets

As the individual budgets move up the organizational hierarchy in the negotiation process, they must be examined in relation to each other. Some budgets may be out of balance with others and need modifying so that they will be compatible with other conditions, constraints and plans that are beyond a manager's knowledge or control. During the co-ordination process, a budgeted profit and loss account, balance sheet and a cash flow statement should be prepared to ensure that all the parts combine to produce an acceptable whole.

Final acceptance of budgets

When all the budgets are in harmony with each other, they are summarized into a master budget that is then passed down through the organization to the appropriate responsibility centres for the managers to carry out the plans contained in each budget.

Ongoing review of budgets

The budget process should not stop when the budgets have been agreed. Periodically, the actual results should be compared with the budgeted results. These comparisons should be used to motivate staff and to detect areas where items are not going according to plan and to investigate the reasons for the differences. The process of variance analysis, explained below, is based on this approach.

Hierarchy of budgets

The relationship between budgets can be seen by drawing a diagram of the budget hierarchy, such as that in Figure 20.2. This hierarchy starts with a sales budget. In practice it will start with forecasts of sales volumes for each product for each month, which are then converted into sales values and added to form the budgeted sales value for each month.

The process of budget formation closely follows that of manufacturing planning, covered in Chapter 13 (see Figure 13.2). Just as forecast sales volumes are translated into planned work and material purchase schedules, so sales value is translated into budgeted labour, material and other costs. These are produced for monthly periods and are split into the different sections of the organization and different types of cost for each section. These then provide the financial objectives of the managers of each section, i.e. each cost centre. The hierarchy of budgets will therefore reflect the hierarchy of the organization structure.

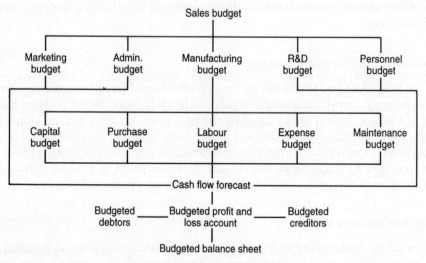

Figure 20.2 *Hierarchy of budgets.*

BUDGETARY CONTROL

Budgetary control is the management process of using budgets to control the performance of the organization. It is based on the use of standard costs and variance analysis. Standard costs were referred to in Chapter 18. A standard cost is the expected cost of a unit of output or input. It can now be seen to be a budgeted unit cost. Standard costs will be derived for the cost of direct labour, direct material and overheads at component and product level. These will be multiplied by the volume to generate the total cost, which is used in the budget. Standard costs are derived historically but are used predictively. For purposes of analysis it is important to use both total and unit costs since this helps to explain causes of differences between actual and budgeted values.

VARIANCE ANALYSIS

In order to achieve the forecast financial results, the plans and actions assumed in the preparation of the budgets must be implemented. As mentioned above, the actual performance will normally be compared with the budgeted performance, usually at monthly intervals, as a means of monitoring progress towards the achievement of the financial objectives. The technique used to do this is called variance analysis. Variance as used here is not the same as variance as used in statistics. A variance in accounting is the difference between actual and budgeted revenue or cost:

cost variance = actual cost − budgeted cost   **(20.16)**

or

cost variance = Δ cost   **(20.17)**

where Δ is the Greek 'delta' and is used here to mean difference between actual and budgeted values.

The order of these terms in the equation above can be easily remembered because A precedes B in the alphabet. It should be noted, however, that some books define variances as budget − actual. This does not matter so long as use is consistent. What matters is whether the variance is beneficial or not.

It would be insufficient merely to note the difference between actual and budgeted *total* cost since that would offer no explanation of the cause of the difference. Therefore variances are measured for each item of revenue and cost in the budget for a cost centre in order to identify the causes of a higher-level variance.

Variances can be positive or negative. However, this is not very useful since whether a variance is beneficial or not depends not only on its sign but also whether it is a revenue (inflow) or a cost (outflow) variance. It is common practice, therefore, to refer instead to variances as favourable or adverse (unfavourable). Table 20.1 shows how to determine whether a variance is favourable or adverse using equation (20.16) above.

Table 20.1   Favourable/adverse variances

Item	Variance	Effect
Revenue	Positive	Favourable
Revenue	Negative	Adverse
Cost	Positive	Adverse
Cost	Negative	Favourable

Favourable means better off which means either:

more money received = a +ve revenue variance

or

less money spent   = a −ve cost variance

Adverse means worse off, which means either:

less money received = a −ve revenue variance

or

more money spent   = a +ve cost variance

The letters (F) or (A) are sometimes attached to variances to indicate whether they are favourable or adverse.

Notice that in traditional accounting, profit is seen as an inflow but wages are seen as an outflow, so a positive profit variance is favourable but a positive wage variance is adverse. This is because the organization is seen in accounting terms from the shareholders' perspective only.

Use of variances

By analyzing variances the reasons why actual performance differs from planned performance can be uncovered. When the variance is favourable it is important to understand why, so that the conditions can be sustained or re-created in the future. When the variance is unfavourable, locating the cause may enable the management to avoid or minimize such effects in future. Variances can be measured at two levels: the total values for a period (year or month) and values per unit. Total values are used in budgetary control; unit values are used in standard costing, which was referred to in Chapter 18.

Variances will normally be calculated monthly and be notified directly to the managers responsible for the work areas in which they occur.

Revenue variance

Revenue variance is the difference in an income, usually sales and profit. Most variances can be split into two or more parts, giving greater detail of the cause of the total variance. Here the sales revenue variance is examined in detail but the same detailed analysis can be made of the cost variances.

Variance in the budgeted revenue may be due to more or fewer of the items being sold than expected (this is the volume variance). It may also be due to a reduction or increase in the selling price made necessary by market conditions (this is the price variance). The total sales revenue variance will be the sum of the volume and price variances.

This concept can be seen visually in a graph plotting unit revenue (i.e. price) against volume. This graph is shown in Figure 20.3 for the example explained below. The budgeted and actual sales revenues are seen to be areas on the graph. Price and volume variances are also seen to be areas. The total sales variance is an area – the sum of the areas of the price and volume variances. The total sales variance is the difference in area between the actual and budgeted sales revenue areas. It can therefore be seen that the total variance comprises two parts. In this way the total variance is split into two assignable causes.

The exact form of this graph depends on whether actual values are higher or lower than the budgeted values. It is drawn for both price and volume being lower than budget.

An example of price and volume variance

If, for example, a company budgeted to sell 10,000 units at £50 each, but because

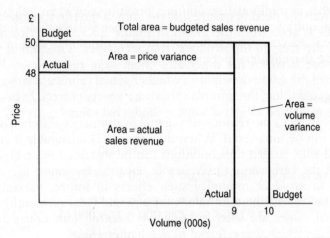

Figure 20.3 *Analysis of sales variance.*

of market forces had to reduce the price to £48 and even so were only able to sell 9000 at this price. Then:

$$\text{actual sales revenue} = 9000 \times £48 = £432,000 \qquad (20.18)$$

$$\text{budgeted sales revenue} = 10,000 \times £50 = £500,000 \qquad (20.19)$$

$$\begin{aligned}\text{total sales variance} &= £432,000 - £500,000 \\ &= -£68,000 = £68,000\,(A) \end{aligned} \qquad (20.20)$$

It can be seen from the graph that the actual and sales revenues are the inner and outer large rectangles, and that the total variance is the difference between these two areas.

This variance is negative, which means that less revenue was obtained than expected; this is therefore adverse and is marked with an (A). This variance is made up of two elements. The reduction in volume of 1000 units and the reduction in selling price of £2 per unit. The total variance is the sum of its two parts:

$$\text{price variance} = \text{change in price} \times \text{actual volume} \qquad (20.21)$$

or

$$\text{price variance} = \Delta\,\text{price} \times \text{actual volume} \qquad (20.22)$$

$$\text{volume variance} = \text{change in volume} \times \text{budgeted price} \qquad (20.23)$$

or

$$\text{volume variance} = \Delta\,\text{volume} \times \text{budgeted price} \qquad (20.24)$$

The use of the actual value in one formula and the budgeted value in the other arises from the need to ensure that the sum of the two variances is the total variance. This is seen from the graph in Figure 20.3. The price and volume variances are the areas of the two long, thin rectangles. The sum of these areas is the difference between the inner and outer large rectangles. Although the detailed form of the graph depends on whether actual values are larger or smaller than the budgeted value, the formulae above are always correct. In both formulae the change is measured as actual value − budgeted value.

For the example:

$$\text{price variance} = (48 - 50) \times 9000 = -\pounds18,000 \tag{20.25}$$

$$\text{volume variance} = (9000 - 10,000) \times \pounds50 = -\pounds50,000 \tag{20.26}$$

$$\begin{aligned}\text{sales variance} &= \text{volume variance} + \text{price variance} \\ &= -\pounds50,000 + -\pounds18,000 = \pounds68,000 \, (A)\end{aligned} \tag{20.27}$$

This is the same value as calculated in equation (20.20).

The main loss of revenue, £50,000, can be seen from this analysis to have been caused by the reduction in volume, with only a smaller amount caused by the change in price.

The breakdown of sales revenue variance into two components – those due to price change and those due to volume change – can be done also on the cost variances.

Cost variances

Cost variances are differences in outflows or costs. These would include manufacturing, sales expenditure and administrative costs.

Manufacturing variances

The biggest single expenditure area in a company is normally the manufacturing or operations function. The most significant cost variances are therefore usually related to manufacturing cost areas such as raw materials and components, labour costs and manufacturing overheads.

Material variance

$$\text{TMV} = \text{actual material cost} - \text{budgeted material cost} \tag{20.28}$$

This is the variance in total material cost over the accounting period. It can be split into volume and cost variances, with the later further split into price and usage variances. The split into volume and cost variance is on the same basis as that for revenue explained above. At the level of price and usage variance the analysis is initially at a unit level, the result being multiplied by the actual manufacturing volume to obtain the annual or monthly variance. This explains the need for three figures in the calculation of variance at this level. This is seen in Table 20.4 below.

MATERIAL VOLUME VARIANCE

$$MVV = \Delta \text{ volume} \times \text{budgeted cost/unit} \tag{20.29}$$

This is the variance in material cost due to the difference in manufactured volume. It does not refer to any difference in volume of material used in making the actual output, which is the usage variance below.

MATERIAL COST VARIANCE

$$MCV = \Delta \text{ cost/unit} \times \text{actual volume} \tag{20.30}$$

This is the variance in the cost of materials used in the actual output. It may be due to differences in price or volume used per unit so this is further split into:

MATERIAL PRICE VARIANCE

$$MPV = \Delta \text{ price/kilo} \times \text{actual kilo/unit} \times \text{actual volume made} \tag{20.31}$$

The actual price paid for raw materials or components is different from that budgeted for. This may be due to unexpected price increases. Kilograms are used in the formula above but it could be square metres, cubic metres, litres – whatever units the material is purchased in.

MATERIAL USAGE VARIANCE

$$MUV = \Delta \text{ kilo/unit} \times \text{budgeted price/unit} \times \text{actual volume made} \tag{20.32}$$

When manufacturing the item, more or less material was used than that originally intended – perhaps due to the use of different tooling or wastage. This is the volume variance at the level of the material used to make the output.

Labour variance

$$TLV = \text{actual labour cost} - \text{budgeted labour cost} \tag{20.33}$$

This is the variance in the total labour cost over the accounting period. It can be split in the same way as material variance.

LABOUR VOLUME VARIANCE

$$LVV = \Delta \text{ volume} \times \text{budgeted labour cost/unit} \tag{20.34}$$

This the variance in total labour cost due to the difference in manufactured volume. This does not refer to the volume of labour used for a actual output which is the efficiency variance below.

LABOUR COST VARIANCE

$$LCV = \Delta \text{ labour cost/unit} \times \text{actual volume} \tag{20.35}$$

This is the variance in the cost of labour for the actual output. This variance may be due to changes in cost or price of labour or the volume of labour used per unit of output, so this can be further split, in the same way as material cost variance, into:

LABOUR RATE VARIANCE

$$LRV = \Delta \text{ labour cost/hr} \times \text{actual hr/unit} \times \text{actual volume made} \tag{20.36}$$

This is the labour price variance at the level of the pay rate per hour. It arises from a difference in pay rates compared with the standard. This could be due to a greater than expected pay rise or the unexpected use of a higher grade of labour.

LABOUR EFFICIENCY VARIANCE

$$LEV = \Delta \text{ hours/unit} \times \text{budgeted labour cost/hour} \times \text{actual volume made} \tag{20.37}$$

The difference between actual and budgeted (standard) time per person per unit of output. This is the labour volume variance at the level of the volume of labour used to produce the output. This may be the result of lack of skill, effort or motivation. Training and involvement may improve this factor.

Overhead variance

Variances can be measured and analyzed as above for each of the overhead costs. In this book overhead variances are only analyzed down by one level into cost and volume elements. It is possible to split them further into price and efficiency elements by examining the overhead absorption rates and the amount of over- or underabsorption, but this is left for more advanced studies of the subject.

Manufacturing overheads or expenses are a part of the total manufacturing variance.

Overhead volume variance is measured as:

$$OVV = \Delta \text{ volume made} \times \text{budgeted overhead/unit} \tag{20.38}$$

Overhead cost variance is measured as:

$$OCV = \Delta \text{ overhead/unit} \times \text{actual volume made} \tag{20.39}$$

The manufacturing variances all sum together to produce the manufacturing variance.

Total cost variance

Administrative and sales expenditure can be calculated as above and added to the manufacturing variance to give the total cost variance. This plus the sales revenue variance is the *profit variance*.

In this way the variance of profit is analyzed and attributed to its separate causes. This enables management to identify, understand and solve problems as they occur and before they become irreversible.

Example of variance analysis

The data in Table 20.2 shows manufacturing and cost data, both budgeted and actual figures, which can be used for variance analysis. This basic data gives unit costs for materials and overhead and the labour rate and work content (standard time) for labour. These must be converted into the total costs for the given volume manufactured. From this the budgeted and actual profit can be calculated. Table 20.3 shows these calculations. On each row the total variance for each factor is shown as actual − budgeted value.

Table 20.2 Manufacturing cost data

Item	Unit	Budget	Actual
Sales volume	year	15,000	16,500
Price	unit	75	72
Material price	per kg	10	8
Material use	kg/year	45,000	66,000
Labour cost	hour	5.0	5.3
Mfg overhead	year	75,000	79,200
Standard time	min	35	34
Admin. costs	year	150,000	146,500
Sales costs	year	75,000	80,000

Table 20.3 Revenue, costs and profit

Item	Unit	Budget	Actual	Variance
Sales volume	year	15,000	16,500	1,500 F
Price	unit R	75	72	−3 A
Material price	per kg	10	8	−2 F
Material use	kg/year	45,000	66,000	21,000 A
Material use	kg/unit	3	4	1 A
Labour cost	hour C	5.0	5.3	0.3 A
Standard time	min C	35	34	−1 F
Overhead cost	year C	75,000	79,200	4,200 A
Admin. costs	year C	150,000	146,500	−3,500 F
Sales costs	year C	75,000	80,000	5,000 A
Standard time	hour C	0.5833	0.5667	−0.1667 F
Labour cost	unit C	2.9167	3.0033	0.0867 A
Material cost	unit C	30	32	2 A
Total variance cost	unit C	37.9167	39.8033	1.8867 A
Material cost	year C	450,000	528,000	78,000 A
Labour cost	year C	43,750	49,555	5,805 A
Overhead cost	unit C	5.0	4.8	−0.2 F
Manufacturing cost	year C	568,750	656,755	88,005 A
Total cost	year C	793,750	883,255	89,505 A
Sales revenue	year R	1,125,000	1,188,000	63,000 F
Net profit	year R	331,250	304,745	−26,505 A

Table 20.4 shows both the breakdown of these total variances into price and volume variances, and how each variance is calculated.

Table 20.4 Variance analysis

Item		Difference ×	Value =	Variance
Sales price	variance R	−3 ×	16,500 =	−49,500 A
Sales volume	variance R	1,500 ×	75 =	112,500 F
Material volume	variance C	1,500 ×	30 =	45,000 A
Material cost	variance C	2 ×	16,500 =	33,000 A
Material price	variance C	−2 × 4 ×	16,500 =	−132,000 F
Material usage	variance C	1 × 10 ×	16,500 =	165,000 A
Overhead cost	variance C	−0.2 ×	16,500 =	−3,300 F
Overhead volume	variance C	1,500 ×	5 =	7,500 A
Labour volume	variance C	1,500 ×	2.9167 =	4,375 A
Labour cost	variance C	0.0867 ×	16,500 =	1,430 A
Labour payrate	variance C	0.3 × 0.567 ×	16,500 =	2,805 A
Labour efficiency	variance C	−0.016 × 5 ×	16,500 =	−1,375 F

Hierarchy of variances

Just as a hierarchy of budgets is compiled, so a hierarchy of variances can be constructed that reflects the organization's hierarchical structure of departments and areas of responsibility. The profit variance is the head of a tree of variances that cascade down through the organization in increasing detail of explanation. Figure 20.4 shows a hierarchy of variances, with the values derived for the

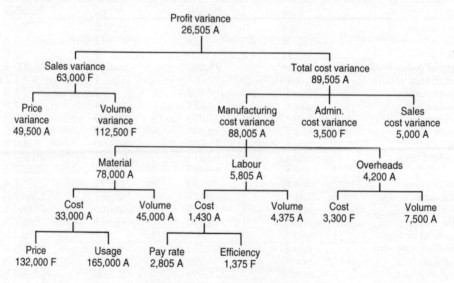

Figure 20.4 *Hierarchy of variances.*

example above. Notice that the variances at a lower level add to give the variance at the higher level. Care must be taken of the favourable or adverse nature of each variance when they are added.

From this hierarchy of variances the causes of low profit and means of improving profit can be seen. The total profit variance has been broken down into several constituent parts, thus explaining the cause of the profit variance. This assists managers in identifying problems areas. Comparing variances from one period to another provides a measure of improvement (or failure!).

Causes of low profit

A tree diagram, similar in structure the hierarchy of variances, can be used to identify the causes of low profit of an organization as a starting point for improvement of performance. Figure 20.5 shows a tree diagram of possible causes of low profit. A similar diagram changing 'low' to 'high' would show the paths to productivity, performance and profit improvement.

There are only four ways of replanning to achieve a better return on sales: reduce production costs per unit, reduce overhead costs, increase selling price and increase sales volume. Because of market forces these last two options may be difficult to achieve simultaneously. The usual method is to target the costs for reduction.

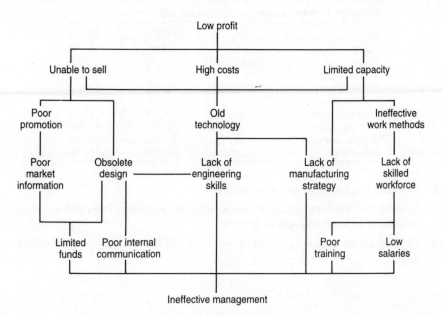

Figure 20.5   *Causes of low profit.*

INFORMATION TECHNOLOGY IN MANAGEMENT ACCOUNTING

The volume of data to be processed for budgetary control makes it sensible to use databases or spreadsheets for the purpose. The processing of the data uses standard formulae repetitively, a feature for which spreadsheets were designed.

A computer-based accounting application will be centred on cost recording and will be used for both financial and management accounting purposes. From the financial accounts, the various ratios explained above can be easily calculated and presented with earlier values for comparison. All that is needed is to ensure that the cost and financial data, which will already exist within the organization, are made available to managers to aid their performance-monitoring and decision-making functions.

Task sheet 20   Management accounting

20.1 Use the data from the profit and loss account and balance sheet in Chapter 19 to calculate the financial ratios derived in this chapter.

Table 20.5 gives some manufacturing and cost data similar to that in Table 20.2. Apply the processes described in this chapter to this data.

Table 20.5   Manufacturing cost data for task

Item	Unit	Budget	Actual
Sales volume	year	5,000	4,760
Price	unit	34	32.6
Material price	m^2	2.5	3
Material use	m^2/year	16,000	14,280
Labour cost	hour	3.9	4
Standard time	mins	15	13.6
Mfg overhead	year	60,000	47,600
Admin. costs	year	30,000	32,500
Sales costs	year	18,000	16,750

20.2 Calculate the budgeted and actual profits.

20.3 Calculate the variances shown in the hierarchy of variances in Figure 20.4, indicating whether they are favourable or adverse.

20.4 Graph these variances as a hierarchy and check that they add up correctly at each level.

CHAPTER 21

PROJECT APPRAISAL

INVESTMENT APPRAISAL

The need for investment appraisal

As far as budgeting is concerned, an organization is looking to the short-term future and its recurrent revenues and costs. It is concerned with the ongoing day-to-day activities of the organization – its steady-state operation. Investment is longer term and is concerned with capital costs. The following is a precise definition of investment:

> An expenditure in cash or its equivalent in one time period or periods in the anticipation or hope of obtaining a net inflow of cash or its equivalent in some future time period or periods. It is speculative by nature.

Investment in a new product or a new factory is likely to involve most departments within the organization. Figure 1.1 in Chapter 1 will assist the reader in visualizing how each department is likely to be involved in a major change project. Various departments will spend time and money on the project.

An organization may be considering expansion. Should they build or acquire additional manufacturing facilities? Management would need to be able to evaluate the return available from selling products made with the new facilities. Another option to consider might be the replacement of some or all of their old equipment with new and more efficient equipment, perhaps involving more automation. Such an investment will incur high initial capital costs but may be cheaper to run in the future since less labour would be used. How can such investment decisions be evaluated?

This type of analysis applies to a change in the organization. It was shown in Chapter 14 that a change involves a set of activities which together form a project. Chapter 14 covered the management and organization of projects. The techniques of budgetary control in Chapter 20 can be applied to projects. This chapter covers the analysis of investment decisions using approaches of investment appraisal. These techniques enable management to evaluate alternative investment proposals in terms of their financial viability. Where there is an existing facility, the proposals can be compared with the performance of the current system. Since proposals are evaluated prior to implementation, this analysis uses budgeted values. It is also possible to use the technique historically, after the investment, to monitor whether actual performance achieved budgeted objectives.

Economic life of a project

This is the time between the acquisition of the facility and the end of the earnings stream generated by it. This could be one of the following:

- When the cost of major replacements or the renovation of facilities is unlikely to be economical in relation to the benefits likely to result.
- When market obsolescence renders continuation of the project no longer viable.
- When rising maintenance costs exceed the estimated disposal value.
- When the availability of new, upgraded plant may justify new investment.

In general, the longer the life expectancy of a project, the better the financial return will be.

Accounting period

Cash will be flowing in and out at various times during the economic life of the project. For most projects annual budgeting is sufficient and fits in with the requirement for production of formal accounts for the shareholders and the Registrar of Companies. Conventionally the time up to the start of the project is known as year 0; any investment made at the very start of the project will be in year 0. Year 1 is the year from starting the project and includes revenues and costs in that year. The project year may not be the same as the organization's normal financial year.

ANALYTICAL TECHNIQUES OF INVESTMENT APPRAISAL

Data for investment appraisal

Data is needed for an analysis. Several different analytical techniques are available. These differ in what they measure but also in the basic underlying concept.

The data needed is shown in Table 21.1. Different techniques use different parts of the data. Some techniques are based on cash flows, others on profit. Calculations from this data are shown in Table 21.2. It was shown in Chapter 18 that cash and profit are not the same thing and must be treated separately.

Table 21.1 Project appraisal data

Year	0	1	2	3	4	5
Investment	−20,000					
Revenue		13,000	16,000	20,000	23,000	20,000
Costs		−8,000	−10,000	−12,000	−14,000	−12,000
Depreciation		−4,000	−4,000	−4,000	−4,000	−4,000

Table 21.2 Calculations from the project financial data

Year	0	1	2	3	4	5
Net cash inflow	−20,000	5,000	6,000	8,000	9,000	8,000
Cumulative cash	−20,000	−15,000	−9,000	−1,000	8,000	16,000
Profit		1,000	2,000	4,000	5,000	4,000
Discount factor	1	0.909	0.826	0.751	0.683	0.621
Discounted cash	−20,000	4,545.4	4,958.7	6,010.5	6,147.1	4,967.4

Cash outflow

There is likely to be an initial investment of capital in plant and machinery. There will also be the subsequent recurrent costs of running the system. Therefore cash will be flowing out of the organization over the entire economic life of the project. In the early stages these cash outflows will be very high. For a startup operation the delay in receiving revenue from customers means a cash outflow to cover early recurrent costs. This is the working capital referred to in Chapter 19. At the end of the life of a manufacturing system it may be necessary to pay for decommissioning of obsolete plant. With opencast mining it is usually a requirement to restore and landscape the site after use; in the case of nuclear power stations the decommissioning costs are enormous.

Cash inflow

Once the new equipment or process starts to operate, some cash should start to flow into the company from new or extra sales. This should continue until the end of the economic life of the manufacturing system. If a capital asset has a scrap value at the end of the life of a project, it will be shown in the final year as a cash inflow. In these calculations the recurrent costs include the direct costs and the fixed costs of services, sales and administration but do not include loan interest or repayment since these are part of the return to an investor.

Net cash inflow

Net cash inflow is the difference between cash inflow and cash outflow in a given time period:

$$\text{net cash inflow} = \text{cash inflow} - \text{cash outflow} \tag{21.1}$$

To avoid confusion net cash *inflow* is always used. If more cash has gone out than has come in, the net cash inflow will be negative. The sign +ve or −ve is therefore very important. Accountants use brackets instead of a negative sign, but a negative sign should be used if data is being processed in a spreadsheet.

Cumulative net cash inflow

This is a running total of the net cash inflows to date. This is required, as shown in Chapter 19, to ensure that there is always enough money available to pay the bills. It is required here to calculate the payback period, explained below.

Profit

Profit from a project in a period is the net cash inflow minus the depreciation of the capital asset during the period.

Discount factor

This is a ratio used to discount or reduce the value of a net cash inflow. It is used in the discounted cash flow (DCF) technique explained below.

Measures of project performance

Accounting rate of return (ARR)

This method of evaluation is based on a measure of the profit generated by the project. The initial cost of the capital investment is depreciated over the life of the project in the same way that it would be in the company accounts. The average annual profit, after tax and depreciation have been deducted, is then expressed as a percentage of the initial investment. Working capital requirements are ignored in all years, since they do not affect the resulting trading profits. The objective is to budget the figures that will appear in the final accounts in future years. One of the main disadvantages is that the technique ignores the timing of the cash flows altogether. Thus two projects with the same total profit will be evaluated as equal even though one may have most of that profit in year 1 and the other may have its profit only in year 5.

$$\text{ARR} = \text{average net annual profit/initial investment}$$

or

$$\text{ARR} = \text{total profit/initial investment} \times \text{no. of years} \qquad (21.2)$$

For the example data:

$$\text{ARR} = £16,000/(£20,000 \times 5) = 0.16 \text{ or } 16 \text{ per cent} \qquad (21.3)$$

Whether this is a good rate of return would depend on the current rates available from other forms of investment. The organization will probably have criteria for minimum acceptable returns. This and the risk of the project would have to be considered before the investment was approved.

Payback period (PBP)

This is the period that it takes for the anticipated net cash inflows to add up to the initial cash investment in the project. The net cash inflows for each period are calculated and then the cumulative net cash inflow is calculated year on year until it moves from a negative value to a positive. During that period the cumulative net cash inflow must have been equal to the initial investment. The exact time at which this occurs will be found by interpolation.

The time at which the cumulative net cash inflow equals zero is known as the payback period. This is sometimes called the break-even point on the grounds that at this point of time the project breaks even, meaning that its revenues equal its investment costs. However, this term is confusing since the term break even is also applied (see Chapter 18) to the concept of a volume of production at which revenue equals costs. To avoid any confusion the term 'payback period' should be used in this situation. This measure is useful because it takes some account of when the money flows in, but only up to the payback period.

PBP = time at which cumulative net cash inflow = capital investment   (21.4)

By inspection of the data in Table 21.2 it can be seen that the cumulative net cash inflow was minus £1000 at the end of year 3 and plus £8000 at the end of year 4. Therefore the cumulative net cash inflow must have equalled zero at some point during that year. The actual point can be determined graphically or by interpolation.

Figure 21.1 shows graphically the changeover from a cumulative cash outflow, A, at time a to a cumulative inflow, B, at time b. The time at which the cumulative inflow is zero, x, can be derived from a scaled graph or from the following formula:

$$x = a + [(b - a) \times A/(A - B)]$$ (21.5)

This formula is correct for interpolation and extrapolation (where x is outside the range between a and b). Care must be taken of + and − signs. In this case A is negative, so a negative value for A must be used. The value of $(A - B)$ is numerically the sum of A and B if signs are ignored. The value $A/(A - B)$ is positive because the signs for both As are negative.

In the example the payback period can be calculated as follows:

$$PBP = 3 + (4 - 3) \times -1/(-1 - 8) = 3\frac{1}{9} = 3.11 \text{ years}$$ (21.6)

Payback calculations are based on cash, not profit.

It is possible to identify from the graph the point at which the value of cumulative negative cash inflow is a maximum and the depth of the 'cash trough'.

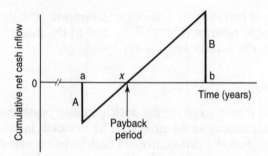

Figure 21.1 *Interpolation to find payback period.*

This is the point at which the cumulative negative cash inflow is a maximum. It is used to determine the size of the investment required to cover initial capital and working capital. This was shown in Figure 19.2 in Chapter 19.

The payback period itself does not give any information about the performance of the project after the payback period. Projects with identical payback periods could perform very differently.

There is a tendency for the project with the minimum payback period to be accepted in preference to more profitable projects with longer payback period. This tendency is fostered by the British banking system and the need of ambitious managers for good short-term results.

The time value of money

Modern approaches to investment appraisal take account of the time value of money, which is ignored in the two techniques explained above.

The time value of money is a concept based on an acceptance that money has a different value at different points in time. If this is the case, then money at different points in time cannot be added. That would be like adding apples to bananas! The two techniques below are based on this concept.

If a person were offered £100 now or £100 in a year's time, most people would prefer to have £100 now. This is because if they had it now it could be invested in a building society and would be worth, say, £110 at the end of the year. When offered £100 now or £120 at the end of a year, £120 would be preferred. This concept is shown in Figure 21.2.

If the rate of interest, or cost of capital was 10 per cent per year, then people would say that £100 now had the same value as £110 at the end of the year. Thus the same amount of money has different values at different points of time, and different amounts of money at different points of time may have the same value. This is the concept of the time value of money. Notice that if £100 becomes £110 after one year, it becomes £121 after two years. This is because the increase in value in the second year is based on the £110 available at the beginning of the second year.

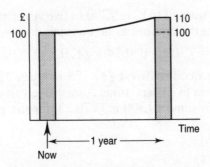

Figure 21.2 *The time value of money.*

This increase in value of money going forward in time is based on the interest received. Similarly the value reduces when going backward in time. It does not matter whether the cash flow is in or out, or whether money is borrowed or available capital used – the process is the same. If available capital is used, then, as an alternative to investing in a project, it could be put into a building society and earn interest. Clearly a project should offer a greater return than a building society for it to be worth investing in.

Notice that these differences in value are due to the return or cost of capital. They are not due to changes in the value of money due to inflation.

The formula for the value of money at different points of time is as follows:

$$S_n = P(1 + r)^n \tag{21.7}$$

P = the principal or present value
S = is the sum of money at the end of year n
r = is the rate of interest as a proportion
n = is the year at which a sum is calculated

$(1 + r)^n$ is called the compound interest factor, which when multiplied by the principal, P, increases it to its future value, S_n.

For example, if the principal or initial value is £100, the rate is 10 per cent and the period of the project is five years, then the future value would be:

$$£100 (1 + 0.10)^5 = £100 \times 1.61051 = £161.05 \tag{21.8}$$

In investment appraisal the need is to bring future value back to the present time, and so the reverse, discounting process is used.

Discounting uses the formula $(1 + r)^{-n}$. This is called the discount factor by which a future sum, S_n, is multiplied to obtain its present value, P. Thus:

$$P = S_n (1 + r)^{-n} \tag{21.9}$$

If an organization wanted to find out how much cash it would need to invest today to receive a certain sum at a future time, it would have to use the formula for discounting.

If an organization wanted to have £300 in five years' time, given an interest rate of 12 per cent, then the amount to be invested now would be:

$$£300 (1 + 0.12)^{-5} = £300 \times 0.5674 = £170.23 \qquad (21.10)$$

It would therefore need to invest £170.23 today at 12 per cent compound interest to receive £300 in five years' time. Alternatively, it could be said that the present value of a future sum of £300 is £170.23, for the given time period and cost of capital.

Discounted cash flow (DCF)

This technique is based on the time value of money. It uses cash flows, and discounts future net cash inflows to the present time to obtain their present value. Since all cash flows, including the initial capital investment, are now at the same time, they can be added. The result is the net present value (NPV) of the project.

Net present value (NPV)

The NPV can be calculated by determining the net cash inflow for each year, discounting each one to its present value and then adding them together. Notice that these values can be added together after discounting as they are then all at the same time, i.e. the present time.

In the example data a discount rate of 10 per cent is used. The NPV is shown to be:

$$NPV = £26,629 - £20,000 = £6629 \qquad (21.11)$$

This means that when discounted at 10 per cent the project will generate income equivalent to £6629 in today's cash terms. If the organization were to invest that sum in the bank at a rate of 10 per cent, it would generate the same cash inflows. If the rate of interest at a building society were greater than 10 per cent, it would make more money than the project with less risk.

The disadvantage of the NPV calculation is that it gives a money value, but its meaning is not well understood. What is a good or bad NPV? This can be explained as follows.

If the NPV is positive, then the return from the project is greater than the cost of capital; if it is negative, then the return is less than the cost of capital. A project is only worth investing in if its NPV is positive.

This explanation is of only slight value, however, since the magnitude of the NPV is not taken into account, only its sign. Clearly a higher NPV would have a greater return, so in comparing two projects, the one with the higher NPV would be selected. This is a deficient criterion, however, since it does not take account of the relative size of the investment.

What is needed is a measure of the financial return on the project, but one that takes account of the time value of money. The internal rate of return has been devised to provide this measure.

Internal rate of return (IRR)

It was stated above that with a positive NPV the return on a project was greater than the cost of capital. If the cost of capital increases, then the NPV for a given project would decrease because more money would be used to pay for the capital and less would be available as a return to the investor.

If, hypothetically, a high cost of capital were considered, then a value of cost of capital would be reached at which the NPV would fall to zero. At this cost of capital the return on the project would equal the cost of capital and there would be no return to the investors.

From this it can be seen that the rate of return on a project is equal to the cost of capital at which the NPV is zero. This provides a way of measuring the return on the project.

The IRR cannot easily be calculated from a formula, so the easiest method is by trial and error, substituting different values into the discounting formula until the NPV = zero. By plotting NPV against discount rate the IRR can be estimated from where the NPV crosses the zero line. The variation of NPV with discount rate is not linear, but may be assumed to be linear over a small range. Therefore different values of r should be tried until small positive and negative values of NPV are obtained. Between these values the NPV value can be obtained by interpolation using the formula given in equation (21.5) above. Many spreadsheets have an IRR function which calculates the IRR automatically.

For the example data the IRR is:

$$IRR = 0.2106 \text{ or } 21.1 \text{ per cent} \tag{21.12}$$

This means that the project will provide a return to the investor of 21.1 per cent. The investor must judge whether the project is worth investing in on the basis of this return and some perception of the risk involved in the project. This return would be compared with the return that could be obtained by putting the money in a building society or some other investment.

Sensitivity and risk

Investment appraisal is based on budgeted values which are derived from estimates and forecasts and hence are subject to inaccuracy and uncertainty. Sales volumes may not be as planned, the selling price may have to be reduced, the hoped-for savings on personnel may not be achieved, interest rates on loans may increase, delays may result in reduced income, etc. All of the variables used to build up the budget are subject to uncertainty. Some factors may have more effect than others: a small reduction in selling price may have a dramatic effect on cash inflow, whereas a doubling of maintenance costs may hardly affect the project at all. The project can then be said to be sensitive to selling price fluctuations and insensitive to changes in maintenance costs. Sensitivity analysis examines the effects on profit and the IRR separately of each input variable; risk analysis examines the total effect of the interactions of uncertainty. Sensitivity and risk analysis are beyond the scope of this book and are left for a more advanced study of the subject.

Project selection

There is no one single measure that can be considered the best way to appraise an investment – they all have advantages and disadvantages. They all use a single criterion and do not provide a continuous cash flow profile, which is essential during periods of negative net cash inflow if solvency is to be maintained. The payback period used on its own discourages long-term investment; however, if the cumulative net cash inflow curve is plotted it will reveal the timings of maximum rates of cash flow and the maximum negative cumulative net cash inflow that the company must finance before the project starts to generate cash and then decline.

The preferred approach is to use all available methods and to weigh up the pros and cons of each measure against any standards imposed by the company and the sources of finance. An insistence on high short-term gains may prevent the organization from developing the new products and using the new technologies that may genuinely be needed for long-term survival.

Task sheet 21   Project appraisal

Table 21.3 provides data for a project. Two alternatives are being considered.

Table 21.3   Financial data for two projects

	Year					
	0	1	2	3	4	5
Project A						
Investment	−10,000					
Revenue		10,000	12,000	12,000	12,000	12,000
Costs		−5,000	−6,000	−6,000	−6,000	−6,000
Depreciation		−2,000	−2,000	−2,000	−2,000	−2,000
Project B						
Investment	−17,500					
Revenue		10,000	12,000	12,000	12,000	12,000
Costs		−2,000	−2,400	−2,400	−2,400	−2,400
Depreciation		−3,500	−3,500	−3,500	−3,500	−3,500

Analyze the data provided and for each project alternative calculate the following:

21.1 The average rate of return.

21.2 The payback period.

21.3 The net present value.

21.4 The internal rate of return.

21.5 Which project, if any, would you approve?

PART 7

MANAGING HUMAN RESOURCES

LEARNING OBJECTIVES

7.1 Identify the role and function of personnel management.

7.2 Identify the stages of the recruitment process.

7.3 Prepare a job description and job specification.

7.4 Identify the data needs for manpower planning.

7.5 Be aware of UK employment legislation.

7.6 Prepare a statement of terms and conditions of employment.

7.7 Recognize the legal requirements to avoid discrimination at work.

7.8 Recognize the legal requirements regarding redundancy and dismissal.

7.9 Be aware of the legal requirements for health and safety at work.

7.10 Be aware of UK legislation on industrial relations and trade unions.

7.11 Describe the structure and decision-making processes of a trade union.

7.12 Define and differentiate the recognition, procedural and substantive types of agreement.

7.13 Describe the processes of consultation, conciliation, mediation and arbitration.

7.14 Explain the concepts of group behaviour.

7.15 Be aware of the reasons for resistance to change and means of overcoming them.

7.16 Understand the significance of Maslow's Hierarchy of Needs to motivation.

7.17 List the factors in Herzberg's Motivator–Hygiene theory and explain them.

7.18 Identify the characteristics of McGregor's Theory X and Theory Y behaviour.

7.19 Relate motivation to productivity.

7.20 List the management systems relevant to human resource management.

7.21 Identify and describe the concepts and techniques of job design.

7.22 Explain the need for and objectives and techniques of job evaluation.

7.23 Explain and evaluate different forms of payment system.

7.24 Explain and evaluate different forms of employee participation.

CHAPTER 22

PERSONNEL MANAGEMENT

PERSONNEL MANAGEMENT

The scope of personnel management

Personnel management is the set of activities carried out by an organization to manage effectively the skills and resources of the people who work within the organization.

The Institute of Personnel Development defines the function as follows:

> Personnel Management is that part of management concerned with people at work and with their relationships within an enterprise. Its aim is to bring together and develop into an effective organisation the men and women who make up an enterprise and, having regard for the well-being of the individual and of working groups, to enable them to make their best contribution to its success.

In many organizations the personnel manager is seen as a specialist supporting the management team in matters relating to the management of people. Depending upon the size of the company, the personnel manager may be an individual specialist providing the total range of personnel services, or in larger organizations a departmental manager controlling individual functional managers.

The functions undertaken either by the individual specialist or by the personnel function can be expressed in two clearly distinctive areas: those activities relating to the interaction of the employees and their performance, and those activities for which there is legal requirement. The former is internal to the organization and involves the design of management systems; the latter is a response to externally set legislation.

The effective interaction of the people in the organization requires the recruitment of capable people, the training of those people in the specific skills the organization requires and the provision of an environment in which employees will willingly exchange their skills for reward. This is crucial to the success of the organization.

This chapter was co-written with Trevor Pye.

Supporting that endeavour is the legal constraint within which the company operates. This requires the firm to recruit fairly, to abide by health and safety regulations, not to discriminate on grounds of sex, or race, and to deal with recognized trade unions in contractual or grievance procedures.

Organization of personnel management

Figure 1.1 in Chapter 1 shows how personnel management fits with other functions of management within an organization. Figures 4.1–4.6 show alternative forms of organization structure within which personnel management fits. The organization structure chart in Figure 22.1 shows a typical personnel department and the associated activities for the individual functions. These subsections of the personnel department can be traced back to the history and traditions of personnel management. These functional divisions also portray the changing values of society itself. A change in values has seen the personnel profession rise from welfare officers dealing with problems caused by employment, to the advent of personnel directors involved in policy and strategic decisions.

THE WELFARE STAGE

Historical background to personnel management

This changing approach to dealing with people at work has seen the role of the personnel function change considerably. The discipline has accordingly grown in status and become more influential. Until the First World War the job was mainly the domain of the well-intentioned, liberally minded, property-owning classes. These individuals were largely involved with improving working conditions and visiting the sick.

Until the advent of factories and large-scale mining during the Industrial

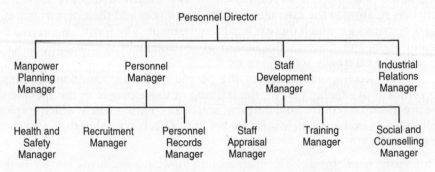

Figure 22.1 *Organization of personnel management.*

Revolution, employment had been largely an activity involving small employers. In these enterprises, merchants, guild masters or farmers had employed only small numbers of labourers paid by the day when work was available. For the skilled artisan the notion of regular employment was also unknown and they would be employed on a casual job basis – a situation not unlike many plumbers, electricians and builders are today where individuals requiring work to be done seek a tradesperson to carry out a specific job.

The main regulation on employment was through common law and the notion of 'master' and 'servant'. For many small employers, and for all those employed, this meant little effective control over employment.

The advent of the Industrial Revolution, in the early 19th century, brought about the need for large-scale organizations. The rapid adoption of new technologies and the consequent growth in markets started the movement towards industrialization. These new technologies and the forms of organization based on them principally brought about the economic concepts of economy of scale and the division of labour, which radically changed the pattern of life.

Initially the expansion of the factory system proceeded unchecked. Output increased, and increasing numbers of people were attracted to the towns to seek employment in these newly expanding industries.

For nearly the first quarter of the 19th century the expansion of the industrial system continued with no change to the legal system. In 1824 the Combination Acts, which prevented employees from forming trade unions, were repealed. This followed pressure from the 'Benthamites', a radical, liberal-thinking group in Britain. More details of trade union legislation are given in Chapter 23. This radical and liberal view continued to grow and in 1832 a Reform Bill was passed that redefined parliamentary constituencies, and expanded the right to vote to a larger group of property-owning men.

These changes principally benefited the newly emerging industrialists but can be seen as the foundations for a more general change. The Trade Union Act of 1871 permitted trade unions, and the Reform Act of 1884 extended the right to vote to male agricultural workers. The trade unions, the embryonic Labour Party and active individuals such as the Rowntree, Fry and Cadbury families demanded radical change.

Many see the first stirrings of personnel management as an attempt by the employers to respond to this demand for improved conditions within the factories and mines. These earliest attempts to consider employees as other than slaves or an economic resource saw the introduction of welfare policies. Initially these policies were adopted by only a few organizations and were generally in industries with a large concentration of women employees.

Welfare officers

This early work by welfare officers was viewed with suspicion by both the employers and their employees. As many of the welfare officers had the same social background and upbringing as the employers, employees were unsure of their motives. Equally, the employers were concerned that profits would be eroded if criteria other than narrowly defined economic ones were used in the management of industrial enterprises. The political orientation of employers led them to an adversarial style of management. This caused an alienative response from employees, who formed trade unions to defend themselves from exploitation.

Was the introduction of welfare policies by Quaker families such as the Cadburys, Frys and Rowntrees purely a matter of social conscience? Clearly their religious and social beliefs were predominant in determining their responses to the growing significance of the factory system. Their concern for the morals and working conditions of their employees was based on a desire to improve the lot of those working for them. There is also a need to recognize that the rules associated with that concern also had an element of regulation and control of the workforce.

With the rise of industrialization had come an increase in alcohol-related problems. Alcohol had become a major consumer product as it was the only commodity that had developed an extensive network of outlets, public houses, and had created a demand for an addictive product. The Quaker families, with their belief in sobriety, had initially developed the chocolate house as an alternative to the attraction of public houses. It is therefore not surprising that as they developed their companies into the manufacture of chocolate products, they should still have this concern with the adverse effects of alcohol and the welfare of their employees. This concern was not just in order to treat employees as human, but was also based on the recognition that productivity would rise if employees were healthy, well fed and well housed. Giving employees more wages could therefore increase rather than reduce profit.

Many of the rules adopted by the companies and used by the welfare officers were couched in the framework of protecting morals and encouraging good citizenship. These rules clearly also had a role in controlling and developing the behaviour necessary for effective working practices. The rules shown in Table 22.1 are those in existence in the Fry's chocolate factory in Bristol during the 1880s. The Frys were contemporaries of the Cadbury and Rowntree families and were taken over by Cadbury in the early 1920s. It must be remembered these were office staff and that Fry's were benevolent employees; the conditions of manual workers and in other organizations were worse.

This conflict of interests – between the need to offer help to individuals within the organization and the requirement to be the custodians of a policy for the way that the organization interacts with its employees – was very apparent to the early welfare officers and is still a dilemma that present-day personnel practitioners have to face.

Table 22.1 Conditions of work in the 1880s

1 Office employees will daily, sweep the floors, dust the furniture, shelves and show-cases.
2 Each day fill lamps, trim wicks and clean chimneys and wash the windows once a week.
3 Each clerk will bring a bucket of water and a scuttle of coal for the day's business.
4 Make your pens carefully, nibs may be whittled to your individual taste.
5 This office will open at 7 a.m. and close at 8 p.m. daily, except on the Sabbath, on which day it will remain closed. Each employee is expected to spend the Sabbath by attending at church and contributing liberally to the cause of the Lord.
6 Men employees will be given an evening off each week for courting purposes, or two evenings if they go regularly to church.
7 After an employee has spent 13 hours of labour in the office he should spend the time reading the Bible and other good books while contemplating the Glories of God and the building up of the Kingdom.
8 Every employee should lay aside from each week's pay a goodly sum of his earnings for his benefit during his declining years, so that he will not become a burden upon the charity of his betters.
9 Any employee who smokes Spanish cigars, gets shaved at a barber's shop or frequents pool and public halls, will give good reason to suspect his worth, intentions integrity and honesty.
10 The employee who has performed his labours faithfully and without fault for a period of five years in my service and who has been thrifty and attentive to his religious duties and is looked upon by his fellow men as a substantial and law abiding citizen, will be given an increase of 5 pence per day in his pay, providing a just return in profits from the business permits it.

THE INDUSTRIAL RELATIONS STAGE

The circumstances of the Second World War required a planned economy in Britain to organize manufacturing and the distribution of food. This changed fundamentally the relationships between employers and employees. Both government and trade unions recognized that, in the short term at least, productivity could be improved through joint consultation. The result of this was that between 1945 and the late 1970s increased consultation took place between successive governments and the Trades Union Congress (TUC).

This recognition of the role of centralized bargaining by the trade unions and government took place at a time when the shop stewards movement was becoming increasingly powerful. Industrial relations often only refers to the recognition and trade disputes conducted between trade union officers and the representatives of the company and employers' associations. The concept of national agreements was that the presidents of trade unions and the industrial relation directors of employers' associations would agree recognition procedures, pay increases or conditions of employment. The role of the shop stewards was then to police their members to ensure the agreements were upheld. In situations where employers did not uphold the bargain, the trade union officers would evoke the procedural process to ensure the employer came into line. Shop stewards found these delays unacceptable and reasoned that work-to-rules and strikes could resolve the situation more quickly. Increasingly this informal trade union activity became the focus for the membership rather than the centralized formal relationships. The development of industrial relations is covered in more detail in Chapter 23.

Against this background an increasing amount of legislation was enacted, bringing the law into the employment arena. Following the traditions of the welfare approach, the initial legislation was in the area of employment regulations (contracts, recruitment practices and safety). Later legislation amended many of these initial regulations and extended them to cover new areas such as discrimination but within a framework that governed the relationships between trade unions and employers.

The personnel professional therefore needs to be aware of the laws that provide the background to the formal relationships. This knowledge combined with negotiating skills enables the practitioner to offer advice in new situations as they arise. The major areas of the law that affect the day-to-day work of the personnel department are explained below.

Contracts of employment

The legislation relating to contracts of employment is based on the civil law concept of an arrangement between two parties of equal standing. The legislation enacted in the 1963 Contract of Employment Act clarifies the relationship between the employer and employee. The Act and subsequent amendments regulate the terms and conditions of the employment to eliminate bad practices.

Employees are employed under a contract *of* service which may be for a fixed term or permanent. This differs from a contract *for* service under which a contractor, consultant or casual labourer would be used to undertake a specific and usually short-term task.

Employees have a duty to take 'reasonable care' but are indemnified in their actions at work except in matters of negligence and harassment. Employees also have a duty of 'fidelity'. This is an implied obligation of trust and loyalty and requires an employee *not* to divulge company information, set up in competition or patent ideas learned at work.

For employment in teaching and in hospitals, job applicants must disclose any previous sexual or violent offences. Otherwise applicants can ignore previous offences after a stated period of time under the Rehabilitation of Offenders Act.

Terms and conditions of work

Terms of employment are those matters, which may include the level of pay, that are agreed and can only be changed by agreement. 'Express terms' are specifically stated but 'implied terms' may be deemed to exist on the basis of custom and practice. Conditions such as working hours are imposed and are accepted in accepting the employment. They can be changed unilaterally.

The requirements of the law are to provide a written statement covering the major aspects of the contract. Within thirteen weeks of starting a job, every employee must have been given or have easy access to this statement. These terms can be individual or be covered by collective agreements for the organization or the industry. The statement itself does not have to include the details of the terms, but if it does not, it must identify the relevant documents and where they can be located. These may include works rule books, grievance procedures, pensions handbooks and codes of practice.

The information required to be given either directly or indirectly to an employee in a written statement of terms and conditions of employment is shown in Table 22.2.

Table 22.2   Statement of terms and conditions of work

1   The name of the employer.
2   The name of the employee.
3   The title of the job the person is employed to carry out.
4   The date when the employment started.
5   Whether employment with a previous employer counts as part of the period of continuous employment.
6   The scale or rates of pay or the method of calculation.
7   The interval at which payment is made.
8   Terms and conditions relating to hours of work.
9   Terms and conditions relating to holidays, holiday pay, leave for sickness, sickness pay and pension schemes.
10 The length of notice of termination of employment which the employee is obliged to give and entitled to receive, or the date of the end of the contract if it is for a fixed period.

As individuals progress through the organization, either through changing jobs or promotion, their conditions of employment will also change. The Act requires information relating to these changes to be updated as and when required.

Other rights of employees

The Employment Protection Act of 1975 and subsequent legislation gives employees certain rights in their contract of employment.

Employees have a right to Statutory Sick Pay, which is initially paid by the employer but then claimed from the government. They also have a right to some payment if 'laid off' for a period or for a part of each week because there is no work to do. Historically salaried employees got this but waged employees did not. Wages are expressed as a pay rate per hour. Employees get paid for the hours they work and lose time if they go to the dentist or take time off. Salaries are expressed as an annual amount and are usually paid monthly. Salaried employees do not lose pay if they go to the dentist but can be disciplined for excessive unauthorized time off. The word 'salary' derives from the Latin for 'salt' which was used as a form of currency in Roman times.

Female employees have rights to maternity pay, time off, right to return and a right to non-dismissal on grounds of pregnancy.

Employees have rights to time off without pay for public duties (magistrates and jury service) and to time off with pay for antenatal care, for a safety representative, for Territorial Army camps and time to look for other work if being made redundant.

Discrimination in the workplace

The changing nature of society and of the working population, together with a growing recognition that discrimination did exist, brought about the need for direct intervention in employment practices to prevent discrimination. The two main areas of discrimination that caused concern were on grounds of sex and race. From the 1960s legislation was enacted to make discrimination illegal.

Sex discrimination

The Equal Pay Act of 1970 and the Sex Discrimination Act of 1975 sought to prevent discrimination in the payment and other treatment of men and women. Prior to this date women doing the same job as men were in some cases paid less. The Act insists on equal pay for the same or similar work, but defining 'similar' is not easy. The Act assists by referring to job evaluation techniques, which evaluate the level of responsibility of jobs. This topic is explained in Chapter 25. Where job evaluation has shown a similar level of responsibility between two jobs in the organization, then employees should receive the same basic level of pay and the same rules for other elements of pay.

Historically women tended to do less skilled jobs and to receive less pay. It was rare in the past for women to enter some professions. These days opportunities have widened but because of the time lag the proportion of women who are, for example, senior judges is still limited. Men do not become senior judges until they are over sixty years old and have about forty years of experience. The proportion of women who joined the legal profession forty years ago was small, so this is a cause of the low current proportion of women in this job. It does not mean that there is any discrimination against women.

These Acts also prevent discrimination against men. The second reason for the low proportion of senior women judges is that women can retire at sixty and receive a state pension, but men cannot receive such a pension until the age of sixty-five. Women have therefore retired before they become senior judges. European Union legislation permits discrimination against men in matters of retirement and pension.

Sexual harassment is an offence at work and elsewhere. Employers have a duty to prevent it occurring at work and can be sued for the actions of an employee.

Job titles must be gender-free. Postmen no longer exist but 'postal workers' do; waiters are 'table staff'; dustmen are 'environmental cleansing operatives'. In reverse, air hostesses are now 'cabin crew'. The title 'manageress' was used in the past for female managers, particularly where female employees were predominant. The title manager is now used for all managers. This is not a gender-biased word since it does not derive from the word *man*, which is of Germanic origin, but from the Italian *menaggio* which derives from the Latin *manus*, meaning hand. The word originally described people who worked with their hands. Over time this came to be applied to skilled people such as stonemasons who, while working with their hands, managed the work of other people who were unskilled.

Advertising for employees '1.8 m or taller' would be an indirect discrimination against women since women are on average shorter than men. Unless a tall person was genuinely required for the job such wording would be seen as an attempt to prevent women from applying for the job.

Racial discrimination

The Race Relations Act of 1976 prevents discrimination on grounds of race or ethnic origin generally but has specific implications for employment. The Act covers race, colour, ethnic origin and nationality but does not cover religion. It is still permitted for a government to discriminate on grounds of nationality on matters of work permits and residence in a country. Within the European Union, however, all nationals have rights of entry and employment in all countries in the Union.

Segregation at work was sometimes practised. For example, some ethnic groups were put onto anti-social hours or allowed to work only in certain areas or departments. This is rendered illegal by the Act.

It is illegal to advertise for 'English' employees since this is seen as an attempt to discriminate against employing people of non-English ethnic origin. Interestingly, it is not illegal to advertise for 'Cornish' or 'Yorkshire' people.

It is permitted to advertise with a requirement for an ability to speak a particular ethnic language where the job requires such an ability, e.g. for speaking with speakers of that language in the community. This is not discriminating against other groups, including the majority language group, since persons from such groups could be appointed if they were able to speak the required language. Those not able to speak the language would be incapable of doing the job.

Discrimination on grounds of religion is not covered by the Act and as such is not illegal. Rights available in Britain to Christians do not extend to other religions in modern multicultural Britain, and this is not seen as discriminatory.

Sikhs are a religious group but they have acquired the right to not wear hard hats on construction sites, or crash helmets on motorbikes, due to a religious requirement to wear a turban. This is permitted because case law has defined Sikhs as a racial group rather than as a religious group. The law requiring safety headgear was thus seen to be discriminatory.

Genuine occupational qualification

Within the laws of non-discrimination on the grounds of sex and race, discrimination is permitted where it can be shown that there is a special need for a person of a given sex or race. In such cases a 'genuine occupational qualification' (GOQ) needs to be shown. Discrimination is permitted only for the following reasons:

- Physiology. In the performing arts the choice of a man to play the part of a man, or a white person to play the part of a white person is not discriminating against others groups.
- Decency, sanitation and sleeping accommodation. Where employees have to undress or sleep, and facilities are available for only one sex, then employment of only that sex is permitted. In practice employers would try to provide separate facilities for each sex.
- A job in a private household, e.g. as a nanny, cook or cleaner.
- Overseas duties in countries which practice sex discrimination.
- A job is one of a pair for a married couple. Residential jobs sometimes provide accommodation for a married couple where both are employed.

Termination of employment

Various Acts have prescribed practices to protect employees from 'unfair' dismissal and to give employees rights not to be unfairly dismissed. Dismissal means the termination of employment by an employer. Employees who are dismissed must be told the reason. Dismissal for being a member, or for not being a member, of a trade union is 'unfair dismissal' and gives an employee a right to reinstatement or to compensation for unfair dismissal.

Employees have a right to a period of notice of dismissal or to a statutory minimum period which depends on the length of service. However, summary, i.e. immediate, dismissal is permitted for gross misconduct or negligence.

The following are the only fair grounds for dismissal:

- Incapability
- Misconduct
- Redundancy
- Non-renewal of a fixed-term contract.
- Reaching 'normal' retirement age

What exactly constitutes incapability and misconduct is not easy to define. Legal cases establish precedents. Failure to obey an instruction, excessive levels of absenteeism and permanent illness are fair grounds for dismissal. Falsely pretending to have qualifications or experience, or failing to disclose a criminal record where it is legally required to do so also constitute fair grounds for dismissal.

Where an employee resigns from a job and then shows that this was caused by unfair pressure or misconduct by someone else (e.g. sexual harassment), then the employee can claim that they were 'constructively dismissed'. This is treated as unfair dismissal even though there was no actual act of dismissal by the employer.

Redundancy

Redundancy is a form of dismissal or termination of employment by the employer on the grounds that there is no work for the employee to do or that the job done is no longer required. It is fair grounds for dismissal. 'Short-time working' (working, say, only three days a week instead of five) and 'laying-off' for a short period are not termed as redundancy since continued employment is envisaged. Redundancy is complete and permanent.

Employees who work for more than twenty hours per week and have worked for at least two years are entitled to financial compensation for loss of employment. The amount of compensation is that stated in the terms and conditions of employment or a statutory minimum which depends on age and length of service.

Employees are not entitled to compensation if they unreasonably refuse a reasonable alternative offer of employment, but may reasonably refuse an unreasonable alternative offer.

A redundancy agreement may be made between an organization and a trade union representing employees. This would cover both the procedures and amounts of financial compensation. Clearly the loss of a job is a difficulty for most employees, so redundancies will be resisted and conflict will occur over who is to be made redundant. A redundancy agreement can minimize this conflict.

A redundancy agreement will often lead to offers of compensation for early retirement and voluntary redundancy as means to reduce staffing without any compulsion. This may not reduce staff in the areas of required reduction or provide sufficient reduction. When compulsory redundancy is required, rules will be applied, based perhaps on length of service, to ensure that the procedure for selection for redundancy is fair. Employees may claim unfair dismissal if they believe that they were unfairly treated. Simply being made redundant is not unfair dismissal, but if all the people being made redundant were of one ethnic minority, then a claim for unfair dismissal may exist under sex discrimination legislation.

Enforcement of employee rights

There is little value in employees having rights to fair treatment if there is no way for those rights to be upheld. In Britain *industrial tribunals* have been established as law courts for the enforcement of industrial legislation.

The Equal Opportunities Commission and Race Relations Commission are independent, but government-funded, bodies which seek to ensure equal opportunities and non-discrimination. They often take cases involving employment to industrial tribunals but individuals can also take their own case. An industrial tribunal may order the reinstatement of an employee or compensation to an employee or former employee.

Employees may be able to take a case to the European Court of Human Rights if they are aggrieved by a decision of an industrial tribunal.

HEALTH AND SAFETY AT WORK

Historical background to industrial safety

The first legislation to deal with safety in places of work came into existence early in the 19th century. The Health and Morals of Apprentices Act of 1802 sought to improve the health and safety of children at work in Britain and gave Britain a lead in matters of safety. This is no surprise considering that at the time the Industrial Revolution was more advanced in Britain than in any other economy. The Factories Act of 1833 was the first act to allow for enforcement with the appointment of four factories inspectors to challenge the working practices of the day.

The regulations covering safety at work expanded on an *ad hoc* basis over the next 120 years, culminating in the Factories Act of 1961 and the Offices, Shops and Railway Premises Act of 1963. Over the years other laws specific to individual industries had been passed relating to agriculture, mining and quarrying, and nuclear power. This legislation was mainly concerned with placing responsibility on the employer to provide a safe place of work. Regulations concerning the requirements for machine guards, certificates for operation of cranes and hoists, procedures for dangerous working areas, and hours of work for young people and women were covered by these acts.

By the mid-1960s the limitations of this legislation were becoming clear and change was recognized as being necessary. A Committee of Inquiry into Health and Safety at Work was set up, by the then Labour government, and reported in 1972. The committee, chaired by Lord Robens, criticized existing legislation, which excluded some 5 million people, including lecturers and teachers, the clergy, market research interviewers and many local government employees. The Robens Report also the criticized the complexity of the legislation: there was too much law and what existed required simplification and co-ordination. The criticism that caused the most concern, however, was the lack of a positive attitude to health and safety. The prevailing attitude was one of compensation and prosecution rather than prevention.

The Health and Safety at Work Act 1974

In 1974 The Health and Safety Act (HASAWA) was introduced based on a proposal in the Robens Report. The Act has been considered to be a watershed in safety as it forced a change in the attitudes of both employers and employees. Employers were forced to consider prevention rather than just risk prosecution and to involve employees in safety by the election of safety representatives to a Safety Committee.

Due to the specialist nature of health and safety, the Act is what is described as an 'enabling act'. It provided for government to issue subsequent regulations and codes of practice without having to pass a new Act. The Act introduced new and radical concepts into the arena of health and safety, namely the following:

- The concept that health and safety is everybody's responsibility.
- That control of working practices and the monitoring of the effectiveness should be exercised by those directly involved.
- The requirement to set up an organization within the firm to make policy and take action.
- All employees to be covered by the Act.

To ensure these concepts were implemented five categories of people who were affected by and could influence safety were identified:

1. Employers
2. Employees
3 Suppliers of goods and services
4 Owners of sites, plant and equipment
5 The self-employed

The Act lays down the duties and responsibilities of each of these groups.

Employers have a duty to ensure, so far as is reasonably practical, the health, safety and welfare at work of all their employees. Employers are also responsible for the provision and maintenance of plant and systems of work that are, so far as is reasonably practicable, safe and without risks to the physical or mental health of employees. They must therefore ensure that machinery is properly guarded, floors kept free of obstruction and sufficient staff are employed to ensure that employees are not overloaded.

Employees have a duty to take reasonable care for the health and safety of themselves and of any other person who may be affected by their acts or emissions at work. Similarly employees have a responsibility not to intentionally or recklessly interfere with or misuse anything provided in the interest of health, safety or welfare.

In the case of suppliers, duties are placed on individuals or firms who design, manufacture, import, supply or install equipment to ensure its safety.

The Health and Safety Commission

The operation of the HASAWA is the responsibility of the Health and Safety Commission and the Health and Safety Executive. As the HASAWA is an enabling act it relies on regulations drafted by the Commission to give it the power of the law. These regulations are formulated after consultations with interested parties and given over to the Secretary of State to present to Parliament. They may be general in scope (e.g. dealing with environmental standards or the reporting of accidents) or be of a more specific nature (e.g. concerning the use of abrasive wheels or protective equipment).

In addition to regulations, which have the full force of law, codes of practice are used to promulgate good practice and to provide the standards and technical requirements which should be met. The Health and Safety Executive also draw up and issue guidance notes. These have no legal power but indicate good practice for carrying out specific tasks. HSE inspectors have powers to visit factories in order to enforce safety. More details of this aspect of safety were given in Chapter 10.

THE EMPLOYEE DEVELOPMENT STAGE

Manpower planning

Manpower planning derives from an organization's strategy in the area of human resource management. It seeks to provide the right people in the right jobs with the motivation and commitment to achieve high productivity for the organization. Its role is to improve the ability of the organization to achieve its corporate objectives by enhancing the contribution and commitment of employees. Without manpower planning an organization is likely to operate at low productivity.

Manpower planning seeks to balance the current and future demand for employees of given skills with the supply of such employees within the organization. This can be seen in Figure 22.2. The means of achieving the balance can be seen to be through recruitment, training and, on occasion, redundancy.

Manpower planning requires data about employees. This is needed to identify the number of employees in each job, of each skill, their age and length of service and in which department they work.

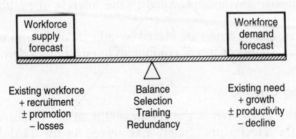

Figure 22.2 *Manpower planning.*

The current and future requirements can be measured and estimated to show any changes needed. This shows where more employees are needed and what training is needed. A company needs a model of its current and future staffing needs. Figure 22.3 illustrates a simple staffing model. It shows the employees in different job categories and how recruitment, promotion and leavers cause changes in the number of people in each category. By knowing the ages of employees, the numbers retiring at given times can be estimated. The rate of labour turnover (proportion of workforce leaving per year) can be measured historically and used as an estimate for the future. A policy of compensation for early retirement can increase the rate of retirement if required.

Recruitment

Recruitment is not a 'new' function of management but in the days of welfare officers 'hiring and firing' was the prerogative of the senior manager rather than a personnel manager. Recruitment is not the simple matter of hiring and firing; it is important to ensure the recruitment of the right people into the right jobs if the organization is to be effective. Quality is therefore an important part of the recruitment process. The objective of recruitment is to obtain *and* retain good-quality employees.

Recruitment is an expensive process, more so if it is unsuccessful. The costs of advertising and interviewing are not insignificant and can be high if incurred more often than necessary. Successful recruitment will provide a stable, skilled, committed and motivated workforce. Recruitment therefore has an important role in achieving the productivity of the organization.

The following stages are needed in the recruitment process:

1. Job analysis
2. Job description
3. Job specification
4. Job advertising

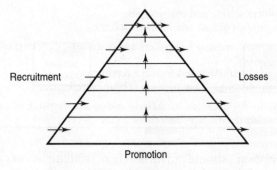

Figure 22.3 *Manpower planning model.*

5. Candidate short-listing
6. Interviewing of candidates
7. Selection of appointee
8. Negotiation of terms
9. Notification
10. Induction

Job analysis

Job analysis is the process of collecting and organizing data about the jobs within an organization. It provides data about the characteristics of the population of jobs. This is in addition to data about the people employed to do the jobs. Data is usually held in a job description.

Job description

A job description identifies the main tasks of the job and the resources for which the job holder is responsible. Table 22.3 shows a typical job description.

Table 22.3 A job description

Job title:	Office Manager	Job no.:	101
Department:	Commercial	No. of jobs:	1
Company:	Superior Office Equipment	Job grade:	4

JOB OBJECTIVE
   To manage the office and administrative functions

MAIN TASKS
   Supervision of office staff, payroll, accounts, purchase and sales order processing.

RESPONSIBILITIES

Factor	Responsibilities of the job
Responsibility	Responsible for buildings and office equipment to value of £200,000. Responsible for sales revenue and costs to value of £800,000 per year.
Supervision	Supervision and organization of work of two office staff, driver and salesperson.
Complexity	Problem solving and cost analysis. Preparation of cost and financial accounts.
Communication	Communication with suppliers and customers, directors, sales and factory employees. Prepare marketing and financial reports. Chair meetings. Take minutes at board meetings.
Decision span	Day-to-day decisions on material orders and despatch of goods. Planning decisions by month for a year.

A job description should give a clear statement of objectives and responsibilities but must not be so prescriptive that it reduces opportunity for initiative and development. It should be written in such a way that if an employee follows it, then work is done effectively. It must not be written in such a way that if the employee does the tasks described then output is very limited and the employee does not do what is intended. Employees who do exactly what is in their job description, but nothing else, are 'working-to-rule'. The format and detail in a job description should match the form of job evaluation, since the job description is the prime source of information used in job evaluation. Job evaluation is referred to in Chapter 25. Job descriptions have other purposes, e.g. as the basis of skills analysis referred to later in this chapter.

Job specification

A job specification is a statement of the requirements needed by a person to do the job. It provides the information needed against which to assess the suitability of applicants for the job. The job specification will be in similar format to the job description. Table 22.4 shows the job specification for the same job as that for which the job description is shown in Table 22.3 above.

Table 22.4   A job specification

Job title:	Office Manager	**Job no.:**	101
Department:	Commercial	**No. of jobs:**	1
Company:	Superior Office Equipment	**Job grade:**	4

JOB OBJECTIVE
   To manage the office and administrative functions

Factor	**Requirements for the job**
Qualification	Undergraduate degree in a business subject or degree in any subject plus postgraduate certificate in management or business studies.
Experience	8 years experience in an office function including at least 3 years as a manager preferably in a manufacturing organization.
Responsibility	Previous responsibility for buildings and office equipment.
Supervision	Previous responsibility for a small team.
Complexity	Able to handle and solve complex problems associated with sales order processing and accounts.
Communication	Able to develop good relationships with suppliers and customers. Able to work with directors and liaise with the manufacturing supervisor. Meet weekly with directors.
Decision span	Able to work on own initiative. Plan work by month for up to a year.
Physical needs and environment	Able to cope with stress. Own office in pleasant office environment.

Organization of recruitment

An organization must decide whether to handle this stage themselves or to subcontract it to an agency. Job Centres as well as recruitment consultancies could be used. They have interviewing skills and access to the labour market of job seekers. For some senior jobs 'headhunters' may be used to approach suitable people who may not actually be looking for a new job. A recruitment consultancy is used in some cases where the name of the organization is not stated in the advertisement and may not even be given in further written details. This information would only be disclosed at interview stage to those candidates who get through to this stage. Occasionally a person will apply for such a job only to discover that it is his/her own job that is being advertised! This can occur when an organization is dissatisfied with an employee and use a job advertisement to see if they can find anyone who may be better. If they discover, because they receive an application for the advertised job, that the particular employee is applying for other jobs then they may find grounds for dismissing the employee.

Job advertising

A decision is needed on where to advertise. In some organizations jobs are first advertised internally and only later externally. Fairness dictates that external applicants should have an equal opportunity. Some, but not all, public sector organizations are required to advertise externally. Where current employees are being made redundant they may be offered a new job before it is advertised externally. Advertising can be costly, so the size, number and channels of advertising must be chosen carefully. Since newspapers tend to attract different types of reader, one should be chosen that is likely to be read by the type of person sought. This will increase the pool of possible applicants and may lead to a better appointment.

The wording of a job advertisement must be an accurate reflection of the job, otherwise unsuitable people will apply. Wording must be non-discriminatory and give potential applicants the information they need to assess their suitability for the job.

The advertisement will normally indicate the location of the job, the extent of travel which would be required and much of the information from the job specification regarding qualifications and experience required. Often a salary is stated or the expected current salary of applicants to give an indication of the experience requirement. Some organizations prefer not to state a salary, in order to keep such information confidential.

The advertisement should indicate whether a curriculum vitae or an application form should be submitted, and must indicate the date and to whom it should be submitted. Normally candidates will be invited to apply for a document of further details about the job or to contact a senior manager in the organization for an informal discussion. In many cases potential applicants will be invited to apply for further details of the job, so these need to be written in advance.

After receiving the job applications, these must be analyzed. With a large number this is time consuming. The use of a standard application form enables the systematic presentation of data and makes easier the collation of information from a large number of applications.

Job interviewing

With a large number of applicants it would be usual to interview only a small number of them. Applications must be analyzed to select a suitable shortlist of those to be interviewed. The line manager as well as a personnel manager should be involved in this process.

Interviews cost a lot of money. Several managers are likely to be involved. Each candidate will normally be paid their travel expenses, which can be high if they travel any distance. Overnight hotels may be needed.

Interviewing requires some skills if it is to be useful in selecting suitable people. The interview must be structured and the interviewers must know what they are looking for. They may wish to clarify ambiguities found in the application form.

The purpose of the interview is to enable the organization to decide who is the most suitable candidate for the job. Question and answer is the most common form of interview, but increasingly organizations are using other methods to improve the process. In assessing the experience of an applicant it is important to decide whether a person with twenty years stated experience actually has twenty years of experience or merely one years experience repeated nineteen times! Evidence should be sought of the candidate's previous career growth and development.

Sometimes candidates are required to submit a report or make a presentation to a wider audience in the organization. This could include the work team within which the appointee would work. The view of the team may be taken into account in the appointment since it is important for the group to accept the new member if it is to work effectively.

Where several candidates are being interviewed for a job, they are sometimes invited to discuss a problem of relevance to the industry. The interviewers observe the discussion. This exercise not only discloses how much each candidate knows but also their interpersonal communication and group working skills. The interviewers will discover the leadership skills of the applicants.

Some organizations use psychometric tests to determine the personality of applicants. Assessing personality and potential attitude and commitment is not easy in conventional questioning. The personalities of the candidates can then be taken into account in selection. Other tests, e.g. of keyboard skills or of knowledge in a relevant topic, may be used where appropriate.

Clearly the capability of the person to do the job is an important factor. Relevant qualifications and experience are paper indicators of this, but must be supplemented and tested by interview.

The views of previous employers may be sought by asking for references. There are dangers in this approach since if an employer is keen to get rid of a poor employee they may give a good reference, and similarly give a poor reference to a good employee they want to retain.

Selection, negotiation and induction

Following interview the job will be offered to one of the applicants, assuming that a suitable person has been found. The start date and salary may need to be negotiated. When the offer is accepted all the unsuccessful candidates must be advised. At this stage the appointee's National Insurance number, next of kin and tax details will be required.

On starting the job, the person appointed will need an induction programme to gain familiarity with the organization, its objectives and procedures and with the people within it with whom the appointee is likely to interact.

Equal opportunity

Recruitment must be based on capability to do the job as the sole criterion for appointment. The same criterion applies to promotion, which is a form of internal 'recruitment' of a current employee into a different job. Many organizations state in job advertisements that they are 'equal opportunity' employers or that they welcome applicants from underrepresented groups. Race, age, gender or membership of particular clubs or political parties should not enter the consideration of suitability of a person for a job. Unsuccessful applicants have no redress under law even if they think they are more capable than the appointee. The concept of 'unfair' dismissal of employees does not extend to 'unfair' non-appointment or non-promotion. The laws of non-discrimination do, however, apply, so a case could be brought on these grounds.

Training

Training is an important means of enhancing the skills of current employees. It may be cheaper, better and involve less risk to 'grow' a known current employee to a higher level of skill and experience than to recruit an unknown outsider to a given job. The significance of training can be seen in the quotation shown in Table 22.5 by Kuan Fung Tzu (known in Europe as Confucious) which dates from the 5th century BC.

Skills analysis

An important part of manpower planning is to determine the skills available

Table 22.5 Pay increments, grades and ratios

If you wish to plan for a year – sow seeds.
If you wish to plan for ten years – plant trees.
If you wish to plan for a lifetime – develop people.
Kwan Fung Tzu, about 500 BC

among current employees and the skills needed to do the jobs in the organization. The skills needed and those available in the organization may not match, especially where jobs are changing. Data about the manual and mental skills needed to do a job can be collected and form part of a job description. This assists in recruitment of employees capable of doing the job. Some jobs require the jobholder to have good colour vision. This is therefore a requirement for those jobs. If applicants are not asked or tested for colour vision, then a person with poor colour vision may be inadvertently recruited to a job for which they are subsequently found incapable of doing.

Training needs analysis

Where differences are found between the skills available in the current employees and the skills required for the jobs in the organization, then these indicate the need for training and or recruitment.

Training can be done in-house, by consultants, by universities or by other training providers. Training is expensive, so the costs and benefits must be identified and analyzed.

Organizations may be reluctant to pay for training for fear that trained employees would leave and go to another employer. Some employers in the past had a reputation for not paying for training; they simply poached employees trained by other employers. By not paying for training they could afford to pay higher salaries. This has happened in the health service and in airlines. To overcome these difficulties governments may provide some funds or tax relief for training in order to encourage its provision.

In Britain a training levy was introduced in the Industrial Training Act of 1964 in order to spread the cost of training across all employers in an industry. Those employers undertaking training could get it funded from this levy. By the 1990s that system has been much modified but some elements of it still exist. CITB is the training board for the construction industry and EnTrA is the engineering training authority. They exist to provide training in particular industries. Training and Enterprise Councils, Industrial Training Organizations and Industrial Training Authorities offer industrial training with some element of financial support from government and employers.

Personal appraisal

Having recruited and trained employees, the next role of management is to ensure the motivation and commitment of staff and to monitor the performance of employees. Motivation is explained in Chapter 24, and the overall approach of modern human resource management is developed in Chapter 25. Personnel management may not have a direct role in personal appraisal, since it is usually a matter involving all managers, but will organize it and file the results.

Personal appraisal usually takes the form of an annual interview between an employee and his/her immediate manager. This will review past performance at agreed objectives for the future. In this way employees are involved in setting their own objectives; this should increase commitment to achieving them. Personal appraisal is a key feature of management by objectives which is explained in more detail in Chapter 24.

Reasons for non-achievement of objectives agreed the previous year will be discussed. Care should be taken to avoid this becoming just a complaint or 'buck-passing' session. Positive lessons should be learned so future failures can be prevented. Failures may be due to organizational problems rather than individual inadequacies.

In many organizations the personal appraisal system is used to give a discretionary increase in pay to those employees judged to have performed well. This is sometimes called 'merit' or 'performance related' pay and is covered in Chapter 25.

MODERN PERSONNEL MANAGEMENT

Managing an organization's human resources is a crucially important task since the effectiveness of the organization depends on it being done well. This chapter has introduced the concepts of personnel management. Chapter 23 covers the management of conflict and industrial relations. Chapter 24 provides an understanding of the basics of human behaviour, particularly behaviour in groups, motivation and the style of management or leadership. Chapter 25 then builds on this understanding and covers the design of management systems such as the organization structure, payment structure and means of involving employees in decision making in the organization.

Together the four chapters in this part of the book explain the historical background, which can cause mistrust and conflict, and provide an understanding on which to develop modern management systems necessary for the success of an organization.

INFORMATION TECHNOLOGY IN PERSONNEL MANAGEMENT

Information technology has a long history in personnel management as it was used from the outset to pay wages and salaries. The modern proactive approach to personnel management requires rapid access to manpower data and the construction of a manpower model for an organization as referred to above.

A database of jobs will identify jobs, and record job descriptions and salary levels. A database of employees will record their personal data, including date of birth, date they started employment, address, current salary, tax code, National Insurance number, and their employment and training history.

A spreadsheet can be used to form the manpower model. This enables personnel managers to estimate the effects and costs of proposed changes in the number and pay levels of employees in particular job categories. It is also used to show the effects of estimated future changes in the organization's need for people in manpower planning.

The standard letters for job applicants and other personnel tasks can be created on wordprocessors.

Task sheet 22   Personnel management

22.1 Examine several job advertisements in the national newspapers. Evaluate whether they provide the information necessary to enable an applicant to assess the type of job it is and whether s/he would be suited to apply for it.

22.2 Prepare a job description for one of the jobs which you examined in task 22.1.

22.3 From the job description you prepared for task 22.2, prepare a job specification.

22.4 List the factors you would expect to be covered in a job interview. Prepare a curriculum vitae which includes these factors.

22.5 If you work, assess your personal training needs. If you are a student, assess your likely training needs for an intended future career. Identify which of these needs is likely to be satisfied within your study programme and which may need enhancement after graduation.

CHAPTER 23

MANAGING CONFLICT

INDUSTRIAL RELATIONS

Scope and objectives of industrial relations

Industrial relations is the term given to the relationships that exist between management and non-managerial employees in an organization. The total relationship is covered in all chapters in this part of this book, but the term industrial relations applies specifically to those aspects involving conflict and in which employees are members of trade unions. Figure 1.1 in Chapter 1 shows how the different parts of an organization work together. Although 'employees' are shown only as the direct operators, employees work in all parts of the organization. The general approach of this book is based on creating common objectives and win–win results. This might be taken to imply that conflict can be eliminated and is therefore unnecessary. However, on matters such as pay, working conditions and redundancy, there will be conflicts of interest between the objectives of an organization and its employees.

As well as conflicts of interest there are conflicts of values or beliefs between managers and other employees. These arise from the political beliefs of individuals and indicate the political nature of industrial conflict. Nevertheless much can be done within an organization to minimize conflict through the development of a suitable industrial relations strategy. 'Win–win' strategies, referred to in several chapters in this book, have a significant part to play in managing industrial relations and minimizing conflict.

However in the areas of pay, working conditions and redundancy some conflict may occur and needs to be managed. This is the role of the industrial relations manager. In the same way that quality is best achieved by preventing poor quality, so good industrial relations is best achieved by preventing rather that just managing conflict. This requires a proactive approach by management.

The industrial relations role exists in parallel with that of personnel management, and in a small organization both functions may have the same manager. In larger organizations, however, there is likely to be a separate section or department. The managerial skills and management style for managing industrial relations needs to be different from those for personnel management. There are benefits in separating the management of conflict from the management of matters about which common objectives can be created.

Comments have been made elsewhere in this book on traditional management approaches, which are based on raising profit by reducing costs. This applies not only to material costs but also to employee or labour costs. The term 'labour' is usually applied narrowly to include only the manual workers within an organization but this chapter addresses relationships with all employees. Although traditional approaches by management had the objective of profit, they failed to achieve profit because the approach was superficial and based on a narrow fixed-sum view of wealth, referred to in Chapter 2. A modern approach is based on viewing the organization as a system and treating people as people. Such conflicts that do occur are easier to resolve in a climate of good management and mutual trust.

TRADE UNIONS

The development of trade unions

With traditional styles of management, employees feel threatened and powerless and that they are not valued by the organization. In this situation they react by forming or joining a trade union as a means of being collectively represented and so enhancing their power in negotiations with management over matters such as pay.

Inflation occurs in most societies, which means that prices rise over a period of time. This means that a given wage or salary will be able to buy fewer goods at the end of a year than it did at the beginning. In this situation the standard of living of employees will fall, they feel aggrieved and will seek an increase in their pay.

The earliest recorded strike occurred when tomb builders in Egypt went on strike in 1170 BC. This was due to an erosion of their standard of living because of inflation caused by the loss by Egypt of territories in West Asia following a war with the 'Peoples of the Sea'. Low-cost food from occupied areas was no longer available, so prices rose. This pattern of conflict has repeated itself many times in history.

A pay rise only equal to inflation will be seen by employees as no rise at all but merely as compensation for inflation, i.e. the loss of value of the currency in terms of what it will buy, and a restoration of a previous standard of living. Hence pay increases higher than the rate of inflation will be sought. To the employer and the accountant, however, a pay rise means paying more money to employees even if it is less than the rate of inflation. Unless a pay rise is financed by higher productivity, it will cause prices to rise. Thus a positive feedback loop is created. Pay rises cause price rises which cause pay rises. This relationship between pay, productivity and inflation was introduced in Chapter 2. This shows the significance of productivity since it is the only way in which pay can rise without causing prices to rise. Productivity is therefore a common objective between an organization and its employees and one on which a win–win strategy can be built.

Given the conflict arising from the conflicting objectives of management and employees over pay, it is clear that industrial relations is a political matter at the level of the organization. It is also a political matter at the national level as the view of a party in power will influence the legislation they propose and pass in Parliament and the policies they develop to control inflation. In Britain the two largest political parties, Conservative and Labour, have very different views on industrial relations as the two parties have members and views reflecting the two sides of the conflict in industry. This political structure is at least partly the cause of industrial conflict and this conflict is needed by the political parties for their survival. The Labour Party was originally created by trade unions in order to provide a voice in Parliament for working people. It was not until 1910 that all adult males in Britain were entitled to vote, and women not until 1928.

In Britain there has been a long political battle over the rights of employees, particularly whether employees should be able to organize themselves into trade unions and what actions they could take legally in pursuing a dispute about pay or some similar matter. It is perhaps no surprise to find that similar conflicts and laws fixing the levels of wages and prices were in operation in the ancient Egyptian, Assyrian and Roman societies. The Code of Hammurabi of 1762 BC and the Edict of Diocletian of AD 301 fixed both wages and prices in attempts to prevent the damaging effect of industrial conflict. In Britain the Statute of Labourers of 1349 fixed maximum wages as a means of preventing wages from rising during a period of high inflation due to the Black Death. This led to a century-long period of zero wage inflation. Legislation on industrial relations and trade unions has continued to the present day, with new legislation being passed every year or so.

Historical background

Prior to the Industrial Revolution, most of the population were engaged in agriculture and were self-sufficient. With the growth of industry people moved to towns and, having no land, became dependent on having a job as a means of survival. Managers used fear as a form of control, and used their power to hold down wages and to exploit employees.

There were no social security benefits in those days, so if people had no job they begged or starved to death. Being in a feudal society and having no vote meant that employees had no power. While the wealth of Britain was increasing in the 18th century due to colonization, slavery and industrialization, most of this wealth went to only a small proportion of the population. The standard of living of employees in Britain fell by 1800 to a half of its 1750 value during this period of industrialization.

Employees were in many cases not paid a wage but given food and housing by their employer. In other cases employees were paid in tokens that could only be exchanged at the local shop, which was owned by the employer and charged high prices. Employees could shop nowhere else. This malpractice, known as 'trucking' was made illegal by the Truck Act of 1831 which gave employees the right to be paid in coins. (In modern times this has prevented employers from paying employees by bank transfer. The Truck Act was repealed in 1986 to enable payment of wages in this way.)

Employees in this situation tried to combine together to form trade unions to give themselves some protection from exploitation. Legislation was passed by Parliament – which then comprised only wealthy citizens elected by the 2 per cent of the population entitled to vote – to prevent these 'combinations' of workers. In Britain a Combination Act was passed in 1799 which prevented workers from combining to form trade unions. To do so was a criminal conspiracy for which the penalty was death or transportation to the penal colony of Australia – a fate many feared more than death.

The Combination Acts were repealed in 1824 and some freedoms granted. However, the actions of workers in pursuing disputes were then deemed to be contrary to other laws. This swing between permission and prohibition of trade union activity has continued ever since. The Trade Union Act of 1871 allowed trade unionists to take industrial actions legally for the first time. Further legislation in the 20th century redefined some actions as illegal. There have been many laws since enhancing or reducing the ability of employees to take action in pursuit of a trade dispute, depending on which political party was in power.

Employees, having become dependent on a job, feared unemployment due to the mechanization which was developing in the 19th century. In 1811 workers in the textile trade in Nottingham organized themselves under the leadership of Ned Ludd. They were known as Luddites and sabotaged some of the machines which threatened their jobs. For this many were hung or transported to Australia for life. The term Luddite is still used to refer to people who cling to traditional practices and resist change.

In the midst of these conflicts and the neanderthal (Luddite!) style of management used at the time, there were some employers who took a different view. In Birmingham in 1792 Matthew Boulton established his Soho Manufactory (from which the word factory evolved as an abbreviation). Boulton established standards of factory management and welfare for employees unusual for the time.

Also prominent was Robert Owen, who in 1817 set up a factory in Lanarkshire, Scotland. Owen took care of his employees, providing housing with adequate sanitation, child-care facilities and canteens. He recognized that hungry workers could not work effectively and saw benefit to his organization in having healthy and well fed employees. He also advocated co-operatives and self-governing workshops in which all of the financial benefits of the activities of the organization would be distributed among those who worked there rather than go to external capitalists. Owen saw the dangers of excessive mechanization and insisted that people should come before machines. By 1834 Owen retired from being an employer and set up a Grand National Consolidated Trade Union as a means of assisting employees to protect themselves from powerful employers.

In 1834 agricultural workers in Dorset were facing a large reduction in their wages and formed a branch of this trade union. For this they were transported to Australia. A public outcry followed this sentence, and the six people involved became known as the Tolpuddle Martyrs after the village in Dorset from which they came.

Other employers who expressed concern at the way employees were being treated established factories on different principles. The Quaker families – Cadbury, Rowntree and Fry – were also concerned about the adverse effects of alcoholism among manual workers. They set up chocolate factories and provided for the social welfare needs of their employees in a style not greatly different from that of Japanese companies.

It is clear from this brief review that there is a long history of conflict in industrial relations which has set the climate for the recent past. Entrenched attitudes of managers and trade unionists have arisen from this history, the mutual mistrust of each other and from their personal political views.

The role of trade unions

The primary role of a trade union is to work on behalf of its members in their relationships with their employers on matters such as pay and working conditions. The objective of a union is to protect employees from exploitation, to improve their pay and safety at work and to reduce working hours. It was not until 1880 that the standard of living in Britain rose to the level it had been in 1450. At the end of the 19th century office employees, who were usually better treated than manual workers, worked a 78-hour week. Wages for unskilled workers were about 2 pence an hour or £1.50 per week.

Given the attitudes on both sides, conflict over pay was inevitable. Trade unions developed as a reaction to the adversarial style of management at the time and the only action they could take – threatening to strike – was itself adversarial.

The role of trade unions in negotiating pay and conditions, although destructive, did provide a mechanism for resolving disputes. Without trade unions, management actions were dictatorial, as they still are in some countries.

Trade unions were also created to provide a second function, that of welfare of their members. Members paid an amount to the union which was a form of private health insurance and for a pension. Workers at the time received no state pension and those off sick or on holiday received no pay. The trade union provided these services for their members in the days before the state took over these functions.

Trade unions were also active in promoting education, which otherwise was available only to the wealthy. In 1833, at the time of many reforms in Britain and the repeal of the Combination Acts, a Factories Act was passed which enforced half-time education for employees aged 9–12 years. This had to be provided by employers. The state was not involved in education until the 1880s.

Trade union structure

Trade unions evolved to meet the situation described above. Different patterns of organization evolved in different industries. These led to the creation of craft, general or industrial unions. The number of trade unions in Britain is large and many of them are very small with less than 1000 members. The number of unions is falling as industries disappear and unions merge for reasons of economic survival.

Craft unions

In the engineering industry, manual workers were apprenticed and trained in specific skills such as tool setting, machining, fitting, welding, wiring and plumbing. Workers formed trade unions comprising people with a single craft or skill. These are referred to as craft unions. They sought better pay only for their own members and were not interested in workers with different skills. In fact conflict between different craft unions occurred over such matters as which union had which members and which jobs each should do. The textile industry was one of the first to develop in the Industrial Revolution and many of its different processes were done in different towns in the north of England. This led to trade unions being formed with membership limited to a single town.

The British shipbuilding industry was renowned for its demarcation disputes – disputes about who does what. Separate employees would be used for different tasks. This was very inefficient but employees feared they would lose their jobs if other workers were given tasks they themselves were doing. Unions, while protecting the jobs of their members, were seen therefore as holding down productivity and resisting the changes that management were trying to bring in to raise productivity.

General unions

Unskilled workers such as lorry drivers, labourers and street cleaners did not fit into a craft union, so general unions evolved to cover them. This type of union therefore has members doing a wide variety of jobs in a wide range of industries in both the public and private sectors of the economy.

Industrial unions

In some other industries, particularly previously nationalized utilities, a single union evolved to cover all workers in a given industry. Coal mining, railways, shipping and power generation are examples of this type of union. In practice it is rare for there to be a single union covering all employees. Office employees are generally in a different union, if at all, from that covering manual workers in the same industry. Even among manual workers there are often two or more unions. For example, the National Union of Railwaymen (NUR, now part of the Rail, Maritime and Transport union, RMT) covers train drivers, but some drivers are in a different union. The Association of Locomotive Engineers and Footplatemen (ASLEF) existed to cover those who stoked the fire on steam locomotives. British Rail has not used steam locomotives since the 1960s but there is still a trade union for stokers.

Company unions

In Japan the culture and company structure is such that a trade union will exist for employees of a single company only. This form of union does not exist else-where and such a union would not be seen as being independent of the employer.

Trade union structure in Europe

In 1947 a British government report, the Mant Report, recommended reorganizing trade unions. The report recommended the industrial union, one per industry, as the most suitable form on the grounds that it would enable an easier resolution of industrial conflict and an avoidance of demarcation disputes and other inter-union disputes. The British government did not force unions to reorganize and the unions did not want to, so nothing happened. However, Britain, as an occupying power in Germany following World War Two, imposed this system on Germany. Sweden decided to introduce such reforms at their own initiative. The very low levels of industrial conflict in Germany and Sweden can be attributed, at least to some extent, to the existence of a single union per industry.

Trade union organization

Trade unions in Britain are formed as Friendly Societies and registered with the Trade Union Certification Office. This body ensures that the trade union rules and procedures conform with legislation. Procedures for voting and calling strikes are now controlled by legislation.

Members of a union pay a subscription which covers the costs of running the union. Separate amounts may be paid into sickness benefit and strike benefit funds. A 'political levy' may be paid to support the political activities of a union. Political activities are anything other than direct negotiations with an employer, and include lobbying Parliament for higher pay for their members and sponsoring Members of Parliament, usually Labour Members. Trade Union subscriptions can be deducted from pay and passed over to the union.

Trade unions normally have full-time officials who manage the union. These officials are elected by the members. The union may also have other employees.

The main representative of employees at the place of work is a shop steward. The shop steward is an employee of the employing organization and has a job there. Some time-off from the job would be given for the steward to engage in union activities such as discussions with management on problems as they arise, on matters such as safety and working conditions. Apart from these problem-solving activities, a more formal role occurs in negotiation with management over changes in pay and working conditions. In large organizations the time taken by stewards can be so large that they become in effect full-time stewards. They are, however, still paid by the employer. In Britain there may be one such steward per thousand employees. The employer pays for this person and does so on the grounds that it is beneficial in providing a channel of communication with employees and can lead to problems being brought to their attention and resolved at less cost than the conflict and alienation which may occur without such a channel of communication.

An analogous position occurs in the running of student unions where the students involved take a year off from their studies and become full-time sabbatical officers.

Trade unions are organized into branches managed by an elected branch committee. Several branches form a district. At this level full-time district officials are employed to manage the union, give advice to shop stewards and in some cases to participate in negotiations with employers. At national level there is usually a national executive committee chaired by the union general secretary who is in effect the chief executive officer of the union. A national conference, comprising delegates sent by branches, usually meets annually to debate issues, pass resolutions and set union policy. Figure 23.1 shows the structure of a typical trade union.

Membership of trade unions

Trade union membership was predominant among male manual workers who saw the need for and benefit in membership. Female and office workers were less likely to be members, perhaps because office employees felt that management would look after them without the need for a union. The 1960s saw a growth in membership of 'white-collar' unions for office workers, particularly in the public sector.

Membership of unions is higher in public sector, government-funded, industries because government has used its power to hold down funding and hence pay for some employees. The pay of university lecturers, for example, has fallen by more than 30 per cent relative to average earnings in Britain during the 1980s. Manual workers in nationalized industries have usually maintained their wage levels but only by taking strike action which has caused difficulty for the

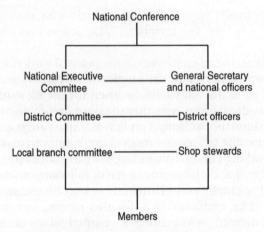

Figure 23.1   *Trade union organization structure.*

public. They were able to do this because of the essential utility nature of their industries. This gave them power to disrupt daily life by strike action in the electricity, coal and rail industries.

In Britain membership of unions has never exceeded 50 per cent of all employees but has formed the majority among manual workers in industry and a large majority of both manual and office employees in the public sector. Membership peaked in the late 1970s. With the decline of manufacturing and rising unemployment in Britain, union membership has been falling.

Membership levels are similar in France, lower in America, and higher in Germany and Sweden. The level of conflict is not positively related to union membership since the levels of conflict in Germany and Sweden are very low compared with other countries.

COLLECTIVE BARGAINING

Collective bargaining is the name given to the processes of joint consultation between an employer and its employees, represented collectively by a shop steward or full-time union official. In most private sector industries this consultation is at the level of the plant or at company level. In the engineering industry, and some others, this consultation is at national level and involves an employers' association representing the employers in the industry. In the public sector consultation is at national level since there is in fact only one employer, ultimately the government.

Collective bargaining involves several processes which may be used at different stages of resolution, or non-resolution, of an industrial dispute. Different types of agreement may result from joint consultation.

Agreements and procedures in industrial relations

Recognition agreement

A recognition agreement is an agreement between an employer and union that the union is recognized by the employer for the purpose of joint consultation. Such an agreement enables the consultation to take place. Without it there is no basis for any consultation or collective representation to occur. An employer would recognize a union if a substantial proportion of employees were in a union and the employer saw benefit in consulting with the union, representing its employees collectively, as a useful mechanism for resolving matters of conflict. Small employers and organizations in which few employees are members of a union would not recognize a union.

Problems arise when employees are in several unions, a situation known as multi-unionism. Traditionally employers have recognized several unions (more than thirty in the motor industry). It would be usual to recognize different unions for manual and office employees. However, there are difficulties for an employer if the organization has to negotiate, separately or together, with different unions.

Endless conflict would occur if a pay rise were granted to each group separately as each group would seek to catch up with other groups in what are called 'leap-frog' wage claims. For this reason employers prefer to deal with, and hence recognize, a single union. The multiplicity of unions in Britain, and the conflicts between them, has caused unions to resist a single-union agreement in which only one union is recognized.

Single-union agreements are becoming more common as unions merge and recognize benefits to themselves of having all union members at a place of work in the same union. Conflicts, of course, occur as to which union should be the single union to be recognized. Japanese companies that have set up in Britain have insisted on single-union agreements.

Where a union is recognized, employees sought in the past to enhance their bargaining power by making closed-shop agreements with employers. These agreements required that all relevant employees were members of the union and sometimes that jobs were offered only to members of the union. The latter agreement was a way of giving priority in employment to unemployed members of the union and also to ensure that new employees were suitably skilled. In the engineering industry there was a fear that less skilled people would be employed at less pay and threaten the pay of existing employees. In the performing arts the nature of the profession leads to intermittent work and periods of unemployment. Equity, the union for actors, had agreements ensuring that Equity members were offered jobs rather than non-members. By the 1990s closed-shop agreements were made illegal in Britain. Similar practices ensure that only professionally trained doctors and solicitors can act as doctors and solicitors, so this is a form of closed shop. However, this is done by different legislation in order to protect the public by ensuring that only qualified people act in these jobs.

Procedural agreements

Procedural agreements are agreements on procedures. Procedures on how a dispute is to be handled can be agreed in advance of such disputes. This should lead to a better managed resolution of the conflict. A disputes procedure may or may not include an agreement on arbitration or a no-strike agreement (details below). A redundancy procedural agreement would agree a procedure in advance for handling redundancy. It would agree a timescale of notification and a priority order in which employees would be selected for redundancy. It may or may not include an agreement that redundancies will be voluntary but not compulsory.

Substantive agreements

Substantive agreements are agreements on substantive matters such as pay, working hours and the amount of compensation for those made redundant. These agreements would set the levels of pay and hours of work of employees covered by the agreement.

Clearly these are matters about which conflict is likely to occur. Making an agreement through collective bargaining and negotiation is a means of resolving these matters of conflict.

Resolving disputes

Conflicts in industry are seen by both sides as a win–lose situation. Each side will attempt to obtain a result favourable to themselves. To do this they will use what bargaining power is available to them. Being represented collectively is one means of enhancing the bargaining power of employees. Employers also have bargaining power and can resist claims for higher pay in a recession or if their levels of pay are reasonable compared with other employers. Employers have power in matters of redundancy where they can dismiss employees, so long as the correct contractual procedures are followed.

Going on strike, or threatening to do so, is a further means of applying bargaining pressure with a view to obtaining a more favourable result for employees. Strikes result in a loss of pay for employees in the short term, which is accepted as a means of obtaining higher pay in the longer term. The loss of production causes losses to the employer and may lead to pay being raised as a means of resuming production. The result is lose–lose: employees are taking actions to achieve their objectives in ways that are destructive. A strike over pay lessens the ability of the employer to afford higher pay. Employers, too, can take action. They can threaten to or actually 'lock out' their employees and not pay them. This was done in the past as a means of forcing employees to accept changes or reductions in wages. Such practices do not occur these days.

While conflicts of this nature cannot be eliminated, it is clear that a proper management of industrial relations should lead to a reduction of conflict and an easier resolution of such conflict. The general theme of this book is the creation of win–win situations. This can be achieved in many areas of human resource management. An industrial relations strategy is needed to minimize conflict and maximize benefits for both the organization and its employees. In the area of safety, for example, there should be a common objective between management and employees, so there can be consultation and agreement. In matters such as pay, there are conflicting objectives, so negotiation is needed to resolve such conflicts. These approaches are developed in Chapter 25.

Failure to agree

In the event of a failure by management and employees to agree after negotiation, employees may consider taking strike action. Also employers may consider refusing their employees entry to the place of work (known as a 'lockout') or dismissing them. To avoid such drastic actions, alternative procedures may be used to resolve a dispute. These include conciliation, mediation, arbitration and no-strike agreements.

Conciliation

Conciliation involves bringing in an independent consultant as an advisor who will try to bring the two sides together and get them to an agreement that they cannot get themselves to because of the history of conflict and mistrust. The role of the consultant is that of a catalyst. The process is similar to that of a marriage guidance counsellor.

In Britain the Advisory, Conciliation and Arbitration Service (ACAS) is an independent but government-funded body of industrial relations advisors available to assist in the resolution of disputes.

Mediation

Mediation is similar to conciliation but goes further in that the consultant suggests a solution, perhaps a level of pay which is in between that demanded and that offered. Mediation would be used if conciliation did not succeed.

Arbitration

Arbitration goes further than mediation in that the arbitrator imposes a solution. This is the end of the road since the solution is imposed. Arbitration requires that both sides agree in advance that they will accept the decision of the arbitrator. Naturally the party to the conflict which believes that it has a strong case will favour going to arbitration in the belief that the arbitrator will impose a solution that is near to their own offer or demand. The other party will resist arbitration. Sometimes it is the employer who resists arbitration, sometimes it is the union.

Employers generally do not favour the idea that an external consultant will, in effect, determine the level of pay. The government, as employer in the public sector, does not favour arbitration since the level of pay affects public expenditure. However, Independent Pay Review bodies are sometimes established to set the level of pay for the military, police, firefighters, judges, doctors and Members of Parliament and other groups in 'essential' industries for whom going on strike is deemed inappropriate. This is a form of arbitration. In this case arbitration is used to prevent a dispute rather than resolving one after a strike has occurred.

No-strike agreement

A no-strike agreement is a procedural agreement which could form part of a dispute agreement. It is of recent origin, being favoured by Japanese employers in Britain. Employees agree to not take strike action. Since strike action was seen in the past as a necessary part of the bargaining power of a union, unions may be reluctant to agree to this. However, as a part of a no-strike agreement there would be an agreement on how to resolve disputes. This would include negotiation and would probably include an agreement to use arbitration in the event of a failure to agree by negotiation.

In this way a resolution of the dispute is assured, so strike action becomes unnecessary. Since both parties would have agreed in advance to go to arbitration if necessary, they will ensure that their demand or offer is reasonable since an unreasonable view would not be accepted by an arbitrator. Such procedures do a lot to remove fear and mistrust, and significantly improve the climate of industrial relations. In such a climate attitudes are more positive and productivity is likely to rise. The rise in productivity enables the organization to afford higher pay for its employees.

Although arbitration is a part of a no-strike agreement, it is rarely used. The better climate of industrial relations tends to moderate demands on both sides, enabling a satisfactory agreement to be made without resort to arbitration.

Strikes and other actions by employees

If all procedures fail to resolve a dispute and there is no no-strike agreement, then employees may take strike action. In going on strike they withdraw their labour, i.e. they refuse to work. It is accepted legally that in so doing they are not resigning but temporarily refusing to work. They remain employees but of course do not get paid. The purpose of strike action is to put pressure on employers as a strike causes employers to lose money from loss of output.

The decision to go on strike is taken by a vote by members. Such decisions used to be taken by a show of hands at a union meeting, but in Britain recent legislation has regulated the procedures by which strikes may be called. Strikes are still legal but procedures must be followed. The right of employees to withdraw their labour is accepted in civilized countries. British legislation now requires a postal ballot of members before a strike can be called. This prevents the instantaneous calling of a strike by a shop steward (a so-called 'wildcat' strike), ensures a time delay for reflection and that a majority of the membership agree with strike action before it takes place.

Picketing

Since only trade union members go on strike, and in many cases these will be only part of the total workforce, strikers will stand outside the factory gate to persuade non-strikers not to go into work. This process is referred to as picketing. Picketing is supposed to be a process of peaceful persuasion but the emotion of the situation has led in the past to conflict and fighting on the picket line. Pickets would lie down in front of lorries trying to deliver materials to the factory in an attempt to prevent the factory from working. Police would become involved because of the disorder and threat of violence. While the police were neutral in the dispute, they would be seen as acting in the interests of the employer when they took action against strikers.

Sometimes pickets would travel to the factories of suppliers and attempt to close them. This was referred to as secondary picketing since it affected companies not directly involved in the dispute. By the 1990s secondary picketing was made illegal in Britain and the number of pickets limited to six. Pickets have a right to hold placards and to peacefully persuade non-strikers not to work, but must not physically prevent anyone from entering the factory.

Go-slow

There are other actions which employees can take apart from going on strike. A 'go-slow' involves working, but only very slowly. Since they are working, employees are paid, but output falls. This puts pressure on employers without employees losing pay. A point may arise at which the employer decides that the failure to work normally is a breach of the contract of employment which could lead to dismissal. An employee can be disciplined and dismissed for refusing to work or to comply with an instruction. However, it is difficult to determine in law at which point this occurs, so such actions are not easily resolved.

Work-to-rule

A further action is a 'work-to-rule'. In this situation employees closely follow their work instructions. In many cases these written procedures are so rigid that if followed closely, work is done very inefficiently. Only by ignoring some procedures is work done efficiently. Working to rule therefore leads to a loss of output without employees losing pay. Employees are not in breach of their contracts of employment since they are working as instructed. They may be interpreting their instructions in such a way that no work gets done. Job descriptions and work instructions should be written in such a way that following them does not lead to this type of problem.

Lost time due to strikes and other causes

A strike is the result of a failure to agree. It is a lose–lose result since both sides lose. Not only do employers and employees lose but the nation loses too. The employer's output and profit is reduced, employees earnings are lost so they buy less goods and pay less tax. The whole economy is affected.

It used to be said by the press that going on strike was the 'English disease', implying that lost time due to strikes was worse in Britain than elsewhere. In fact this was never the case. The rate of lost time per employee has always been higher in the United States, Canada, Australia, Italy, Ireland and India. Lost time in Germany and Scandinavia has usually been less than in Britain.

Table 23.1 shows figures for lost time due to strikes for 1975–79 and 1992. Lost time due to strikes has fallen because of economic recession, which reduces employee bargaining power. These figures must be viewed with caution since they do vary from year to year and definitions vary from one country to another.

Table 23.1 Lost time due to strikes

Country	Lost time/employee/year	
	1975–79	1992
India	1.4 days	2.0 days
Italy	1.8 days	NA
Ireland	1.6 days	1.0 day
Spain	2.3 days	4 hours
Canada	2.0 days	3.6 hours
Australia	1.2 days	2.8 hours
South Korea	NA	2.0 hours
Norway	34 min	1.6 hours
Japan	68 min	NA
USA	1.0 day	50 min
Denmark	76 min	36 min
Germany	44 min	34 min
France	3.0 hours	30 min
Netherlands	18 min	12 min
Britain	1.0 day	10 min

There are other causes of lost time at work than that due to strikes. Lost time due to accidents, illness and absenteeism is much higher than that due to strikes yet receives less media attention. Accidents at work lose an average of 1.4 days per employee per year, while accidents at home and on the road lose a further 0.7 days. Illness causes an average loss of nine days per employee per year, but this does vary between industries. The time lost due to particular illnesses is shown in Table 23.2. Absenteeism is said to cost British employers £9 billion per

Table 23.2 Lost time due to illness and accidents

Cause	Days lost/year (millions)
Absenteeism	300
Stress	131
Dysmenorrhoea	75
Headaches	64
Smoking	50
Road accidents	50
Ulcers	45
Rheumatism	44
Work accidents	30
Lung cancer	17
Backache	15
Home accidents	15
Tooth decay	12
Food poisoning	8

year, i.e. £1 billion for each of the average number of days off. Lost time data is not systematically recorded. The data presented here comes from a variety of sources at different times and so must be treated with caution.

Absenteeism may be due to a lack of interest in the job or a perceived need to take time off to deal with family matters. Employees were reluctant to take holiday time for such purposes when holidays were limited to two weeks, but in modern times many employees get four or more weeks holiday per year, so it could be expected that they would take holiday to deal with such matters. High levels of absenteeism are a symptom of poor industrial relations and an unmotivated workforce. Absenteeism can significantly affect productivity and profit. A modern approach to human resource management, developed in Chapter 25, will reduce absenteeism by involving employees more within the organization and gaining their motivation and commitment.

Productivity and technology agreements

Much of the action of trade unions is seen to be negative, a reaction to a neanderthal style of management. Nevertheless there is scope for a more positive involvement of employees and trade unions within organizations. Since safety is, or should be, a common objective, trade unions can be included on safety committees involved in improving safety at work. Safety Committees are a requirement in Britain under the 1974 Health and Safety at Work Act.

Productivity can be a common objective if suitably measured at group or factory level. Where this is the case, significant improvements in productivity can be achieved by the involvement of employees. A productivity agreement is a substantive agreement on working methods, staffing levels and pay. It is a framework for agreeing to an improvement in productivity for mutual benefit.

Productivity agreements avoid the conflict which prevents the achievement of increasing productivity. They also link pay to productivity and so ensure that pay rises are financed by productivity and are not inflationary.

The introduction of new technology, or any other form of change, may be resisted by employees if they believe their jobs are threatened. Little has changed, it seems, since the days of the Luddites referred to above. In a climate of hostility employees may use the introduction of change as a bargaining lever, agreeing to the change in exchange for a pay rise.

Trade union members may resist change in order to preserve their jobs, but if productivity is held down, so is the ability of the organization to pay wages. This can lead to job losses. If there is no opportunity to increase sales, then an increase in productivity will cause a loss of jobs. This can be avoided by the creation of new products and new markets so that sales can grow. With larger sales from productivity, productivity can rise without causing job losses. This is a 'win–win' strategy and shows that productivity is a common objective.

With a productivity agreement, employees will more readily accept change – indeed they may welcome it and even pressure management to introduce more changes. They will accept some loss of jobs since they will benefit from the resulting increase in productivity.

The introduction of change can be of mutual benefit. It can enable employees to achieve greater output with less physical effort. It increases productivity by employees working smarter, whereas in the past employers were seen to be forcing their staff to work harder to increase productivity and profit. Where employees are involved and can be assured that they will benefit from the change, they will accept change. It is up to management to sell the change to employees rather than force it on them.

Wage Councils

In some industries such as agriculture, hairdressing, retail selling, home working in textiles and much of the toy industry, employees work in very small organizations and do not meet colleagues in the same industry with other employers. In such situations of fragmentation it is not possible for employees to form trade unions. There is a possibility that such employees will be exploited and be paid wages much lower than employees with a similar skill level in other industries. In Britain in 1909 Trade Boards, later called Wage Councils, were established to set a minimum wage for such employees. Similar systems exist in other countries. In Britain in the early 1990s the minimum wage set by Wage Councils was around £3.00 per hour for an adult doing a full-time 40-hour week. This equates to a salary of £6000 per year, a similar level to that of junior clerical employees in public sector organizations. This was about 33 per cent of average earnings or 50 per cent of average manual earnings. In 1993 Wage Councils were abolished in Britain. The concept of a minimum wage is that it prevents employees falling below the poverty level and prevents exploitation of vulnerable employees in 'sweat shops'. It also ensures that people in work get more money than they would if they were unemployed and so avoids the reluctance to work which would otherwise occur among the unemployed.

If employees' wages are below the poverty level, then, in developed nations, social security payments would be made to supplement the wage. This would be a government subsidy of wages of private sector organizations.

A minimum wage exists in all countries of the European Union except Britain.

MODERN PRACTICE IN INDUSTRIAL RELATIONS

The growth of competition from East Asia and the recession in Europe and other parts of the world has changed the economic climate significantly. This has had a major effect on the bargaining power of employees and hence on industrial relations. There has been an increasing recognition that people are important to the achievement of productivity and that traditional styles of management have led to conflict rather than to productivity and profit.

Employees have also recognized that there are limits to what can be achieved by going on strike and that such action reduces the ability of an employer to pay higher wages. There is now a recognition that there are mutual benefits in working together to resolve problems rather than shop stewards 'thumping on the desk' for a pay rise and managers trying to hold wages down.

A modern approach to the management of industrial relations is based on a systems view of the organization and the plurality of objectives of people within it. It works to create a win–win situation in the relationship between managers and other employees. The role of management is no longer to force employees to work harder as a means of raising profit: that approach has been shown by history to be ineffective. The modern role of management is to create the climate within which employees want to work effectively. Higher productivity will result, and from it both higher profits and higher wages.

The emphasis is now on preventing conflicts rather than dealing with them when they occur. The damage due to strikes extends to customer confidence and can affect future sales. In the past the actions of both management and trade unions were reactive; they are now becoming proactive for both parties. This requires a positive industrial relations strategy.

By changing the culture and style of management an organization can gain the motivation and commitment of its employees. Attitudes will change, conflict will reduce and productivity will rise substantially. This approach to human resource management is developed in Chapters 24 and 25.

In some organizations it may be the view that with good industrial relations, trade unions are unnecessary. This view has led to a few cases of 'de-recognition' of unions. In most organizations, however, trade unions are seen to provide a useful channel for communication, procedures for preventing conflict and a structure for managing industrial relations.

Task sheet 23   Industrial relations

Read about a recent industrial dispute.

23.1   What is the basis of the claim by employees?

23.2   What is the basis of the claim by the employer?

23.3   Analyze the progress of the dispute and the process used to resolve it.

You are a shop steward at a mail order company which employs 300 people in processing orders and in packing and dispatching them. Most employees are unskilled and female. You have just been told that the company is moving to a new site 10 km away and introducing modern computer information systems.

Some of your friends are concerned about these changes and ask you to arrange a meeting of the union which you will chair.

23.4   How will you structure the meeting?

23.5   List the issues which are likely to be raised.

23.6   What strategy should the union adopt to the changes?

23.7   What issues will you raise with management?

CHAPTER 24

MOTIVATION AND BEHAVIOUR

MOTIVATION AND BEHAVIOUR

Human behaviour

Organizations are complex sociotechnical systems in which the behaviour of the system depends significantly on the people within the organization. The role of management is to guide the organization towards the objectives set for it in terms of productivity, profit or cost-effectiveness. This can only be done if management understand something of human behaviour since this affects themselves as managers and the other employees in the organization. It is increasingly being recognized that in order to be successful, organizations must fully harness the skills and talents of their employees and create a climate in which all employees are motivated and committed to the organization. In this situation productivity will be high. Most of the problems of low productivity are due to poor motivation and low commitment of people at work.

Many people have tried to explain human behaviour. However, human behaviour is too complex to be explained by theories. Individuals vary in their behaviour and in their response to situations. The way people behave in groups is quite different from their behaviour as individuals. Nevertheless, in an attempt to achieve some partial understanding, many theories have been developed which explain some aspects of human behaviour. These theories cannot be applied with the accuracy or precision of a scientific theory, such as gravity, but they do form a framework for understanding how people generally behave in certain situations. As long as this limitation is accepted, then it is useful to consider some theories of motivation and behaviour which are relevant to understanding the behaviour of people in organizations.

Chapter 4 considered some aspects and theories of organization; this chapter examines the human aspects of organization. Within the scope of this book aspects of human behaviour in organizations are addressed only at an introductory level; a more detailed treatment is left for a more advanced study of the subject. This chapter covers group behaviour, the motivation of individuals and the style of management – i.e. the style of relationship between managers and non-managerial employees and organizational culture. Chapter 25 uses this understanding of human behaviour and develops a modern approach to the design of management systems in the area of human resources. It is conventional and relevant to consider these by referring to the name of the originating authors. The bibliography at the end of this book provides references to the original works of these authors.

GROUP BEHAVIOUR

Humans are not solitary animals: they live, work and play in groups. Modern management is based far more than in the past on people working in groups. Effectiveness of groups and of their members is crucial for the effectiveness of the organization. It is important, therefore, to understand how people behave in groups. Large amounts have been written on group behaviour and group dynamics, most of which is left to a further study of the subject. Here only the basic concepts are introduced.

Group development

Several stages of development of groups have been identified by Tuckman.

Forming

When a group is first set up, people allocated to it will be unfamiliar with each other and perhaps unclear as to the objectives of the group and their role within it. The formation stage will attempt to clarify these points, and relationships and procedures will begin to be developed.

Storming

As procedures are being developed and people get a clearer view of objectives and roles, conflicts will occur over objectives and procedures. Some individuals will strive for or attempt to impose leadership on the group. As people gain confidence they become more assertive. These conflicts can, however, be constructive in that they lead to workable and acceptable procedures that enable the task to be done. At this stage some individuals in voluntary groups may decide to leave the group if it does not proceed in a way that is acceptable to them.

Norming

As procedures are agreed, norms develop. Norms are shared views and beliefs and expectations about forms and standards of behaviour, in this case of members in the group. Norms may develop for the pace of work, the relationships between members and even for dress and language. Group cohesion and harmony replace the conflict of the storming stage.

Performing

Only after a group has been through the previous stages will it be able to perform affectively. Mutual trust and co-operation have replaced fear and hostility. Group members have become motivated through involvement and committed to the group and its objectives. These are prerequisites for effective individual and group performance.

Dorming

This stage is not in the original work but has been identified later as something to avoid. There is the danger that once the performing stage is reached, the further stage of maintaining commitment will be neglected. If this happens performance will fall. The group is 'dorming' or 'falling asleep'. This shows the need to involve people continually in decisions about what they do and how they do it.

Groupthink

This too is something to be avoided. Janis has identified a behaviour in group decision making which is called 'groupthink'. A situation can occur in which the group becomes so impressed by its own perception of its abilities that it makes disastrous decisions. Feelings of infallibility can cause a rejection of information that does not reinforce preconceived views. Individuals acquiesce, believing that the decision is right, and are amazed when it is found later to be completely wrong.

Roles in groups

Benne and Sheats did some early work in the 1940s on the different roles that people play in groups. They identified twenty seven different roles which are too detailed to include here. In the 1980s Belbin simplified this to a set of nine roles listed below.

- *Co-ordinator* goal focused, manages interactions.
- *Shaper* task focused, pushes the group.
- *Plant* innovative, original and creative thinker.
- *Investigator* collects information, negotiates.
- *Monitor-evaluator* analyzes and evaluates.
- *Team worker* creates harmony and team spirit.
- *Completer* checks details, ensures completion.
- *Implementer* does basic tasks and gets others to.
- *Technical specialist* uses expert knowledge.

Individuals may adopt more than one role and switch between them. Their personality may direct them to develop a particular role in which they feel happiest.

The chairperson in some groups is appointed, in others elected, but in many informal groups a chairperson may emerge by a dominant personality or by acceptance of a person who is seen to be a good organizer.

Involvement of members in groups

Elton May – Australian-born American psychologist of the 1920s

Mayo did some early work on group behaviour and showed how people behave in groups and how their behaviour and motivation are affected by their membership of a group. Mayo did some of his work at the Hawthorne Factory of Western Electric in Chicago. He observed small groups of assemblers, changed their working conditions and observed the effect in terms of productivity. He introduced different patterns of work and rest periods, introduced 'music-while-you-work' and improved lighting levels. He found that after every change there was an improvement in productivity – even when the conditions were changed back to the original conditions. This effect, of a transient improvement, is still referred to today as the 'Hawthorne Effect'.

Mayo reached the following conclusions:

- People behave differently when they are being observed.
- The increase in productivity was due to employees having attention paid to them and being involved.
- The increase in productivity resulting from a change was transient.
- People behave differently in groups to the way they behave as individuals.
- Groups developed norms of behaviour to which members conformed.
- Attachment to group norms was stronger than to financial incentive systems offered by management.

Mayo found that even where a financial incentive scheme was in operation, employees would reduce their output, rather than, as expected, increasing their output in order to receive a financial bonus. What actually happened was that the group set themselves an agreed level of output which was at a comfortable pace, without exerting themselves. Although individuals could easily exceed this level of output, they would limit their effort and achieve only the agreed output. They wanted to be a member of the group and to be accepted by the group. If their actions went against group norms they might be rejected by the group.

The implication of this early work is that managers must understand employees as people. The prevalent view at the time was that manual workers were there to do what they were told to do, to obey the commands of management. This reflected the class-based autocratic society of the time. Mayo showed that employees were people with thoughts who organized their work in ways that suited themselves and not always in ways that management wanted and expected them to do.

Modern manufacturing organizations use groups and teams for problem solving, for projects, and for design and marketing activities. It is important to understand group working and to manage groups so that they are effective.

Human aspects of managing change

Problems of managing change have already been referred to in Chapter 14. Attitudes to change had been mentioned in Chapter 23. The focus here is on understanding the human aspects of change.

Kurt Lewin – German-born American sociologist of the 1930s

Lewin did some early work on the dynamics of group behaviour. In the area of managing change he analyzed the 'forces' in favour and against change and identified the natural reluctance of people to change. He advocated four stages in the implementation of change:

1. Identify the forces for and against the change.
2. Unfreeze the attitudes (overcome resistance).
3. Implement the change.
4. Refreeze the attitudes (reinforce acceptance).

Clearly managers must 'sell' the advantages of the change. In this way a common objective is established and the 'for' forces are strengthened. Resistance must be understood and ways found to minimize any adverse effects, reducing but not ignoring the 'against' forces.

Rosbeth Moss Kanter – American sociologist of the 1980s

Kanter writes generally about the changes needed in organizations in the changing economic situations. She concludes that change is going to be the norm and that organizations need to be 'lean', shallow, flexible and responsive. Traditional forms of management are seen to be bureaucratic and prevent the organization from being effective.

All staff are seen as a source of value (in contrast to the traditional view which sees staff as a non-value-adding overhead). All employees are empowered by delegating decisions and creating autonomous work groups. The role of management is supportive and facilitative rather than authoritarian.

Relationships between groups within the organization, and between the organization and its suppliers and customers, are based of them being PALs. PAL stands for:

Pool resources
Ally to exploit opportunity
Link systems in partnerships

Kanter goes on to identify the new skills needed by managers in managing change, which is the 'normal' situation for now and the future. These are based on a supportive role and participative style of management.

MOTIVATION

Human motivation

As indicated above, the behaviour of people is too complex to be described by theories. Many writers have written about the motivation of individuals, though only a brief selection is referred to here – for a more detailed study, specialist books in human resource management and behaviourial science should be consulted.

This book on management is concerned with the motivation of people at work. It must be recognized, though, that people are motivated in all areas of their lives. Work is only a small part of a person's life. Some people are far more motivated by their involvement in sport and other social activities, such that motivation at work may not be so important to them. However, it is part of the role of management to consider the motivation of people at work since this significantly affects the effectiveness and productivity of the organization.

It is generally accepted that all people are inherently motivated, but that in some situations they become alienated or demotivated. Management can do a lot to create the conditions in which motivation will be enhanced. Poor management system design will lead to alienation, i.e. negative motivation. The main interface between managers and other employees is the style of management, which is referred to later in this chapter. This has a major effect on attitudes and motivation. The role of management, therefore, is not to create motivation but to design management systems to liberate the motivation that is inherent in everyone. This requires an understanding of human motivation and the design of management systems that take cognisance of the attitudes that such systems generate. This approach to management is developed in Chapter 25.

Abraham Maslow – American psychologist of the 1940s

Maslow analyzed human needs. He suggested that there were five distinct types of need and that these were related hierarchically. This work is of value in marketing in understanding why people buy things – they do so to satisfy their needs. Figure 5.3 in Chapter 5 introduced hierarchy of needs. Here Maslow's work is related to another aspect of human behaviour – motivation.

Maslow's hierarchy of needs

5.   Self-actualization – fulfilment, realization of potential.
4.   Esteem – prestige, recognition and respect by others.

3. Social – belonging, affection, love.
2. Safety – safety, security, stability, lack of threats.
1. Survival – air, food and drink, shelter, sleep, sex.

These needs are numbered from the bottom up as level 1 is seen to be the most basic level of human need. Levels 1 and 2 can be seen to be physiological needs, without which a human being would die. These are important in societies where these basic needs are not guaranteed.

Level 3 are sociological needs, the need of people to be with other people and to feel that they belong to a group, which may be a nation, a town, a religion, a political party, a football team or an employing organization. This aspect correlates with Mayo's work on group behaviour.

Levels 4 and 5 are psychological needs and relate to feelings and attitudes. The sociological and psychological needs are referred to as the higher-level needs.

Having identified these human needs, Maslow went on to relate them to motivation. He concluded that motivation arose as people strove to satisfy their needs. In doing so he explained the source of human motivation. The hierarchical relationship between different types of needs arises because people in different situations attach greatest importance to different types of need.

In a developing society, or one suffering from war or some other catastrophe, food may be difficult to obtain and shelter may not be available. In these situations people's main concern is to get enough food. Hence they are highly motivated to obtain it and in extreme circumstances may kill other people to get it. In order to get money to get food, people will accept work for low pay and in unsafe conditions. Their only interest is food, or money for food.

In developed societies, however, food and shelter can, by most people at least, be taken for granted. That need is therefore already satisfied. In these circumstances people become more interested in the next level of need – safety. They press for better safety and to achieve this they press governments to pass laws to improve safety of products and safety on the roads and at work.

In a similar manner, as each level of need is satisfied, people become more interested in the next higher level of need. Motivation is therefore seen to be due to people striving to achieve the next higher level of unsatisfied need.

The relevance of this work for management is that it shows what is necessary for employees to become motivated. In traditional organizations decisions are centralized, management operate autocratically and jobs are broken down into small, repetitive tasks. Employees are not in any way involved – management make it clear that employees are only there to do what they are told. Employees get a clear message that they must switch off their brains as they walk into the factory. In this situation jobs are dull, boring, monotonous and many involve hard physical work. Management keep telling them to work harder. In this climate there is no motivation because there is no opportunity for motivation. In order to create a climate in which motivation can develop it is necessary to change the organization and decision structures and the style of management so that all employees are involved in decisions. Jobs must be designed which are interesting and challenging. It can be seen that through job design and involvement, employees have the opportunity to satisfy some of their higher-level needs. This creates motivation.

Frederick Herzberg – American psychologist of the 1950s

Herzberg asked people at work what it was that satisfied and dissatisfied them about their work. It might be assumed that money would be a key factor that motivated people at work. The results of Herzberg's work, however, show that psychological factors were more important as satisfiers. These can be seen to derive from the content of the job itself if it is a challenging job. Factors deriving from the context, or organizational environment, of the job were seen as dissatisfiers. Money comes into this category. Dissatisfaction factors do not mean that the factor causes dissatisfaction but rather that its absence causes dissatisfaction. It also means that the provision of such a factor does not cause satisfaction. With money, for example, it means that more money does not increase satisfaction, but inadequate money is a cause of dissatisfaction. Herzberg's satisfiers and dissatisfiers are shown in Figure 24.1.

Herzberg's view correlates with Maslow's. They both suggest that, contrary to popular belief, money is not a primary motivator, but that motivation derives from striving to satisfy higher level needs.

Motivation and productivity

Figure 1.2 in Chapter 1 showed that motivation of employees is a path to productivity. This can now be explained in more detail.

Several writers have sought to relate productivity to motivation. Most of these theories are based on the concept of expectancy and suggest that people will only behave in a particular way if there is a reasonable expectation that such behaviour will lead to an outcome that is favourable to them and provides them with some satisfaction. Satisfaction at work derives primarily from the job and

Satisfiers (motivators)	Dissatisfiers (hygiene factors)
Achievement	Administration
Recognition	Supervision
Advancement	Salary
Responsibility	Relationships
Work contact	Working conditions
Intrinsic factors	Extrinsic factors

Figure 24.1 *Herzberg's satisfiers.*

the activities done as part of the job. The term 'job satisfaction' refers to the satisfaction that people at work derive from their jobs.

Management seek to achieve high productivity within the organization and this includes high productivity of its employees. Traditional management approaches to productivity were based on forcing workers to work harder. This approach is, however, adversarial and generates conflicts which prevent the achievement of productivity. A modern approach is based on working smarter. Chapter 11 referred to the design of work as a means of increasing efficiency and reducing unnecessary activities. This chapter shows the importance of motivation in achieving productivity, and Chapter 25 develops a modern approach to human resource management. All of these approaches are based on increasing the effectiveness of the organization by creating a suitable organizational climate.

Clearly motivation, job satisfaction and productivity are linked, but how are they linked? Which is the cause and which is the result?

Lawler and Porter – American social psychologists of the 1960s

Lawler and Porter developed an expectancy model that related motivation to productivity. In this model, shown in Figure 24.2, people need to be able to make a link or 'path' between their activity and the expectancy that the activity will lead to a favourable outcome. This is known as a path–goal relationship and is seen to be the basis of the link between motivation and productivity.

A circular model of motivation and productivity

Many of the theories or models relating motivation to productivity imply a linear relationship, starting with motivation and ending with productivity. It seems more likely that the relationship is circular, involving feedback. The feeling of pride,

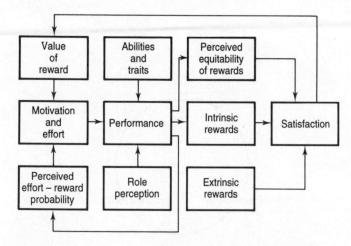

Figure 24.2 *The relationship between motivation and productivity.*

recognition and success that derives from doing a job well and achieving high productivity can be seen to lead to job satisfaction and to motivation. This completes the loop in the original expectancy model in which motivation leads to productivity. This circular relationship is shown in Figure 24.3.

None of these theories or models pretend to be a complete explanation of motivation, but they form a consistent view on which it is possible to develop management systems that enable employees to develop high motivation and thus create high productivity.

MANAGEMENT STYLE

Leadership

Style of management is a term which refers to the nature of the relationship between managers and non-managerial employees. It includes not only personal relationships between people but also the style of communication and the attitudes that managers have towards employees and the attitudes generated in employees.

The term 'leadership' is sometimes used. This refers to the ways in which managers achieve the attitudes and actions of their employees. Usually the actions desired are those which lead to the achievement of organizational objectives. A form of leadership therefore implies a style of management. Some of the styles of management explained below are sometimes referred to as styles of leadership.

The style of management is the whole set of relationships between managers and employees by which a manager leads the team of employees. The different styles of management are sometimes classified within different schools of management.

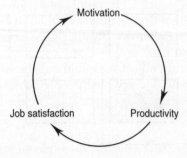

Figure 24.3 *Job satisfaction, motivation and productivity.*

The classical or scientific management school

Adam Smith – Professor of Moral Philosophy at Glasgow in the 1770s

Adam Smith was referred to in Chapter 3 because he saw organizations and economies as systems. In looking at the economics of industrial production he advocated a division of labour and the organization of work methods. Division of labour means that individual employees do only one task, the total set of tasks being divided by management between them. This was done on the grounds that it would take less time to train a worker for a single task and the worker would quickly become efficient in doing it because the single task was being done repetitively. He examined the manufacture of pins and by changing the method of work was able to increase productivity by a factor of 2000.

Charles Babbage – Professor of Mathematics at Cambridge in the 1830s

Apart from inventing the first computer and cowcatcher, Babbage suggested the division of labour and the timing of tasks so that a standard time could be established as a means of forcing workers to a pace of work based on time standards.

Frederick W. Taylor – American manufacturing engineer of the 1880s

Taylor developed a 'scientific' approach to the organization of work. His approach is referred to as 'scientific management' and considered in detail the methods by which work is done. Chapter 11 considered the detail of Taylor's approach to the design of work; here his approach to management is explained. He advocated the separation of the planning and doing parts of a task (planned by managers and done by workers). This was accompanied by a fragmentation of work into small tasks on the grounds that a small task done often would be done efficiently because of practice and experience. It did, however, lead to dull, boring and repetitive tasks. In the society of the time this approach improved productivity considerably. However, in a modern society social disbenefits arise if the psychological aspects of work are ignored as they were by Taylor.

The fields of work study, ergonomics and operational research have all developed from Taylor's work. He addressed some human aspects of organizations, recognizing that organizations would only be effective if managers and employees worked together in 'brotherly co-operation'. He saw that managers wanted low labour costs and that employees wanted higher wages, and recognized that these differing objectives could only be simultaneously achieved by an increase in productivity. He saw that people achieved a maximum output if they worked at brisk but reasonable pace (now defined as a standard rate of work), but that output fell due to tiredness if employees were forced to work faster than this.

Taylor advocated the use of individual payments-by-results incentive payment systems, commonly known as piecework. He did his work in a very different society from that of today. Co-operation between managers and employees was not really possible in a society of the 1880s. Over a period of years his approach was misused as managers tried to reduce labour costs by making employees work harder. Conflict with trade unions grew, indeed trade unions grew, because of the conflict generated by the use of an autocratic style of management attributed to a use of scientific management.

Max Weber – German sociologist of the 1890s

Weber developed a model of organization based on a hierarchy of authority and control in which tasks and procedures were prescribed centrally by management. Employees therefore knew exactly what to do and how to do it. They did not have to think, only to do what they were told. Each person had a clearly defined role and set of tasks. Employees were selected on the basis of qualifications.

Weber considered this form of organization to be the most efficient way for an organization to manage its activities. He called this form of organization 'bureaucracy'. As society has changed, this form of organization was found to be inefficient and the term bureaucracy now refers to an inefficient, centralized organization. This does not, however, invalidate the view that this was an efficient form of organization in the society of that time.

Human relations school

Robert Owen – British industrialist of the 1810s

At this time Britain used slaves to develop agriculture in its colonies. Conditions for agricultural and early industrial workers in Britain were little different. Owen saw that malnourished workers working 84 hours a week were exhausted. He provided meals for his workers and reduced their working hours. He also restricted the use of young children at work. He advocated treating workers with respect and dignity, an unpopular view in society of that time. In doing so he recognized that people are important and that by treating them better they would be able to contribute more to the organization.

Douglas McGregor – American psychologist of the 1960s

McGregor looked at the motivation of employees and sought to relate it to the style of management. He categorized employee behaviour into two types – Theory X and Theory Y.

Theory X employees were lazy and unmotivated; they only worked when forced to and never worked hard. By contrast Theory Y employees seemed to like work, were eager to please and worked hard on their own initiative. Figure 24.4 shows the features of Theory X and Theory Y. Clearly it seems desirable for employees to behave according to Theory Y but in practice some employees behaved according to Theory X.

A superficial view would suggest that if an organization had Theory X

Theory X	Theory Y
1 People hate work	1 People like work
2 People have to be forced to work hard	2 People drive themselves and work effectively
3 People prefer to be told what to do	3 People will take the initiative if given the opportunity
4 People are selfish and have no interest in the organization	4 People will commit themselves to the organization if it is beneficial

Figure 24.4 *Theory X and Theory Y.*

employees, then it would have to use an autocratic style of management in order to force them to work hard. Management power seemed to be the only way to get the employees to work. If, however, an organization had Theory Y employees a more participative style of management seemed to work satisfactorily. Theory X and Y employees were seen as two different types of people by management, who thus had to base their style of management on the type of employee they had. They tried to employ only Theory Y people and blamed trade unions for causing Theory X behaviour.

McGregor rejected these views and suggested that instead of employee behaviour being the *cause* of a management style, it was the *result*. If an autocratic management style is used, Theory X behaviour will be the result. That is, if management treat their employees as idiots, the employees will behave as idiots. If employees behave according to Theory X, then a change in style of management is required in order to create a different organizational climate. If Theory Y behaviour is wanted, then a more participative style of management is needed. That is, if management treat their employees as responsible, the employees will behave responsibly.

Blake and Mouton – American behaviourial scientists of the 1960s

Blake and Mouton looked at how managers created attitudes in employees and saw that managers had concerns for their employees but also concern for the task or output of the organization. Managers were seen to differ in the proportion of concern or importance that they put into people and tasks.

A managerial grid (Figure 24.5) was developed which plotted concern for people and for tasks as the axes of a graph. A particular style of management could therefore be described at a point on the graph, referenced by its co-ordinates on each axis.

No one style was seen as always appropriate: different types of organization needed different styles of management. A social club, for example, created and existing to provide services to its members, would tend to be strong on concern for people but not be pushing them hard (1,9 on the graph). A military

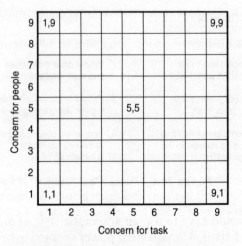

Figure 24.5 *The managerial grid.*

organization would always put the task first and foremost. This would be seen as appropriate for this type of organization (9,1 on the graph). This would also be the style used in an organization operating according to scientific management. A 1,1 style implies very little concern by management for anything, so organizational performance will be poor. A 5,5 style is adequate for an ordinary performance, but the organization would not be operating at its full potential. The implication of this view, however, is that managers need to be concerned for both people and output, so the 9,9 style is seen as the one most likely to lead to organizational effectiveness. Employees would be fully involved and committed, there would be trust between managers and employees, and productivity would be high.

Not only should a style of management be chosen according to the type of organization, but it should be chosen to suit the particular situation in a given organization. For example, the means of dealing with a disciplinary matter will differ from that used in a problem-solving situation. The style of management of routine day-to-day activities will differ from the management of project and design activities. Managers must therefore choose a management style to suit the situation.

Rensis Likert – American social psychologist of the 1960s

Many writers have identified a spectrum of styles of management, ranging from autocratic to democratic. Likert expressed this in terms of four systems or styles of management:

- System 1   Exploitive authoritative: based on centralized commands, no communication and the use of threats and coercion to ensure compliance.

- System 2 Benevolent authoritative: little communication, uses financial 'carrot' rather than 'stick'.
- System 3 Consultative: some trust and involvement, decisions are still centralized but take employees' views into account.
- System 4 Participative: complete trust, employees involved in goal setting, substantial communication and teamwork.

Likert suggested that system 4, a participative style of management, was needed if productivity was to be the result.

Amitai Etzioni – American sociologist of the 1960s

Etzioni identified the same spectrum of management styles and related these to the forms of control implicit in them and the type of response that was likely from employees. Etzioni recognized that for organizations to be effective, employees had to comply with the orders given to them; the alternative would be anarchy from which very little would result. He analyzed how managers sought to achieve compliance and how employees complied.

The autocratic style of management was seen to be based on the use of coercion as a means of control in an attempt to force employees to behave in a particular way. The response of employees to such coercion is seen to be alienative, i.e. they will do as they are told because the alternative may be unemployment, but they are not at all motivated so will only do the absolute minimum required of them to retain their jobs. Clearly productivity in such an organization will not be very high. (Military organizations operate with an autocratic style of management but this is seen as necessary in the situation in which they operate. Alienation should therefore not arise since there is an acceptance of that style of management in that form of organization.) Figure 24.6 relates the form of control and response to the management style.

At the other end of the spectrum a democratic style of management is seen to rely on treating employees as responsible adults. Control is based on trusting employees and delegating decision-making responsibility to them. This form of control relies on employees developing norms of behaviour which are conducive to organizational effectiveness and taking actions which will lead to its achievement. Employee involvement will be moral, i.e. based on intense commitment to the organization arising from a perception of the morality of such action.

Power	Control	Compliance
Coercive	Physical	Alienated
Remunerative	Financial	Calculative
Normative	Symbolic	Identative

Figure 24.6 *Management style, control and response.* Source: A. Etzioni (1964), *Modern Organizations*, Prentice Hall, Englewood Cliffs, NJ.

Tannenbaum and Schmidt – American sociologists of the 1960s

A similar spectrum of management style is seen, varying from centralized authority and decision making to delegated responsibility with considerable freedom of action by work groups.

Seven styles of management are identified as indicated by the numbers in Figure 24.7:

1. A manager makes a decision, tells the group what to do.
2. A manager makes a decision and sells it to the group.
3. A manager makes a decision and invites questions.
4. A manager announces draft decision, consults and decides.
5. A manager explains a problem, invites ideas then decides.
6. A manager explains a problem, the group decides.
7. The group identifies the problem and decides.

Some other styles of management

Management by walking about

This style of management was referred to in Chapter 10 in discussing the Japanese approach to manufacturing management. It seeks to avoid the élitism and isolation of managers who sit all day in their offices issuing orders. It ensures that managers are aware of problems on the shopfloor and sees the role of a manager as problem solver and supporter of the operators. This style of management is similar to that advocated in the 'upside down' organization referred to in Chapter 4. It is used in Japan and known there as *Gemci genbutsu*, meaning literally 'go and see'.

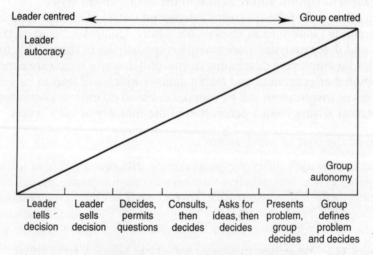

Figure 24.7 *Continuum of leader behaviour.*

Management by objectives

Management by objectives was developed in the 1950s by Humble. It is based on defining objectives for each employee and controlling performance by relating it to the objectives set. People perform well if they have clear objectives, and better if they are involved in setting the objectives. The approach is modern in that it delegates responsibility to individuals and involves them in the setting of the objectives, but has traditional aspects similar to those of Weber. Problems can occur if objectives are set for individuals in isolation. This can lead to incompatibility between the objectives of individuals in different departments who need to work together. Conflict and low productivity result. To be effective, objectives must be compatible and set within interdisciplinary task forces rather than in fragmented departments.

Mushroom management

This is a style of management to be avoided! It is based on keeping employees in the dark and shovelling abuse on them – the way that mushrooms are grown. This style of management is common in traditionally managed organizations.

Seagull management

This is similar to mushroom management and is to be avoided. In this style managers are often absent from the organization. They fly in occasionally, shout at people from a great height and fly off again quickly.

Crisis management

Crisis or firefighting management occurs with incompetent management where there are no strategies or planning. Managers respond to short-term crises and never look beyond the current day or two. This style solves immediate problems but often in ways that cause more problems, which are solved tomorrow. The result is gross inefficiency and low productivity. It is to be avoided but is widely practised.

Implications of the style of management

All of these theories are different, and examine different aspects of leadership and management style. Nevertheless they have a common theme: they all criticize the autocratic style and suggest that such a style will lead to conflict, low motivation and low productivity. They all advocate a more democratic style of management. This does not mean management by committee or making decisions by democratic voting; it means involving people in organizations in some aspects of the running of the organization. The path to productivity is seen to lie in a management style that enables people to satisfy some of their higher-level needs. By so doing, motivation and commitment are enhanced and the organization thus achieves high productivity. Chapter 25 builds on this knowledge and identifies how management systems can be designed to achieve organizational success.

Given the variety of views portrayed in these theories of leadership and management style, it can be seen that a choice of management style must be made. The only safe advice which can be given is that the style of management must 'match' the form of organization and the society in which it operates. There is no one best style. The choice of management style is, however, crucial to the success of the organization. A wrong style of management will generate conflict and alienation – and prevent the organization from achieving its objectives.

ORGANIZATIONAL CULTURE

Handy and others have identified that different organizations have different cultures. Culture is related to the style of management but is a broader concept. Culture refers to a set of shared assumptions, values and beliefs about what is right and how things should be done to set the framework within which people work. Employee actions, language and behaviour all depend on and give an indication of the type of culture in use in a particular organization.

The concept of norms of behaviour was mentioned earlier in this chapter. Norms refer to behaviour at an individual level. Culture is a similar concept but is broader and applies at a higher level of the system – the organizational level. The culture of an organization derives from the sets of norms established within it.

The types of strategy that organizations develop will depend on the culture. Different types of organization have different cultures. The culture in military, autocratic, business, educational, public sector and charitable organizations will differ.

Four types of organizational culture can be identified:

- *Power culture* – centralized, hierarchical and patriarchal.
- *Role culture* – works on rules and procedures, typical of a bureaucracy.
- *Task culture* – analytical, knowledge based, focus on getting the job done.
- *Person culture* – individualism, freedom to operate professionally.

It is important for employees to know what the culture is and for managers to develop an appropriate culture. New employees will not be aware of the culture and this may lead to people not 'fitting in' and not operating effectively. Induction training must make new employees aware of the culture. Culture is not a visible management system to be designed, but is something that evolves in an organization over a period of time and is influenced by the attitudes of its senior members. Managers must be aware of the culture and attempt to modify it if it is not appropriate.

Task sheet 24   Motivation and behaviour

Consider the motivation theories of Maslow and Herzberg.

24.1   Reflect on the extent to which these theories apply to you as an individual. List the aspects of life which you find satisfying.

24.2   If you have work experience, consider whether these theories explain your attitudes and behaviour at work. (If not, try to visualize what your attitudes would be.)

Consider McGregor's Theory X and Theory Y.

24.3   Which style of management do you see in the managers in the organization in which you work or study? What effect does this have on the attitudes and motivation of employees and the effectiveness of the organization?

Consider Handy's concept of organizational culture.

24.4   Which type of culture is used in the organization in which you work or study?

CHAPTER 25

HUMAN RESOURCE MANAGEMENT

HUMAN RESOURCE MANAGEMENT

The scope of human resource management

Human resource management is a modern term, replacing the older term 'personnel management' in order to give a broader scope. Personnel management, as was shown in Chapter 22, is concerned with the recruitment, retention and training of employees, including contracts of employment. Personnel in the past existed at a low level within the organization structure and was seen to have little more than a clerical function. The title 'personnel officer' reflected this low status, the title manager not being used.

Human resource management is a proactive, systems-based approach, based on the development of a human resource strategy for the organization. It recognizes the crucial role of employees in the performance and success of the organization and the need for professional skills in managing people. In a modern organization the higher position of this function within the organization structure reflects the increased importance attached to it.

The term 'human resources' is used to parallel the terms physical and financial resources. This is beneficial in showing that people are resources available to the organization, but can be a problem if used to infer that people are only resources, to be used in the same way as machines, there only for the benefit of the organization.

The objectives of human resource management

The objective of the human resource function of management is to advise the directors of an organization on the design of management systems which enable all employees to satisfy their aspirations at work, and through this to make a more valuable contribution to the effectiveness of the organization. This approach requires the formulation of a human resource strategy for the organization which is based on creating win–win situations. The strategy will identify and design the management systems which the organization needs to achieve high commitment of and harmonious relations with employees as a vehicle for high productivity and profit or cost-effectiveness.

Human resource management involves an understanding of relevant legislation – on employment, industrial relations and safety – together with an understanding of human motivation and the design of a range of management systems. Legislation on safety was referred to in Chapter 10, that on employment in Chapter 22 and on industrial relations in Chapter 23. This chapter builds on the concepts outlined in Chapter 24 and explains how management systems can be designed which are based on an understanding of people and are likely to lead to organizational objectives such as high productivity and profit or cost-effectiveness. The approach taken is a systems approach, based on a holistic view which sees an organization as a system, a concept explained in Chapter 3. The key management systems in the area of human resources are as follows:

- Organization structure
- Decision structure
- Information structure
- Job design
- Job grade structure
- Payment structure

The organization structure, and its related decision and communication structures, are key management systems. It was mentioned in Chapter 4 that these structures often arise by default rather than by design. Positive management action is needed to design structures that are appropriate to an organization since organizational effectiveness depends on them. The basic organization structure will be one of the forms described in Chapter 4 but within it there will be more formal and informal groups and teams. The number of levels in the organization structure is likely to be smaller now than in the past as decisions are delegated to people at lower levels in the organization.

Team building is an important management skill in modern organizations. Some teams will be interdisciplinary and cut across organizational boundaries. The greater use of project-based management to manage change places greater reliance on groups and teams of people being able to work together. Figure 1.1 in Chapter 1 showed a visual representation of an organization, and by implication of the people who work with and within it. Managing teams and the interactions between teams becomes the major task of senior managers.

Management information systems are formal systems for organizing information. It used to be said that employees should be given only the information they need to do their job. The fear being that more information meant more power for employees, which in the past was considered undesirable. The modern approach gives employees access to more information, which enables them to develop their understanding and to make better decisions. Computer-based information systems contain only 'hard', quantitative data. The information needs of managers extend beyond this to 'soft', qualitative information, including, for example, the attitudes and power of customers, suppliers and employees. Communication systems must ensure that all employees have the information they need, not only to do the job but to develop the job and do it more effectively.

The main interface between a person and the organization is the job the person does. This too must be designed if job satisfaction and productivity are to be achieved. Job design was mentioned in Chapter 11 and is further developed below.

Another interface is the pay for the job. Clearly the level of pay is likely to be a matter of conflict, but the payment system and structure are management systems which can influence attitudes and motivation and so need to be designed.

The third interface is the form and extent to which an employee participates and feels involved in the organization. It was mentioned in Chapter 3 that organizations have both multiple and plural objectives and that the behaviour of an organization resulted mainly from the behaviour of the people within the organization. Managing people at work is therefore based on minimizing the conflicts between groups of people and maximizing the common objectives between them. This is the framework within which human resource management systems must be designed if the organization is to be successful. Win–win strategies must be created in this area, as in other areas of the organization.

JOB DESIGN

The principle on which job design is based is that in order to create motivation, jobs must be interesting and challenging. This is seen to provide job satisfaction and to enable high productivity. Without proper job design, jobs are likely to be dull, boring and monotonous. The earlier approach of scientific management deliberately broke jobs down into small elements that a worker would do repetitively during the working day. This was seen to reduce training costs and was thought to lead to faster work, especially if related to an individual financial incentive. This, however, was the approach taken a hundred years ago. It worked in the class-structured society of the time when employee attitudes were not considered to be important.

With an understanding of Chapter 24 it will be seen that jobs need to be designed to be challenging. Job design was referred to in Chapter 11 in the context of the design of work; here the emphasis is on the human aspects of job design.

It could be said that the jobs of managers are interesting and challenging, but how can the jobs of manual workers be made interesting?

The concept of job design is that jobs should contain the following:

- Variety
- Responsibility
- Control

Job design should therefore ensure that all jobs contain variety, some responsibility for decision making and some control over the way the job is done.

There are several approaches to or techniques of job design which seek to provide these characteristics in jobs.

Techniques of job design

Job rotation

Job rotation involves groups of employees rotating between a set of jobs or tasks over the working day. This provides some variety to relieve boredom and monotony. Employees might rotate tasks say, every two hours during the working day. This does require more training time for a range of tasks to be learned. This approach is often taken with manual workers. The change of task may, to some extent, involve using different working postures and reduce tiredness and muscle strain that might occur when a worker does the same short-cycle task all day. However, doing two boring jobs in a day is not much better than doing one, so job rotation may not be very effective. It has been found that little improvement occurs when job rotation is imposed, but that some improvement does occur when employees themselves control the rotation system and decide themselves how and when jobs will be rotated. This is seen to work because it provides some decision responsibility as well as variety.

Job enlargement

Job enlargement involves the grouping together of small tasks into a job with a longer cycle time. This provides variety within rather than between tasks. Instead of ten employees on an assembly line, each doing a single task on every product, each employee may do ten tasks on every tenth product that comes down the line. Not only does this provide variety but it also increases the feeling of responsibility of the employee who is now responsible for a larger part of the work on a product and thus sees the job as being more important. Jobs are enlarged by a process of horizontal integration, the merging of similar tasks requiring similar skills of the level which an employee has.

Job enrichment

Job enrichment goes further than job enlargement in that tasks are integrated vertically. That is, the job now includes some tasks normally associated with a higher grade of employee such as a supervisor. This could include tasks of inspection, recording and reporting. In this way the importance of the job is enhanced. Job enrichment is likely to involve some training of employees in tasks they have not done before. They derive satisfaction from acquiring new skills.

Autonomous work groups

Autonomous work groups are groups of employees given a considerable amount of autonomy within the organization over the total set of jobs which they do. This form of job design embraces all of the concepts and techniques of job design. The group becomes self-managing, and elects a leader rather than having one imposed on them. The group decides who does what and when. This approach is used in the Volvo car company in Sweden. There is an agreement between the organization and the work groups as to the hourly output rates and the level of pay. The groups are then left to organize themselves to achieve that output. This form of organization sees a very different role for management from that in a conventional organization. Instead of centralized authority and control, the role of management is seen to be supportive and facilitative. This form of organization is sometimes referred to as the 'upside down' organization, a concept introduced in Chapter 4. It sees management at the bottom of the organization, supporting and enabling work groups to achieve higher levels of output and productivity. This is not an abdication of managerial responsibility, but an alternative to the traditional view of autocratic management. Management is still responsible for the productivity or profit of the organization, it is just a question of which style of management is more likely to achieve it.

Group working is particularly relevant to project work as used in managing innovation (Chapter 7), and introducing manufacturing change (Chapter 14). Project teams and taskforces are interdisciplinary teams of employees brought together to undertake a set of activities associated with a project. These groups have considerable delegated responsibility and are a form of autonomous workgroup. As change becomes the norm, more work will be organized as projects.

Subcontracting

At the end of the spectrum of delegated responsibility is the abolition of the job as a job within the organization and its replacement by people from outside the organization. The subcontracting of peripheral activities such as catering and of many of the activities of public sector organizations can be seen as passing complete responsibility for such activities to people who are not employees of the main organization but may be self-employed or be employees of the subcontracting organization. This is a form of autonomous workgroup; however, the group members are no longer employees of the main organization but have become suppliers to the organization.

Job design and organization

Clearly job design is related to organization structure and style of management. As management style moves from autocratic to democratic, decisions are delegated and jobs at lower levels absorb this decision-making responsibility. In such an organization groups become more self-managing and fewer managers are required in central positions. The number of managers and levels of management can be reduced in a process known as 'delayering'. This process is happening in Europe and America in the 1990s.

Spans of control can be increased since groups are largely self-managing and do not need the constant supervision associated with the situation of mistrust on which an autocratic style of management is based. More employees are given more challenging jobs, but there are fewer employees in the organization. Productivity increases because of higher motivation, so fewer managers and fewer employees are needed.

JOB EVALUATION

Job evaluation is the process of evaluating the levels of responsibility of jobs as the basis for determining the level of basic pay for a job. It is important to note that job evaluation evaluates jobs – not the people doing them. The basic pay of different jobs in an organization will differ according to the levels of responsibility of the jobs. It would be unusual for every job to be allocated an individual level of pay. It is more likely that jobs would be grouped into job grades in which jobs with a similar level of responsibility would be in the same job grade, and several job grades would cover a population of jobs in the organization. Such a structure is known as a job grade structure, which is explained later in this chapter.

The purpose, then, of job evaluation is to design a job grade structure. This structure will be used to determine the pay of jobs but should be considered as a separate structure. There may be conflict about the levels of pay, but there is a common objective between management and employees about the design of the job grade structure, so this can be designed without the conflict associated with pay. The common objective is the design of a job grade structure which is seen and accepted as being 'fair' and equitable. For this reason job evaluation and job grades are best sorted out before pay is attached since these processes can then be achieved with less conflict.

When a suitable job grade structure has been designed, levels of pay can then be attached to it in order to create a payment structure. Conflict about pay is likely to be less if pay is being attached to an accepted and agreed job grade structure rather than mixing both processes and engaging in conflict about job grades and pay together.

Data needed for job evaluation

In order to evaluate and grade a job some information about the job is required. Evaluation is based on job descriptions (referred to in Chapter 22). It is not necessary to have a different job description for every employee since typically several people may have the same job title and do the same job. Sometimes, however, people with the same job title are doing different jobs, and people with different job titles are doing the same job. It is important, therefore, to ensure that any conflicts about job titles and responsibilities are resolved since such conflicts would lead to non-acceptance of the job grade structure.

Job evaluation techniques

There are many techniques of job evaluation – some informal, others formal. Techniques can be categorized according to whether they are analytical or not and whether they start from an analysis of jobs and end with a job grade structure, or start with a job grade structure and slot jobs into it. Analytical techniques analyze separately several aspects or factors inherent in jobs, such as number of subordinates, financial responsibility or qualifications and experience needed. Non-analytical techniques assess a job as a whole and do not examine factors within a job. Table 25.1 categorizes the main techniques of job evaluation.

Table 25.1 Techniques of job evaluation

Evaluation method	Direction to the job grade structure	
	Jobs to grades	Jobs into grades
Non-analytical	Ranking	Grading
	Simple ranking	Job grading
	Paired comparison	
	Decision banding	
Analytical	Rating	Comparison
	Points rating	Factor comparison
	Job profile	

Ranking

Ranking is the simplest method. This involves a person, or more usually a small group of senior employees, examining each job and using their knowledge of the skills required to do it to place jobs in a list in rank order of responsibility and to create a job grade structure from this list. This method can work in a small business where everyone knows what is involved in each job and centralized decisions are more acceptable, but could cause a lot of conflict if used in a larger organization. This process would be very time consuming in a large organization as a committee would discuss and argue about every job and its value relative to other jobs.

Paired comparison

Paired comparison is a non-analytical, participative technique. It is based on collecting the opinions of a wide range of employees about the relative values of jobs with which they are familiar. The group does not operate as a committee since that would involve conflicts. Instead views are individual, with individuals being asked for their personal judgement. The individual judgements are collated to form a collective view. Instead of asking individuals to rank a set of jobs, they are presented with a list of pairs of jobs. A large number of simple binary decisions are made on the relative value of the two jobs in a pair. One job is seen to 'win' in comparison with the other job which 'loses'. If an individual cannot distinguish between two jobs, they 'draw'. This makes the decision process easier and quicker. The lists of pairs of jobs are organized such that between the total group every job is compared with every other job a given number of times. Table 25.2 shows a paired comparison form. A complete set of forms would compare every job with every other job. The form is completed by ticking the preferred job of the pair in each row.

Table 25.2 A paired comparison form

Job No.	Job title	Decision	Job title	Job No.
1	Skilled operator	☐	Secretary	4
4	Secretary	☐	Maintenance fitter	7
7	Maintenance fitter	☐	Filing clerk	10
10	Filing clerk	☐	Labourer	2
2	Labourer	☐	Semi-skilled operator	5
5	Semi-skilled operator	☐	Office manager	8
8	Office manager	☐	Salesperson	11
11	Salesperson	☐	Manufacturing supervisor	3
3	Manufacturing supervisor	☐	Wordprocessor operator	6
6	Wordprocessor operator	☐	Van driver	9
9	Van driver	☐	Skilled operator	1

Decisions are recorded in the form of two points for a win, one point for a draw and none for a loss. When all decisions have been made, jobs can be listed in points order. The process is similar to that of a football league. At the end of the season every team will have played every other team the same number of times. A football league table then lists teams in order of points, the team winning most often against other teams is top of the league. Similarly, jobs are listed in an order which reflects their number of 'wins' against other jobs.

The paired comparison process achieves a result without conflict since decisions are individual, not argued about in committee. Because a lot of people are involved in the process the result represents the consensus of opinion and will have a high degree of acceptability. It is clear that this process is subjective, but by involving a large number of people in the process the effects of individual bias are minimized.

Decision banding

Several techniques fit into this category. The concept is that the best single factor of relevance to the value of a job is the level of responsibility for decisions. This responsibility can be measured by the financial value of decisions taken by a jobholder or the timespan of discretion (devised by Jaques), which measures the time period during which an employee is left unsupervised and taking decisions without the need to report to a manager.

The benefit suggested for these approaches is that they are measures suitable for all jobs and thus generate a feeling of fairness that all jobs are being evaluated using the same criterion. This avoids conflicts which might otherwise occur.

With these techniques the decision factor is measured for each job and then jobs are listed (ranked) in order of the factor.

Factor comparison

Factor comparison is a complex technique which is rarely used. It assesses the financial value of each of several factors and totals these to give the pay for a job. This technique directly relates job grading to pay. This is likely to lead to conflict about the level of pay allocated to each factor of each job.

Grading

Grading is a technique in which a job grade structure is predetermined. As well as job descriptions, a grade description is needed for each grade to specify the types and levels of responsibility for a grade. Each job description is then assessed against the grade descriptions and slotted into the appropriate grade. This technique is appropriate for office jobs which are reasonably similar across a country. Employers could use this technique in order to create a job grade structure which would be similar to structures used elsewhere for similar jobs. It is also used in government organizations with offices in many different cities. A new job in a local office would be assessed and slotted into a nationally agreed grade structure. In this way similar jobs in different cities would be graded similarly.

Points rating

Points rating is an analytical technique which allocates points to jobs under several different factors. Because it uses numbers, the technique appears to be objective, scientific and precise. However, all decisions made are subjective. A job rating form is used which lists the factors to be considered, the maximum points that can be allocated under each factor and some guidance on the allocation of points within each factor. Table 25.3 shows a job rating form. The choice of factors is a subjective decision. Each job is then assessed under each factor and a number of points allocated to each job under each factor. These points are totalled to give a points score for the job. This results in a list of jobs in points order.

Qualifications are often a factor in job evaluation. It is important to note that it is the qualification *needed* to do the job which is being rated, not the qualifications of the current holder of the job.

Table 25.3   A job rating form

Job Title:					Job no.:
Department:					No. of jobs:

Factor	Extent of application of factor					Max. points
Qualification	Master	Degree	BTEC-H	A-level	5-GCSE	15
needed	NVQ5	NVQ4	NVQ3	NVQ2	NVQ1	
Points	15	12	10	5	2	
Experience	20	15	10	5	2	15
needed (years)						
Points	15	12	10	5	2	
Responsibility	2 points per digit of money value					16
for resources						
Points						
Supervision	100	50	20	10	5	20
No. supervised						
Points	20	15	10	7	3	
Complexity	Value judgement					10
Points						
Communication	Crucial	Important		Normal	Occasional	10
Points	10	7		5	2	
Decision span	5 years	Year	Month	Week	Day	10
Points	10	7	5	2	1	
Physical needs	Heavy	Hot		Moderate	Slight	4
and environment	Dirty	Cold		Discomfort		
Points	4	3		2	1	
					Total	100

Decisions are made by a committee, and so conflict can arise over the number of points to be allocated under each factor for each job. The technique is not suitable in autocratic organizations since conflict will spread to the job evaluation system if it is imposed. It can work in more democratic organizations where trust exists and if a wider group of employees are involved in the evaluation.

Job grade structure

A job grade structure is a set of several job grades across which a population of jobs is spread. Jobs in the same grade will have the same level of basic pay, and jobs in different grades will have different levels of basic pay, the differences reflecting the differing level of responsibility in each grade. All jobs in a grade will not have exactly the same responsibility but will have been evaluated as having similar responsibilities. Where jobs have been listed in points order, jobs with similar points will be allocated to the same job grade. This is done by deciding the number of grades and drawing a set of lines at positions down the list which represent grade boundaries. Table 25.4 shows a typical job grade structure.

Table 25.4 A job grade structure

Grade	Job title
9	Managing director
8	Operations director Marketing director Finance director
7	Manufacturing director Purchasing director Research director Quality director Personnel director
6	Departmental manager
5	Senior engineer Accountant Systems analyst Section leader
4	Design engineer Manufacturing engineer Computer programmer Market analyst
3	Skilled operator Maintenance fitter CAD operator Director's secretary Salesperson
2	Semi-skilled operator Secretary Wordprocessor operator Accounts clerk
1	Labourer Clerk Receptionist

The number of job grades will depend on the size of the organization and whether the grade structure covers all employees or only a particular group. It is unlikely that a job grade structure will have less than four or five grades, and it should not have more than nine or ten grades even for the largest organizations. The number of grades should be kept to a minimum so that when pay is attached a significant increase occurs on being promoted to a higher grade. A larger number of grades will cause such pay increases to be much smaller. A lot of small differences of pay causes a lot of conflict, so a smaller number of job grades is preferred.

The minimum number of grades, however, is the number of levels in the organization structure – otherwise a person could be in the same job and pay grade as their manager. If the number of grades seems too high, the solution may lie in reducing the number of levels in the organization structure rather than in increasing the number of job grades. Grade boundaries should be positioned by examining clusters of jobs with similar points values and placing boundaries where points differences are greater.

The largest organizations in the world, the military forces of major nations, employ millions of people and do so within a structure of no more than ten grades. Commercial and government organizations can have fifty or more grades.

PAYMENT STRUCTURE

The payment structure is the structure created by attaching pay levels to the job grade structure. By looking at the intended ratio of top to bottom pay across the structure it is possible to estimate the level of pay appropriate to each grade by making a constant percentage pay increment between grades.

In some organizations a single level of pay will be attached to a job grade and all employees in that grade will get the same pay. This is based on the concept of the 'rate for the job': everyone doing that job should get the same pay. This approach was common in the past for manual workers. Pay would be expressed as a rate per hour and employees would only be paid for the hours they worked. This form of pay is a wage. There is a modern tendency to pay all employees on the same basis, i.e. a salary instead of a wage. A salary is paid to an employee irrespective of the actual hours worked.

An alternative approach, more common in office and public sector jobs, is for a scale or range of levels of pay to be attached to a job grade. This is based on the concept of 'wage for age'. Employees could be on any point on the scale according to length of service or performance.

An incremental pay structure is likely to have four or five levels of pay within the grade. The top point of a grade may be lower than, equal to or higher than the bottom point of the next higher grade. A simple structure is obtained by devising a single set of pay levels covering the whole set of jobs, and for a particular job grade to be covered by, say, four or five of these pay levels. Such a structure is called a *pay spine*.

A pay spine is developed by considering the ratio of top to bottom salaries, the number of grades, and the pay increments within and between grades. The payment structure shown in Figure 25.1 is based on a 3 per cent increment between scale points and a top to bottom pay ratio of 5. This produces 53 points on the spine and a ratio of 4.8. The annual salary at each point is a multiple of 12 so that the monthly salary is a whole number of pounds.

If nine grades are needed or chosen for a complete set of employees in a large organization, then a 20 per cent increment between grades is available to give a significant pay rise on promotion.

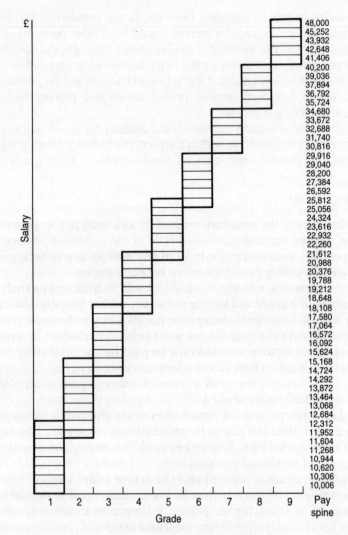

£

Salary

| 48,000 |
| 45,252 |
| 43,932 |
| 42,648 |
| 41,406 |
| 40,200 |
| 39,036 |
| 37,894 |
| 36,792 |
| 35,724 |
| 34,680 |
| 33,672 |
| 32,688 |
| 31,740 |
| 30,816 |
| 29,916 |
| 29,040 |
| 28,200 |
| 27,384 |
| 26,592 |
| 25,812 |
| 25,056 |
| 24,324 |
| 23,616 |
| 22,932 |
| 22,260 |
| 21,612 |
| 20,988 |
| 20,376 |
| 19,788 |
| 19,212 |
| 18,648 |
| 18,108 |
| 17,580 |
| 17,064 |
| 16,572 |
| 16,092 |
| 15,624 |
| 15,168 |
| 14,724 |
| 14,292 |
| 13,872 |
| 13,464 |
| 13,068 |
| 12,684 |
| 12,312 |
| 11,952 |
| 11,604 |
| 11,268 |
| 10,944 |
| 10,620 |
| 10,306 |
| 10,006 |

1 2 3 4 5 6 7 8 9   Pay spine

Grade

Figure 25.1 *A payment structure.*

The number of grades in a payment structure is the number of grades in the job grade structure (assuming the same population of jobs is covered by both). The number of grades and the pay increment between them relates to the ratio of top to bottom pay in the structure as shown in Table 25.5. This shows the nine grades required for a pay ratio of 5 and a 20 per cent increment between grades as shown in Figure 25.1.

Table 25.5 Pay increments, grades and ratios

Ratio	2	3	4	5	8	10	15	20	30
10%	7	12	15						
15%	5	8	10	12	15				
20%	4	6	8	9	11	13	15		
25%		5	6	7	9	10	12	13	15
30%		4	5	6	8	9	10	11	13
40%			4	5	6	7	8	9	10
50%				4	5	6	7	7	8

Pay differentials

There is no such thing as a correct ratio of top to bottom pay for a set of jobs. It would be usual to survey the job market in order to ascertain the market levels for the set of jobs being evaluated.

The ratio or differentials of pay vary from one country to another depending on the culture of the nation. It is not easy to measure the ratio of top to bottom pay in a country since it depends on the pay of a single individual with the highest pay, and such levels of pay are not known publically except for jobs in the public sector. Pay has to be distinguished from income, which may be from sources other than employment. Pay in the private sector is not publicly available but companies are required to publish in their accounts the pay of their highest-paid director. Fringe benefits are not taken into account in these figures.

In Britain the lowest pay for a full-time adult employee was about £4000 in the early 1980s and £6000 in the early 1990s (as fixed by Wage Councils referred to in Chapter 23). Chairpersons of nationalized industries were earning about £60,000 in the early 1980s, but after privatization some were paying themselves up to £600,000 in the early 1990s. The pay ratio for this population of jobs, which exceeds 99 per cent of jobs, has increased from 15:1 in the 1980s to 100:1 in the 1990s. In the private sector a few individuals receive salaries of more than £1 million a year, so the overall ratio is higher. However, these figures must be treated with caution as they are based on incomplete information and the pay of one or two individuals. Excluding the top 20 per cent and bottom 20 per cent of earners gives a more realistic value of 25:1 for Britain in 1991.

Overseas pay comparisons

Pay comparisons between countries are difficult to make. Because management skills are international, good managers can migrate to higher-paying countries, at least to countries in which they can speak the language. Higher pay may not, however, be beneficial if the cost of living is higher as more money is needed for living costs.

In making international pay comparisons it is conventional to use the exchange rate between the currencies to convert a salary in one country to a value in another country. This is deceptive because it does not take account of the cost of living.

Exchange rates are set by governments or central banks and reflect the value of currency for trading purposes. This valuation tends to downvalue the currencies of Eastern European and developing countries, and leads to salaries in these countries being very low when expressed in pounds. However, living costs are low and people can buy more with their pay than they could in Britain.

In order to make more realistic comparisons a comparison of living costs is needed. Since these are not generally available, attempts have been made using the price of a ubiquitous simple food product. Pay comparisons have been made using a conversion rate based on the price of a Mars bar, a can of Coca Cola or a McDonald's hamburger since these products are almost universally available. This approach must be used with caution as the price of such a product in a country may not be typical of costs generally.

The price of a 'basket' of goods can be used to establish purchasing power parity between currencies. This enables a more realistic comparison of wage levels in different countries.

For trading purposes, however, the exchange rate is relevant. The value of the Russian rouble, Turkish lira and several other currencies have fallen relative to the pound and dollar in recent years. Although these nations do not manufacture or sell many goods to Western Europe they are becoming more competitive due to the fall in wage levels relative to the pound. This may attract international manufacturing organizations to transfer manufacture to these nations.

Pay ratios are generally lower in high-productivity, low-conflict nations such as Germany and Japan, but are higher in America, Australia and New Zealand.

PAYMENT SYSTEMS

Apart from a basic level of pay, related to the level of responsibility, it is possible to provide other elements of pay as parts of the total remuneration. These could include the following:

- *Overtime* – paid for additional hours.
- *Service increment* – pay increased each year.
- *Bonus* – performance-related pay.
- *Merit pay* – discretionary increment.
- *Pension* – deferred pay, paid on retirement.
- *Fringe benefits* – company cars or similar.

Elements of pay

Overtime

Employees on wages normally get extra pay for extra hours. The pay rate is usually higher than the basic rate and is called a premium rate. All hours above the normal working week would attract a premium and higher premiums would attach to unsocial hours and weekend working.

Service increment

A service increment would be paid automatically each year until an employee had reached the top of the scale of pay for the job.

Merit pay

A merit element of pay is discretionary and would be allocated by a manager to some of the team working under that manager. An appraisal system would be used, which may or may not involve the employee. A manager would be authorized to allocate a prescribed amount of money for merit increments for those employees who report directly and would normally be prohibited from splitting it equally between all of the team.

Pension

A pension is a part of total remuneration since an employer pays a contribution to it and employees derive benefit from it. An employee would normally also make a contribution. Such contributions are excluded from taxable pay, so employees pay less tax.

Fringe benefits

Fringe benefits include company cars, private health insurance and other benefits in kind. Some organizations pay private school fees or living costs at university for employees' children. At senior levels chauffeurs or domestic servants may be provided at company expense. These may be included as a means of providing benefits and avoiding tax on them, but these days the taxation authority taxes most of them in part.

Choice of a payment system

An organization must design an appropriate payment system if it is to contribute to organizational effectiveness. A particular system would include some of the elements listed above. Although pay is not a prime motivator, feelings of unfairness of the payment system will alienate employees. The payment system must therefore match the total approach of the organization to the management of its human resources. It is not essential that all employees are paid in the same manner – sales staff, for example, are likely to be paid commission, a form of bonus appropriate to their role. However, a modern approach tries to avoid unnecessary differences between employees, since such differences are seen as being unfair, and so a common payment system for all, or most, employees is becoming more common.

Working conditions are not strictly a part of pay but relate closely to it, and could form part of a negotiated agreement on pay and conditions. Working hours for manual workers have gradually reduced over time. At the end of last century working hours were around 80 hours per week, and Saturday was a full working day. By the 1950s hours had fallen to 44 hours per week including Saturday mornings. Hours have gradually fallen to 40 hours per week and in some cases have fallen below this level. It was common in the past for a factory to work 40 hours per week but for an office to work only 35 hours per week. Manual workers lost pay if they had to go to a dentist but office workers did not. These status differentials are now seen as unfair, and there are moves towards reducing or eliminating them in what are called single-status working conditions.

Performance-related pay

A bonus acts as a financial incentive, a reward for productivity as a motivation to productivity. Bonuses were traditionally provided for manual workers using payment-by-results or piecework schemes. Senior managers and directors had share options and in some cases profit-related bonuses, but the majority of employees, and all public sector employees, were on fixed wages or salaries. A share option is the option to buy shares at today's price sometime in the future. An employee would buy if the future price were higher but did not have to if the price were lower. This 'heads you win, tails you do not lose' situation is a financial benefit to selected employees.

A modern approach will seek to include all employees in a bonus system in order to encourage both individual performance and co-operation to gain higher productivity for mutual benefit. This requires the development of appropriate payment systems that incorporate some element of performance-related pay.

The different types of payment system explained below will be seen to relate to different measures of productivity measured at different levels of system. The three main types of productivity measure were introduced in Chapter 11. Figure 25.2 shows how pay would vary with productivity in any payment system with a bonus element of pay.

A basic level of pay for a job would be determined by job evaluation and would apply up to a standard level of productivity. Performance above the standard level would increase pay. The slope of the pay–productivity line must be decided. In Figure 25.2 the relationship is proportional, i.e. an x per cent increase in productivity will give an x per cent increase in pay. This gives 100 per cent of the labour-saving benefit of the productivity increase to the employees.

Payments by results (PBR)

Payment by results or piecework schemes have been around for a century,

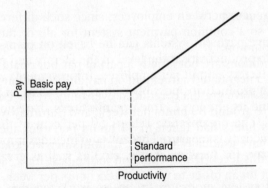

Figure 25.2 *Variation of pay with productivity.*

following their introduction by F. W. Taylor (see Chapter 24). These systems require the standard time for each task to be measured and the actual time taken to do each task to be recorded. The ratio of standard to actual time is a measure of productivity at its most basic level, the level of the task. Pay for individuals can be related to this ratio. These systems relate pay to productivity at the level of the task. The relationship between pay and productivity is usually linear and proportional, beyond a basic pay for a specified performance. This system provides 100 per cent of the benefit of higher productivity to the operator, but only the saving of labour cost is given because productivity is measured narrowly and includes only the speed of work.

In today's society such schemes cause a lot of conflict. Employees put social pressure on the employees responsible for work measurement, with the result that standard times become inaccurate. In this way pay increases, but due to increasing inaccuracy in the measure of productivity, rather than increasing productivity. Some tasks are 'good' from the operator's point of view in that the time is larger than it should be, so a bonus is obtained without any effort. Accurately timed jobs are 'bad' in that a bonus is only achieved by working harder. Employees do only the 'good' jobs and leave the 'bad' jobs to someone else.

Maintenance workers were put onto piecework schemes on the principle that the more work they did, the more pay they got. Maintenance workers, however, have the power to create work. They can maintain a machine in a way that causes it to break down again soon. With maintenance workers on piecework, more maintenance work will be done, but did it need to be done? The work is diagnostic in nature and when the work has been done it is not possible to tell whether it needed to have been undertaken. This creates a situation in which employees will sabotage the machines since by doing so they get more pay.

Employees cannot be blamed if they use a system offered to them by management as a means of increasing their pay. The problem is with the payment system. Such systems clearly motivate employees to actions which are destructive rather than constructive. Such schemes are not widely used today.

Plant-wide incentives (PWI)

Alternative incentive payment systems can be based on plant-wide incentives. These relate pay to productivity at the level of the factory and pay all employees a bonus based on factory output per employee. Improvements in pay are related to improvements in productivity, possibly with a 50/50 share of benefits between the organization and its employees (the organization's share going ultimately to the shareholders). The organization retains 50 per cent of the benefit on the grounds that productivity is measured broadly and includes improvements which the organization pays for (new machines, etc.) as well as greater effort by the employees.

This type of system motivates groups to work together since the bonus depends on people working together. Maintenance workers, for example, are motivated to ensure that machines are available to work all of the time since in that way output will increase and with it their pay. People in offices also receive the bonus, so they too are motivated to ensure that their information-processing and other activities are contributing to the common objective – of getting goods out of the door at the end of the week.

Profit- or performance-related pay (PRP)

A third variant is to relate pay to productivity at the level of the company as a financial system. In profit-making organizations this would be profit, but it could be added value. The same concept can be applied in non-profit-making organizations by relating pay to similar financial measures of performance.

In profit-making organizations, directors have long had profit-related bonuses or the option to buy shares at discounted values. Rarely has the concept been extended to other employees, but there is no reason why it should not.

In non-manufacturing organizations modern approaches to remuneration seek similar benefits. Civil servants, doctors and similar professional employees are now on performance-related pay. This requires the development of suitable measures of financial performance in order that pay may be related to it.

EMPLOYEE PARTICIPATION

The theories of motivation outlined in Chapter 24 have shown that employees need to be involved if they are to be committed. There are several possible forms of involvement at different levels within the organization. Historically there has been considerable political conflict over the involvement or participation of employees in the affairs of the organization. Involvement is likely to include some responsibility for decision making. To an autocratic manager there is no place for any involvement or participation of employees: they are there to do what they are told. Any other style of management sees benefit in involving employees.

Modern approaches to management create win–win situations, and involving employees within an organization is seen to be a way towards creating such situations.

Forms of employee participation

- Worker shareholders
- Worker co-operatives
- Worker directors
- Works councils
- Productivity committees
- Quality circles
- Job design

Worker shareholders

There is no reason why any employee should not own shares in a plc. Many employees would not wish to but they may be encouraged to do so, not only for the financial benefit referred to above, but in order to make employees feel part of the organization and feel they are participating in it. Employees can be offered discounts in share purchases or given some free shares as part of the pay or bonus.

In order to extend the feeling of participation beyond those who feel they can afford to buy shares, a share trust can be established from which the dividends are distributed to all employees. This avoids the necessity for employees having to buy shares and the problems of employees leaving the organization if they were share owners. The employees get the benefits of share ownership and the feeling of participation which that involves without having to pay for shares.

Worker co-operative

In a worker co-operative the benefits of the organization are shared by those who work there. There are no shareholders: any external money requirements are obtained from banks or government loans. In this way all decisions are taken within the organization and all employees are involved. In Britain the Co-operative shops and factories were formed as early as the 1840s as an alternative to the exploitation workers felt in capitalist companies. Some co-operative organizations do exist in Britain today; these were referred to in Chapter 2. In Spain there are co-operatives in the Mondragon region. Israeli kibbutzim are another form of co-operative and so are Russian collective farms. There are few co-operatives in a capitalist society since they represent a political structure that denies capitalism. Banks do not favour lending to such organizations so they find it difficult to get any external finance.

Attitudes have changed recently as some conventional organizations have been sold to their employees in what are termed 'management buy-outs'.

(MBO in this context is not the same as MBO, management by objectives, referred to in Chapter 24.) Some governments have favoured such forms of organization, and so banks have been prepared to make loans to them. These organizations still have shares, so are not co-operatives in the original sense. The shares may be held in a trust from which the benefits go to employees, perhaps in proportion to their salary. The actual amount of involvement of employees may not be significantly different from conventional organizations.

Worker directors

The concept of a worker director is that the policy-making board of directors should include representatives of non-managerial employees (jobholders) as well as senior managers and shareholders. This form of participation is at board level and would enable the board to be appraised of employee views, and employees to be made aware of what was happening at the top of the organization. The analogy is with the city council in local government which comprises councillors (elected representatives of the public) as well as council officers. However, the transfer of the model of local democracy into an organization must be treated with caution. It is more appropriate to voluntary than to business organizations.

This form of participation is politically controversial. It would be suitable within an organization that already had a participative style of management, but with an adversarial style of management any conflict that already existed would be transferred to the board. In a unionized organization it may be the shop steward who is elected as a worker director, so the issues of concern to the shop steward, e.g. pay and working conditions, would be discussed at board meetings. Such matters would normally be negotiated between a human resource manager and the shop stewards. It would not be seen as the function of the board to be involved in such negotiations. Worker directors are used in most countries of the European Union.

Works councils

Works councils are committees of managers and non-managerial employees which meet, perhaps quarterly, to discuss matters of common interest. This form of participation is at senior management level and would enable employee involvement in management decisions at this level. They can be used to jointly solve problems, and suggest and agree ways of making improvements. A works council will only be effective if the different groups of people within the organization share a common objective. Only then can they work together towards the achievement of the common objective. In a climate of conflict a works council will become a forum for conflict and little will be achieved.

In practice there will be some reluctance to allow a works council to take any decisions. Decisions on pay and similar matters would be left to negotiation with trade unions, or centralized without employee involvement. Decisions on financial expenditure would be seen as a management prerogative. There may be very little for a works council to talk about. In some organizations they have been left to arrange cricket matches and other social events. Few companies in Britain have works councils but they are widely used within the European Union.

Productivity committees

A productivity committee is similar to a works council but has a specific task – of identifying and implementing improvements in productivity. With such a focus this form of participation would have more chance of success but is rarely used in Britain. Only when productivity is established as a common objective can such a body be effective.

Quality circles

A quality circle is similar to a productivity committee but operates at workgroup level. As well as quality circles at workgroup level there would be a factory-level quality circle. (Chapter 9 explained quality circles. They are mentioned here because they are a form of employee participation.) They are based on the principle that quality is a common objective, so managers and employees can work together to its achievement. Quality circles are widely used in Japan and are considered to be an effective form of employee participation. They are being introduced in some organizations in Britain. In Japan quality circles have extended their agenda to include safety and productivity since these too are seen to be common objectives.

Job design

Job design has been explained earlier in this chapter. It is restated here because it can be seen to be a form of employee participation. It acts at the job or task level. Job design is perhaps the most powerful and least controversial form of employee participation and likely to be the most effective. It enables participation through the job at a level that is most meaningful to employees.

Participation and effectiveness

The participation of employees in organizations is politically controversial but is seen to be necessary for employee commitment. Some politicians favour selling shares to employees, turning all employees into capitalists and perhaps excluding external shareholders. Others favour co-operatives in which there are no capitalists (in the sense of external shareholders). Despite the extreme polarity of these views, the form of organization advocated is not very different. In both, all of the profit or wealth is distributed among those who work in the organization. This is seen as providing motivation and reward to all employees. What is different is the legal form of organization and the political perception of it.

Management must determine the form and extent of participation which suits the culture of the organization and the style of management used. Since employee motivation and commitment is crucial to the success of the organization, it is important that employee participation is given proper consideration within the organization. This means that it should form part of the human resource strategy of the organization. Employee 'empowerment' is a modern term used to describe approaches to employee participation. The term seems to be less politically controversial than participation.

ORGANIZATION DEVELOPMENT

The modern systems approach to organizations, based on seeing the organization as a system, leads to a consideration of how the performance of an organization can be improved. Each part of this book has examined parts of the organization and developed approaches for analysis and improvement. The interactions between parts have also been stressed as part of the view of the organization as a whole being an interactive system.

In looking in Chapter 22 at employee training and development the benefit was seen of developing individuals within the organization to their full potential, not only for the benefit of the individual but also for the benefit of the organization. A similar concept can be applied to the organization itself. Organization development is the term given to approaches that seek to develop the organization as a whole. Just as individuals learn, so organizations can learn. This requires individual learning, and forms of participation and information sharing that enable the organization to learn from the individuals within it. In doing so the organization is maximizing its potential and its performance.

Organization development is a process of trying to change (develop) an organization from a traditional autocratic, bureaucratic, adversarial style of management to a participative style based on common objectives and a win–win strategy. It is based on changing the attitudes of people, both managers and non-managerial employees, within the organization.

Traditional attitudes and the mistrust between managers and other employees make it difficult for many organizations to make progress on their own towards a more effective organization. A management consultant is often needed to act as a catalyst, getting managers and employees to agree to new systems which they could not agree on their own because of their history of mistrust and the political dimension of the problem.

Organization development is based on involving employees as a means of enabling Theory Y behaviour to develop. This implies a Theory Y style of management, and a range of management systems as described above.

Human resources strategy

This part of the book has shown that the human resources of an organization are crucial to its success. The performance of an organization depends on the motivation and attitudes of its employees. This chapter has introduced some approaches – and they are approaches not techniques – to the design of management systems which enable the organization to become effective.

The application and implementation of these approaches requires a human resource manager with a vision of modern management systems and the skills and authority to introduce them. This can only be done if an organization has a human resources strategy, and one that integrates with the financial, marketing and manufacturing strategies of the organization. A human resources strategy requires the commitment of the board of directors if it is to be successful. It must be seen to be as equally important as the other functional strategies. These functional strategies must be developed with an awareness and acceptance of the strategies in the other functional areas if the organization is to develop a coherent corporate strategy.

Business process re-engineering (BPR)

Business process re-engineering is the name given to a modern approach to management based on viewing the organization as a system and focusing on the processes of the organization and the interactions between them. The various chapters in this book have developed modern approaches to management in each of the functional areas of the organization. The concepts of the lean organization, the internal market, the plant within a plant and the profit centre are all related to a new form of organization in which employees are actively involved and decisions delegated. These changes need changes in the management of human resources. This chapter has shown what changes are possible. Unless employee attitudes are changed by modern management systems, other management initiatives will fail. Many of the changes suggested in this book are designed to improve industrial relations, but they need a positive climate of industrial relations if they are to be effective. Management initiatives in the different functional areas of an organization are interdependent, so an integrated systems approach is needed.

Slavery was abolished 160 years ago but many attitudes of that era remain, with some organizations treating employees virtually as slaves. The task of management is seen as a 'battle' to be won, fighting against employees, who then respond with similar attitudes. Neanderthal styles of management and dinosaur attitudes of some trade union leaders still exist. In this adversarial climate productivity is low and held down by conflict.

This part of the book has shown a little of the historical background, because it conditions present attitudes, but the emphasis has been on a positive way forward. There are some new ideas in this book but most of the approaches recommended are not new.

A 'systems approach' to managing manufacturing organizations has led to the integration and development of new perspectives based on ideas which are themselves not new. The bibliography lists many books from the 1970s and earlier which form the background on which the ideas in this book have been developed. The problem is that these approaches have not been used in many organizations. It will be found that the achievement of world-class standards of productivity is occurring in organizations that are changing their culture and their style of management.

A human resource strategy based on establishing common objectives and win–win outcomes can lead to employee commitment and productivity. Indeed it is the only way to achieve these mutually desired objectives. Organizations that develop modern management systems will succeed; those that do not will fail.

Sir Graham Day, former chief executive of Rover cars, is quoted as saying: 'Survival is not compulsory', implying that survival is something that does not occur naturally but has to be planned, organized and controlled, i.e. needs managing.

Lord Robens, former chairman of the Coal Board, pointed out in the 1950s that: 'In Britain we spend so much time arguing about the *share* of the cake that we fail to realise that the *size* of the cake is getting smaller.' High levels of industrial conflict and autocratic styles of management have contributed to the loss of competitiveness and the closure of a large part of British manufacturing industry.

It is only now being more widely recognized that success depends on the creation of common objectives which enable employees to contribute fully and to achieve the higher productivity that leads to higher wages and higher profit.

Task sheet 25   Human resource management

25.1 Evaluate each of the following jobs in terms of the concepts of job design:

Managing director	Manufacturing manager
Design engineer	Factory supervisor
Sales representative	Maintenance engineer
Machine operator	Factory labourer

25.2 If you are at work, analyze the job grade and payment structures of which you are part. Calculate or estimate the top to bottom pay ratio and determine the pay increment percentage between grades and within grades if appropriate.

25.3 Consider the three measures of productivity used in the three types of performance-related pay referred to in this chapter. You may want to refer back to Chapter 11. What is included in each measure of productivity and how do they differ? What is the likely motivational effect of each type of payment system?

25.4 List and evaluate the likely effectiveness of the forms of employee participation referred to in this chapter.

CASE STUDY

___

SUPERIOR OFFICE EQUIPMENT

OBJECTIVES

This case study has been designed to suit the content of this book in management and any taught course based on it. It provides the opportunity to apply and to integrate the concepts taught in the book. It enables readers to develop their enterprise skills, particularly those of analysis, synthesis and understanding the entrepreneurial environment.

ORGANIZATION AND USE WITH STUDENTS

The case study has twelve stages of work related to the teaching in the book. Case study notes and task sheets are provided for each stage. There are a set of tasks at each stage related to the book and taught course.

Tutorial guidance and feedback can be given in group tutorial sessions where this book is used on a taught course.

Students using this case study will normally submit their reports for each stage (two or three pages of A4, wordprocessed) in the form of recommendations to the board of directors. For numerical stages the use of a spreadsheet is recommended. Information provided or calculated at one stage is needed at later stages.

At some stage during the year students can be required to make a presentation of a stage of analysis.

After each stage has been completed, feedback notes will be provided. These include the 'decisions' made by the board of directors on the recommendations made in students reports.

The reports and presentations can form the assessed coursework for a taught course.

STAGE I   CASE NOTES

Two graduates, who became close friends during their studies, have remained in contact since graduating five years ago. Their current jobs are in the same area, so they continue to meet socially. Both are paying off a mortgage and both of their partners are working. One graduated in engineering, the other in business. They both have experience in a manufacturing organization. When they met recently they discussed the possibility of setting up their own business with themselves as directors.

One graduate recognized that current methods of manufacturing filing cabinets were very inefficient and thought that with better design and manufacture the product could be made for less cost than current manufacturers. Filing cabinets are designed to suit the foolscap size of paper, which is no longer used. A4 paper is smaller than foolscap, so a filing cabinet could be made smaller and save on material cost.

The other graduate recognized that, despite the growth of office computing, there was still a continuing need for filing cabinets to store A4 documents. An A4-sized cabinet would take up less space in the office. There is also a market for the storage of B3-sized computer documents for which no current filing cabinet is suitable, so this might form the basis of a second product.

They are enthusiastic about the idea but realize that there is a lot of work to do to prove its viability. They contact the manufacturing school at the local university and ask for help in developing the manufacturing and management systems for the planned business.

You, as a reader, are invited to do this analysis as part of your study of this book in management.

STAGE 1   TASKS

1.1   Develop a statement of corporate objectives for the planned company.

1.2   Develop some outline ideas for a set of functional strategies.

1.3   Advise the graduates on the legal requirements for setting up a company.

1.4   Advise the graduates whether they should set up a partnership, a private or a public limited company.

1.5   List the main cost headings so that initial estimates of costs and capital can be made.

1.6   Identify the possible sources of funds and the costs and risks of each.

STAGE 2   CASE NOTES

Before committing themselves to setting up the manufacturing facilities, the directors recognize that they must identify the market for their product.

They must also consider what level of sales would be realistic for the initial phase of running a new business, bearing in mind the finance which would be needed.

The Engineering Director has done some initial design drawings. The design will be modular so that new products can be quickly and cost-effectively developed from the design of the initial product.

The main body of the cabinet will be formed from a single sheet of metal. The open front will be braced with a cross-bar between each drawer, and a drawer guide will be fitted at each side of each drawer space. Each drawer will be formed from a single sheet. A runner at each side has two wheels which permit smooth opening of the drawer to its full length.

The Marketing Director has started to think about possible products and market.

STAGE 2   TASKS

2.1   Develop a hierarchical market segmentation model for office filing systems and explain its structure.

2.2   Recommend the best product and market for the company to sell initially. Justify your recommendation.

STAGE 3   CASE NOTES

The Marketing Director decides to commission a marketing consultant to obtain quantitative data about the market for the product which the company intend to produce initially.

The consultant's report makes the following conclusions:

1.   The UK market for the product is 400,000 units per year.
2.   The average retail price of the product is £120.
3.   Retailers add 40 per cent to the trade price.
4.   Market research shows the market share related to the price (expressed as a proportion of the average price). This is shown in Table A1.

Table A1   Market share related to price

Price ratio	0.7	0.8	0.9	1.0
Market share	6.0%	4.4%	3.0%	1.1%

The Engineering Director estimates that 13,750 units per year is a feasible initial capacity of the planned factory. For this factory, fixed costs are estimated to be £300,000 per year and variable costs to be £25 per unit.

STAGE 3   TASKS

3.1   Estimate the potential annual demand for the product depending on the price ratio compared with competitors.

3.2 Analyze revenues and costs for the planned factory for each of the price ratios used in Table A1. Hence determine the profit or loss for the first year of operation on the basis of the estimated costs given.

3.3 Recommend the volume the company should produce annually and the price at which the company should sell its product.

STAGE 4   CASE NOTES

In planning the factory to make 'Superior' filing cabinets an analysis is made of the processes and equipment required. Table A2 shows a process planning chart for a four-drawer cabinet. Operation times are in standard minutes. There are two wheels and two pins per runner, the wheels being held in place by the pins. The runner assembly time is per runner.

Table A2   Process planning data for the four-drawer filing cabinet

Part	Description	No.	Operation		Op. time		Plant reqd
1	Main body	1	(a)	Cut	0.64	(A)	Sheet cutter
			(b)	Bend	6.4	(D)	Bend m/c 3
			(c)	Weld	2.4	(E)	Spot welder
2	Cross-bar	3	(d)	Cut	0.54	(A)	Sheet cutter
			(e)	Bend	1.8	(B)	Bend m/c 1
3	Drawer guide	8	(f)	Cut	0.42	(A)	Sheet cutter
			(g)	Bend	1.6	(B)	Bend m/c 1
4	Body assembly	1	(h)	Weld	1.6	(E)	Spot welder
			(i)	Paint	3.2	(G)	Paint sprayer
			(j)	Fit lock	1.6	(H)	Assembly area
			(k)	Inspect	1.6	(I)	Measuring equip.
5	Drawer body	4	(l)	Cut	0.62	(A)	Sheet cutter
			(m)	Bend	2.4	(C)	Bend m/c 2
			(n)	Weld	2.0	(E)	Spot welder
			(o)	Paint	2.4	(G)	Paint sprayer
			(p)	Fit handle	1.0	(H)	Assembly area
			(q)	Inspect	2.2	(I)	Measuring equip.
6	Runner	8	(r)	Cut	0.46	(A)	Sheet cutter
			(s)	Bend	1.4	(B)	Bend m/c 1
			(t)	Press	0.6	(F)	Small press
			(u)	Assemble	1.3	(H)	Assembly area
7	Final assembly	1	(v)	Assemble	3.4	(H)	Assembly area
			(w)	Final inspect.	2.2	(I)	Measuring equip.
			(x)	Pack	1.8	(J)	Packing equipment

The cost and space requirements for the machine identified in the process planning data are shown in Table A3.

The factory will be planned on the basis of an annual output of 13,750 units per year.

Table A3   Machine cost and space requirements

Machine/facility	Cost (£)	Area (m²)
A Sheet cutter	12,000	10
B Bending machine 1	5,500	10
C Bending machine 2	56,000	20
D Bending machine 3	75,000	20
E Spot welder	12,500	10
F Small press	800	10
G Paint sprayer	6,500	20
H Assembly area	–	10
I Measuring equipment	1,000	10
J Packing equipment	1,200	10
K Handling equipment	8,000	–
L Stores and load area	–	160
M Gangways	–	40

The space allocated allows for working space around the machine

The factory operates 8 hours per day, 5 days per week, 50 weeks per year.

There is some flexibility of labour (which means that labour can move from one machine to another). Flexibility is, however, *not* total. Sets of machines required for processing particular parts of the total process should be located adjacently and labour assumed to be flexible *within* the following groups.

Semi-skilled work groups
I   body, cross-bar and guide manufacture
II   drawer manufacture, drawer assembly
III   runner manufacture, runner assembly
IV   body and final assembly, inspect and pack

Skilled work group
V   welding

STAGE 4   TASKS

4.1   Determine the labour requirements and labour utilization.

4.2   Determine the machine and equipment requirements.

4.3   Determine the capacity of the factory if these machines are acquired.

4.4   Determine the space requirements.

4.5   Draw a product structure diagram.

4.6   Comment on the planned manufacturing methods and machine utilization.

STAGE 5   CASE NOTES

Material costs

Sheet	£1.40 per sq m (6.8 sq m reqd.)
Paint	£2.20 per litre (0.28 litres reqd.)
Lock	£1.25
Wheel	£0.12
Handle	£0.08
Pin	£0.01
Packing	£1.24
Consumables	£0.45

Labour costs

Direct operators are semi-skilled and operate flexibly within separate manufacturing areas. Their wage rate is £5.60 per hour. Welding is done by skilled operators who are paid £6.40 per hour.

Factory sites

In order to avoid the delay associated with building a new factory, an existing factory will be sought. The company would prefer to buy rather than rent, so as to benefit from the (longer-term) appreciation of site value.

Three sites are currently available in the local area. Details are given in Table A4. Each site has a suitable factory with office accommodation.

Table A4   Local factory sites available

Site	Factory size	Cost (£)
1	10 m × 20 m	100,000
2	20 m × 25 m	200,000
3	25 m × 40 m	300,000

STAGE 5   TASKS

5.1   Draw a process chart for the product.

5.2   List the bill of materials for the product.

5.3   Calculate the direct production cost per unit.

5.4   Is this feasible relative to the intended selling price?

5.5   Determine which site to buy.

5.6   Suggest a suitable layout for the factory.

STAGE 6   CASE NOTES

Work is done in batches at each stage of manufacture. A batch is completed before moving on to the next process. This introduces delays in manufacturing. Work will be issued daily to operators, so work will transfer once per day to the next process.

On day 20 of an accounting period the cumulative output at each stage of manufacture is monitored. The data is shown in Table A5.

Table A5   Output at each stage of manufacture

Section	Process	Output	Section	Process	Output
Stores	Main body	1,106	Runner	Cut sheet	7,040
items	Cross-bar	3,153		Bend	6,584
issued	Guide	8,408		Press	6,104
	Drawer body	4,180		Assemble	5,640
	Handle	3,280	Body assembly	Weld	882
	Lock	825		Paint	817
	Runner	7,480		Fit lock	758
	Wheel	12,320		Inspect	703
	Pin	12,320	Drawer	Cut sheet	3,960
	Pack material	605		Bend	3,740
				Weld	3,504
Main body	Cut sheet	1,051		Paint	3,276
	Bend	996		Fit handle	3,048
	Weld	937		Inspection	2,828
Cross-bar	Cut sheet	2,988	Final assembly	Assembly	648
	Bend	2,823		Final inspect.	590
Guide	Cut sheet	7,968		Pack	535
	Bend	7,528		Store	480

Output is the cumulative output from each process from day 1 to day 20

STAGE 6   TASKS

6.1   Draw a time-scaled process chart. (Issue material on the longest path on day 1.)

6.2   Determine the throughput time for the product.

6.3   Construct a 'line of balance' chart showing planned and actual cumulative output for day 20 based on a planned production level of 55 products per day.

6.4   Compare planned and actual output and identify the processes at which delays have occurred.

STAGE 7   CASE NOTES

You have identified the activities needed for the project of setting up the factory. The activities, together with their logical dependencies and durations (in days), are shown in Table A6.

Table A6 Activities for the project

Activity	Days	Dep.	Activity	Days	Dep.
A Technical feasibility	10	–	L Identify site	8	H+D
B Market research	14	–	M Recruit staff	6	F+H+D
C Board approval	2	A+B	N Order material	2	J+K
D Raise capital	15	C	P Wait machines	10	K
E Process analysis	11	C	Q Negotiate site	12	L
F Recruit management	5	C	R Wait materials	20	N
G Identify suppliers	4	E	S Install machines	6	P+Q
H Specify machines	8	E	T Buy office equip.	8	Q
J Negotiate prices	6	G	U Train staff	8	M
K Order machines	5	H+D	V Manufacturing trial	5	R+S+T+U

The dependency shows which activities must be completed before a given activity can be started

STAGE 7 TASKS

7.1 Draw a network diagram for this project.

7.2 Calculate project duration.

7.3 Identify the critical path.

7.4 Redraw the network on a time base.

7.5 Comment and suggest a realistic schedule for the activities.

STAGE 8 CASE NOTES

The directors have estimated that the cost of carrying stock is 20 per cent per year of the purchase value for each item. They further estimate the cost of processing an order to be £20.

They decide that raw materials will be ordered weekly since this fits in with the accounting system and keeps stock levels down. Sheet is ordered by the square metre and paint by the litre. Other items are ordered by unit quantity.

Assume a manufacturing output of 275 products per week.

Note: Consumable material is *not* a single item, so is excluded from the calculations.

STAGE 8 TASKS

8.1 Evaluate the cost of the proposed system for ordering materials.

8.2 Determine the economic order quantities for each material ordered.

8.3 Suggest suitable order quantities and reorder intervals, explaining the logic of your choices.

8.4 Determine the saving to the company if it were to use these order quantities.

8.5 Comment on these calculations and suggest ways in which further savings could be achieved.

STAGES 9, 10 AND 11   CASE NOTES

The directors have made some investigations into sources of funds. The directors will each invest £100,000 share capital. They have approached a finance house (Lending On Advanced New Systems, Help And Resource Capital Society, LOANSHARKS for short) and have arranged permission for company loan capital of up to £500,000 secured on the directors' homes.

The following costs have been estimated:

- Loan interest is 1 per cent of the outstanding balance per month.
- The discount rate for investment appraisal is 14 per cent per year
- The directors will draw fees of £28,800 each annually and do not intend to take any dividend in the first year.
- The directors both lease a car costing £236 per month each.
- A Manufacturing Supervisor will be employed on a salary of £18,600.
- An Office/Sales Manager will be employed at £16,400 and two secretaries at £8350 each.
- A salesperson will be employed on a basic salary of £17,760 plus sales commission of 0.5 per cent of sales.
- A car for the salesperson costs £185 per month.
- Advertising and promotion costs are estimated to be £20,850 per year (spread over twelve months).
- A van driver is employed at £9880.
- A van will be leased for £304 per month.
- Storage and distribution costs are estimated at £1842 per month (spread over twelve months).
- In the factory a maintenance fitter is needed at a wage of £286 per week.
- A storekeeper will be employed at a wage of £173 per week.
- A labourer will be employed at a wage of £146 per week.
- The factory will operate 50 weeks/year and will be shut down for two weeks in August. (Two weeks holiday pay for the direct operators should be included in the factory overhead.)
- Power, fuel and light are estimated at £489 per month (spread over twelve months).
- Business rates (local taxes) are £13,860 per year, payable in November and May.
- Factory insurance is £8880 per year, payable in July and January.
- Building maintenance is 8 per cent per year of the value of the building.
- Machine depreciation is 20 per cent per year of the original value.

- Materials are first ordered at the beginning of April, on two weeks delivery. Materials are paid for one month (four weeks) after delivery.
- Manufacturing starts in May and the first goods are sold in June. Customers pay for their goods one month (four weeks) after delivery.
- Raw material stocks are 3.1 weeks (for all items) and finished goods stocks are 2.6 weeks.
- The number of days of work in progress is the manufacturing throughput time.
- Staff will be recruited and paid from the beginning of April. Prior to production, direct operators will assist in the installation of the equipment and be trained to operate it.
- The company financial year will run from April to March, comprising twelve accounting periods of five, four and four weeks per month for each quarter (but the factory works for only two weeks in August).

Note: For this exercise ignore taxation and wage and price increases. Use manufacturing data from earlier stages.

STAGE 9   TASKS

9.1   Prepare a cash flow forecast for the first year (assume a suitable value of loan capital).

9.2   Recommend the amount of loan capital which the company needs if it is to avoid an overdraft.

9.3   Suggest a schedule for loan repayments.

9.4   Determine the value of debtors, creditors and stocks of raw materials, work in progress and finished goods at the end of the first year.

Note: Value materials at cost, finished goods at 75 per cent of trade sales price and work-in-progress at the average of the raw material and finished goods values.

STAGE 10   TASKS

10.1   Prepare a budgeted manufacturing account for the first year.

10.2   Prepare a budgeted trading account for the first year and hence determine the gross profit.

10.3   Prepare a budgeted profit and loss account for the first year and hence determine the net profit.

10.4   Determine the break-even volume for the company.

STAGE 11   TASKS

11.1   Prepare a budgeted balance sheet for the company at the end of the first year.

11.2 Determine the budgeted added value for the company for the first year.

11.3 Calculate the following measures of return on investment for a five-year period:
   (i) average rate of return;
   (ii) payback period;
   (iii) net present value;
   (iv) internal rate of return.

Note: Second-year figures can be obtained by changing the first-year data to include a *full year* of production (use of a spreadsheet recommended).

   Assume the following: a constant level of output in years 2–5; no inflation in prices or costs; factory assets retain their value (investment = cost of machines only); the machines have no value after five years (exclude loan interest and repayments from costs).

11.4 Calculate the following performance measures for the first year of manufacturing:
   (i) profit/assets and profit/sales;
   (ii) sales/assets and added value/employee;
   (iii) added value/wages and average wages/employee;
   (iv) output/man-hour and output/employee;
   (v) labour utilization.

STAGE 12   CASE NOTES

There are no notes for this stage of the case study.

STAGE 12   TASKS

12.1 Recommend an organization structure for the company and draw the organization chart.

12.2 Prepare a job description for the Manufacturing Supervisor.

12.3 Prepare a job advertisement for the post of Manufacturing Supervisor and recommend where it should be advertised.

12.4 Recommend a job grade structure and suggest how it might be created.

12.5 Recommend a way to relate pay to performance.

BIBLIOGRAPHY

Ackoff, R. L. (1960) 'Systems, organisations and interdisciplinary research', *General Systems Yearbook*, Vol. 5.

Adair, J. (1982) *Effective Time Management*, Pan, London.

Adair, J. (1983) *Effective Leadership*, Pan, London.

Adair, J. (1985) *Effective Decision Making*, Pan, London.

Adair, J. (1986) *Effective Teambuilding*, Pan, London.

Ammer, D. S. (1980) *Materials Management*, Irwin, Homewood, IL.

Ansoff, H. I. (1968) *Corporate Strategy*, Penguin, London.

Ansoff, H. I. (1969) *Business Strategy*, Penguin, London.

Anthony, R. N. and Hekiman, J.S. (1967) *Operations Cost Control*, Irwin, Homewood, IL.

Argyris, C. (1960) *Understanding Organisational Behaviour*, Tavistock, London.

Armstrong, M. and Murlis, H. (1989) *Salary Administration*, Kogan Page, London.

Atkinson, A. B. (1977) *Wealth, Income and Equality*, Penguin, London.

Babbage, C. (1835) 'On the economy of machinery and manufactures' in 1965 *Reprints of Economic Classics*, Augustus Kelly, New York.

Barnard, C. (1938) *The Functions of Executives*, Harvard University Press, Cambridge, MA.

Barnes, R. M. *Motion and Time Study*, Wiley, New York.

Baron, R. A. (1983) *Behaviour in Organisations*, Allyn and Bacon, Newton, MA.

Barrett, B., Rhodes, E. and Beishon, J. (eds) (1975) *Industrial Relations and the Wider Society*, Collier Macmillan, London.

Beach, D. S. (1980) *Personnel*, Macmillan, New York.

Beishon, J. and Peters, G. (1972) *Systems Behaviour*, Open University Press, Milton Keynes.

Belbin, R. M. (1981) *Management Teams: why they succeed or fail*, Heinemann, London.

Benne, K. D. and Sheats, P. (1948) 'Functional roles of group members' *Journal of Social Issues*, vol. 4, no. 2, pp. 41–49.

Bennett, R. (1981) *Managing Personnel and Performance*, Business Books, London.

Berliner, C. and Brimson, J. (eds) (1988) *Cost Management for Today's Advanced Manufacturing: the CAM-I conceptual design*, Harvard Business School Press, Boston, MA.

Bignell, V. (1985) *Manufacturing Systems: context, applications and techniques*, Blackwell, Oxford.

Bignell, V. and Fortune, J. (1984) *Understanding Systems Failures*, Manchester University Press, Manchester.

Blake, R. R. and Mouton, J. S. (1964) *The Managerial Grid*, Gulf, Houston, TX.

Boot, R. L., Cowling, A. G. and Stanworth, M. J. (1977) *Behavioural Sciences for Managers*, Arnold, London.

Bowey, A. and Lupton, T. (1973) *Job and Pay Comparisons*, Gower, Epping.

Branham, J. (1975) *Practical Manpower Planning*, Institute of Personnel Management, London.

Brech, E. F. (1957) *Organisation, the Framework of Management*, Longman, London.

Brown, J. A. C. *Social Psychology of Industry*, Penguin, London.

Brown, S. C. and Martin, J. N. (1977) *Human Aspects of Man-made Systems*, Open University Press, Milton Keynes.

Brown, W. (1965) *Explorations in Management*, Penguin, London.

Brown, W. (1973) *The Earnings Conflict*, Penguin, London.

Buffa, E. S. (1965) 'Sequence analysis for functional layout', *Journal of Industrial Engineering*, vol. 6, no. 4, pp. 12–15.

Burbidge, J. (1975) *The Introduction of Group Technology*, Heinemann, London.

Burns, T. (1963) 'Industry in a new age', *New Society*, 31 Jan. 1963, pp. 17–20.

Burns, T. (1969) *Industrial Man*, Penguin, London.

Burns, T. and Stalker, G. (1961) 'Mechanistic and Organismic systems of management', in *The Management of Innovation*, Tavistock, London.

Buzan, A. (1977) *Make the Most of Your Mind*, Pan, London.

Castles, F. G., Murray, D. G. and Potter, D. C. (1971) *Decisions, Organisation and Society*, Penguin, London.

Checkland, P. (1981) *Systems Thinking, Systems Practice*, Wiley, Chichester.

Child, J. (1977) *Organisation*, Harper and Row, London.

Cleland, D. I. and King, W. R. (1972) *Management: a systems approach*, McGraw-Hill, New York.

Clifford, D. K. and Cavanagh, R. E. (1986) *The Winning Performance*, Sidgwick and Jackson, London.

Cooper, R. (1974) *Job Motivation and Job Design*, Institute of Personnel Management, London.

Cowling, A. G. and Mailer, C. J. (1981) *Managing Human Resources*, Arnold, London.

Craig, C. F. and Harris, R. C. (1973) 'Total productivity measurement at the firm level', *Sloan Management Review*, 1973, pp. 13–29.

Crosby, P. (1979) *Quality is Free: the art of making quality certain*, McGraw-Hill, New York.

Cyert, R. M. and March, J. G. (1963) *A Behavioural Theory of the Firm*, Prentice Hall, Hemel Hempstead.

Davis, L. E. and Taylor, J. L. (1972) *Design of Jobs*, Penguin, London.

de Bono, E. (1967) *Lateral Thinking*, Penguin, London.

de Bono, E. (1968) *The Five Day Course in Thinking*, Penguin, London.

de Bono, E. (1985) *Conflicts: a better way to solve them*, Penguin, London.

de Bono, E. (1985) *Six Thinking Hats*, Harper Collins, London.

de Bono, E. (1991) *Six Action Shoes*, Harper Collins, London.

Deming, W. E. (1982) *Quality, Productivity and Competitive Position*, Massachusetts Institute of Technology, Cambridge, MA.

Deming, W. E. (1986) *Out of the Crisis*, Massachusetts Institute of Technology, Cambridge, MA.

Drucker, P. (1967) *Managing for Results*, Pan, London.

Drucker, P. (1968) *The Practice of Management*, Pan, London.

Drucker, P. (1969) *The Age of Discontinuity*, Heinemann, London.

Drucker, P. (1970) *The Effective Executive*, Harper Row, New York.

Drucker, P. (1986) *Innovation and Entrepreneurship*, Pan, London.

Dubois, P. (1979) *Sabotage in Industry*, Penguin, London.

Dunham, R. B. and Pierce, J. L. (1989) *Management*, Scott Foresman, Glenview, IL.

Economica XXII, vol. 87 (1955).

Economica XXIII, vol. 92 (1956).

Emery, F. E. (1959) *Characteristics of Socio-technical Systems*, Tavistock, London.

Emery, F. E. (1962) *Systems Thinking*, Penguin, London.

Emery, F. E. and Trist, E. L. (1960) 'Socio-technical systems', in Churcham, C. and Verhulst, *Management Science: models and techniques*, vol. 2, Pergamon Press, Oxford.

Encyclopaedia Britannica, 15th ed.

Esland, G. *et al* (1975) *People and Work*, Open University Press, Milton Keynes.

Etzioni, A. (1964) *Modern Organizations*, Prentice Hall, Englewood Cliffs, NJ.

Farnham, D. and Pimlott, J. (1979) *Understanding Industrial Relations*, Cassell, Eastbourne.

Fayol, H. (1916) translated 1949 by Storr, C., *General Industrial Management*, Pitman, London.

Fetter, R. B. (1967) *The Quality Control System*, Irwin, Homewood, IL.

Fiedler, F. E. (1971) *Leadership*, General Learning Press, New York.

Fiedler, F. E. and Chemers, M. M. (1975) *Leadership and Effective Management*, Scott Foresman, Glenview, IL.

Finniston, M. (1980) *The Finniston Report: engineering, manufacturing and the national economic needs*, HMSO, London.

Flanders, A. (1969) *Collective Bargaining*, Penguin, London.

Flanders, A. (1975) *Management and Unions*, Faber and Faber, London.

Follett, M. P., collected papers in Metcalf, H. and Urwick, L. (1941), *Dynamic administration*, Harper, London.

Fox, A. (1985) *Man Mis-management*, Hutchison, London.

Frantzve, J. L. (1983) *Behaving in Organisations*, Allyn and Bacon, Newton, MA.

Galbraith, J. (1967) 'The goals of an industrial system', in *The New Industrial State*, Houghton-Mifflin, Boston, MA.

Gilbert, M. (1972) *Modern Business Enterprise*, Penguin, London.

Goslin, L. N. (1967) *The Product Planning System*, Irwin, Homewood, IL.

Greene, J. H. (1967) *Operations Planning and Control*, Irwin, Homewood, IL.

Greene, T. J. and Sudowski, R. P. (1983) 'Cellular manufacturing control', *Journal of Manufacturing Systems*, vol. 2, no. 2, pp. 137–45.

Guest, D. E. and Fatchet, D. J. (1974) *Worker Participation, Control and Performance*, Institute of Personnel Management, London.

Gyllenhammer, P. G. (1977) *People at Work*, Addison-Wesley, Reading, MA.

Hackman, J. R. and Lawler, E. E. (1971) 'Employee reactions to job characteristics', *Journal of Applied Psychology*, vol. 55, pp. 265–86.

Hackman, J. R. and Oldham, G. R. (1976) 'Motivation through the design of work', *Organisation Behaviour and Human Performance*, vol. 16, pp. 250–79.

Handy, C. (1979) *Gods of Management*, Pan, London.

Handy, C. (1976) *Understanding Organisations*, Penguin, London.

Handy, C. (1984) *The Future of Work*, Blackwell, Oxford.

Harris, T. A. (1973) *I'm OK – You're OK*, Pan, London.

Hawkins, K. (1978) *Management of Industrial Relations*, Penguin, London.

Hayes, R. H. and Schmenner, R. W. (1978) 'How should you organise manufacturing?' *Harvard Business Review*, January–February.

Herzberg, F. (1959) *The Motivation to Work*, Wiley, New York.

Herzberg, F. (1966) *Work and the Nature of Man*, World Publishing, Cleveland, OH.

Herzberg, F. (1968) 'One more time; how do you motivate employees?', *Harvard Business Review*, vol. 46, no. 1, pp. 53–62.

Hickman, C. and Silva, M. (1984) *Creating Excellence*, George Allen and Unwin, London.

Hicks, P. (1977) *Introduction to Industrial Engineering and Management Science*, McGraw-Hill, Kogakusha, Tokyo.

Hill, T. (1993) *Manufacturing Strategy*, Macmillan, Basingstoke.

Hitchins, D. K. (1994) 'World class systems engineering', *IEE Engineering Management Journal*, April, pp. 81–88.

Holt, R. (1977) *Product Innovation*, Newnes Butterworth, Sevenoaks.

Howarth, C. (1984) *The Way People Work*, Oxford University Press, Oxford.

Humble, J. (1971) *Management by Objectives*, McGraw-Hill, Maidenhead.

Hunt, J. (1979) *Managing People at Work*, Pan, London.

Husband, T. M. (1976) *Work Analysis and Payment Structure*, McGraw-Hill, New York.

Ingle, S. (1982) *Quality Circles Master Guide: increasing productivity with people power*, Prentice Hall, Englewood Cliffs, NJ.

Institution of Electrical Engineers (1994) *UK Manufacturing – facing the international change*, Institution of Electrical Engineers, Stevenage.

International Labour Office (1967) *Introduction to Work Study*, International Labour Office, Geneva.

Ishikawa, K. (1985) *What is Total Quality Control? The Japanese Way*, Prentice Hall, Englewood Cliffs, NJ.

Janis, I. L. (1972) *Victims of Groupthink*, Houghton-Mifflin, Boston, MA.

Janis, I. L. and Mann, L. (1977) *Decision Making: a psychological analysis of conflict, choice and commitment*, Free Press, New York.

Jardine, A. K. (1973) *Maintenance, Replacement and Reliability*, Pitman, London.

Jaques, E. (1951) *The Changing Culture of a Factory*, Tavistock, London.

Jaques, E. (1961) *Equitable Payment*, Heinemann, London.

Jay, A. (1972) *Corporation Man*, Professional Library, London.

Jones, G. and Barnes, M. (1967) *Britain on Borrowed Time*, Penguin, London.

Jones, M. (1992) *Are You Managing Purchasing?*, Nicholas Brearley, London.

Juran, J. M. (1951, 3rd edn 1979) *Quality Control Handbook*, McGraw-Hill, New York.

Juran, J. M. (1988) 'Managing for quality', *Journal for Quality and Participation*, vol. 11, no. 1, pp. 8–12.

Kanter, R. M. (1985) *The Change Masters*, Touchstone, New York.

Kanter, R. M. (1989) *When Giants Learn to Dance*, Simon and Schuster, New York.

Kast, F. E. and Rosenzweig, J. E. (1985) *Organisation and Management*, McGraw-Hill, Singapore.

Katz, D. and Kahn, R. L. *The Social Psychology of Organisations*, Wiley, Chichester.

Kay, J. M. (1984) 'The use of modelling and simulation techniques in the design of manufacturing systems', *Conference publication 237, International Conference on the development of flexible automation systems*, Institution of Electrical Engineers, London.

Koontz, H. and O'Donnell, C. (1976) *Management: a systems and contingency analysis of managerial functions*, McGraw-Hill, New York.

Lawler, E. E. (1969) 'Job design and employee motivation', *Personnel Psychology*, vol. 22, pp. 426–35.

Lawler, E. E. and Porter, L. W. (1967) 'Antecedent attitudes of effective managerial performance', *Organisation Behaviour and Human Performance*, vol. 2, pp. 122–42.

Lawrence, P. R. and Lorsh, J. W. (1968) *Organisation and Environment*, Harvard University, Boston, MA.

Leighton, M. (1981) *Men at Work*, Jill Norman, London.

Lewin, K. (1951) *Field Theory and Social Science*, Harper, New York.

Likert, R. (1961) *New Patterns of Management*, McGraw-Hill, New York.

Lock, D. and Farrow, N. (eds) (1988) *The Gower Handbook of Management*, Gower, Aldershot.

Lockett, M . and Spear, R. (1980) *Organisations as Systems*, Open University Press, Milton Keynes.

Lockyer, K. and Gordon, J. (1991) *Critical Path Analysis*, Pitman, London.

Lumsden, N. P. (1972) *Line of Balance Method*, Pergamon Press, Oxford.

Lupton, T. (1971) *Management and the Social Sciences*, Penguin, London.

Lupton, T. (1972) *Payment Systems*, Penguin, London.

Lupton, T. and Bowey, A. (1974) *Wages and Salaries*, Penguin, London.

March, J. L. and Simon, H. A. (1958) *Organisations*, Wiley, Chichester.

Marx, K. (1867) *Capital*.

Marx, K. and Engels, F. (1962) *Selected Works*, vol. 2, Foreign Languages Publishing House, Moscow.

Maslow, A. (1943) 'A theory of human motivation', *Psychological Review*, vol. 50, pp. 370–96.

Maynard, H. B., Stegemerton, G. T. and Schwab, J. L. (1946) *Methods Time Measurement*, McGraw-Hill, London.

Mayo, E. (1937) *The Human Problems of an Industrial Civilisation*, Macmillan, Basingstoke.

Mayo, E. (1949) *The Social Problems of an Industrial Civilisation*, Routledge and Kegan Paul, London.

Meadows, D. H., Meadows, D. L., Randers, J. and Behrens W. (1974) *The Limits to Growth*, Pan, London.

McCarthy, W. E. (1972) *Trade Unions*, Penguin, London.

McDonald, N. and Doyle, M. (1981) *The Stresses of Work*, Nelson, Walton-on-Thames.

McGregor, D. (1960) *The Human Side of Enterprise*, McGraw-Hill, New York.

McKersie, R. B. and Hunter, L. C. (1973) *Pay, Productivity and Collective Bargaining*, Macmillan, Basingstoke.

Mitchell, W. and Corbett, A. R. (1973) *Employee Relations*, Macmillan, Basingstoke.

Mizuno, S. (1986) *Management for Quality Improvement: the 7 new quality control tools*, Productivity Press, Cambridge, MA.

Morley, M. F. (1978) *The Added Value Statement*, Gee, London.

Morris, W. T. (1967) *The Capacity Decision System*, Irwin, Homewood, IL.

Murrell, K. (1965) *Ergonomics*, Chapman and Hall, London.

Muther, R. (1973) *Systematic Layout Planning*, Cahners, Boston, MA.

Nadler, G. (1967) *Work Systems Design: the IDEALS concept*, Irwin, Homewood, IL.

Norman, R. G. and Bahiri, S. (1973) *Productivity Measurement and Incentives*, Butterworth, London.

Oakland, J. S. (1989) *Total Quality Management*, Butterworth Heinemann, Oxford.

Ohmae, K. (1983) *The Mind of the Strategist*, Penguin, London.

Ohmae, K. (1985) *Triad Power: the coming shape of global competition*, Free Press, New York.

Ouchi, W. (1981) *Theory Z: how American business can meet the Japanese challenge*, Addison-Wesley, Reading, MA.

Parkinson, N. (1965) *Parkinson's Law*, Penguin, London.

Parkinson, N. (1965) *The Law and the Profits*, Penguin, London.

Peter, L. J. and Hull, R. (1970) *The Peter Principle*, Pan, London.

Peters, T. (1987) *Thriving on Chaos*, Pan, London.

Peters, T. and Waterman, R. (1982) *In Search of Excellence*, Harper and Row, New York.

Phelps-Brown, H. (1977) *The Inequality of Pay*, Oxford University Press, Oxford.

Pugh, D. (1971) *Organisation Theory*, Penguin, London.

Pugh, D., Hickson, D. J. and Hinings, C. R. (1971) *Writers on Organisations*, Penguin, London.

Reed, R. (1967) *Plant Location, Layout and Maintenance*, Irwin, Homewood, IL.

Rose, M. (1978) *Industrial Behaviour*, Penguin, London.

Salaman, G. and Thompson, K. (1973) *People and Organisations*, Longman, London.

Sasaki, N. (1990) *Management and Industrial Structure in Japan*, Pergamon Press, Oxford.

Schonberger, R. J. (1982) *Japanese Manufacturing Techniques*, Free Press, New York.

Schumacher, E. F. (1973) *Small is Beautiful*, Blond and Briggs, London.

Shanks, M. (1961) *The Stagnant Society*, Penguin, London.

Shaw, A. C. (1952) *The Purpose and Practice of Motion Study*, Harlequin, Manchester.

Shingo, S. (1981) *Study of the Toyota Production System*, Japanese Management Association, Tokyo.

Shingo, S. (1986) *Zero Quality Control: source inspection and the poke-yoke system*, Productivity Press, Cambridge, MA.

Singleton, W. T. (1974) *Man–machine Systems*, Penguin, London.

Skinner, W. (1974) 'The focused factory', *Harvard Business Review*, May/June.

Smith, A. (1776) *The Wealth of Nations*, reprinted 1986, Penguin, London.

Smith, B. M. (1979 and 1981) 'The demise of the British motor cycle industry', *Centre for Urban and Regional Studies*, papers 67 (1979) and 3 (1981), University of Birmingham.

Stainer, G. (1971) *Manpower Planning*, Heinemann, London.

Starbuck, W. H. (1971) *Organisational Growth and Development*, Penguin, London.

Sumanth, D. S. (1985) *Productivity Engineering and Management*, McGraw-Hill, Singapore.

Sutermeister, R. A. (1976) *People and Productivity*, McGraw-Hill, New York.

Taguchi, G. (1974) *Introduction to Off-line Quality Control*, Central Japan Quality Control Association, Magaya, Japan.

Taguchi, G. (1981) *On-line Quality Control During Production*, Japanese Standards Association, Tokyo.

Taguchi, G. (1986) *Introduction to Quality Engineering*, Asia Productivity Association, Tokyo.

Tannenbaum, A. S. (1966) *Social Psychology of the Work Organisation*, Wadsworth, London.

Taylor, F. W. (1911, 1947) *The Principles of Scientific Management*, Harper and Row, New York.

Thomas, J. T. and Bennis, W. G. (1972) *Management of Change and Conflict*, Penguin, London.

Thomason, G. (1975) *Textbook of Personnel Management*, Institute of Personnel Management, London.

Timms, H. L. (1967) *Introduction to Operations Management*, Irwin, Homewood, IL.

Townsend, R. (1971) *Up the Organisation*, Coronet, London.

Tuckman, B. W. (1965) 'Developmental sequence in small groups', *The Psychological Bulletin*.

Vanek, J. (1975) *Self-management*, Penguin, London.

von Bertalanffy, L. (1950) 'The theory of open systems in physics and biology', *Science*, vol. 3, pp. 23–29.

Vroom, V. H. and Deci, E. L. (1970) *Management and Motivation*, Penguin, London.

Walley, B. H. (1973) *Management Services Handbook*, Beckman, London.

Walley, B. H. (1982) *Handbook of Office Management*, Business Books, London.

Warnecke, H. J. and Vettin, G. (1982) 'Technical investment planning of flexible manufacturing systems – the applications of practice-oriented methods', *Journal of Manufacturing Systems*, vol. 1, no. 1.

Weber, M. (1930) *The Protestant Work Ethic and the Rise of Capitalism*, Allen and Unwin, London.

Weber, M. (1922) *The Theory of Social and Economic Organisations*, Free Press, New York. English translation A. Henderson and T. Parsons (1947).

Weir, M. (1976) *Job Satisfaction*, Fontana Collins, Glasgow.

Weir, D. (1973) *Man and Work in Modern Britain*, Fontana Collins, Glasgow.

Welsch, L. A. and Cyert, R.M. (1970) *Management Decision Making*, Penguin, London.

Weston, F. C. (1972) 'Quantitative analysis of plant location', *Industrial Engineering*, vol. 4, no. 4, pp. 22–28.

Whyte, W. H. (1960) *Organisation Man*, Penguin, London.

Wild, R. (1972) *Mass-production Management*, Wiley, New York.

Wild, R. (1975) *Work Organisation*, Wiley, London.

Wild, R. (1984) *Production and Operations Management*, Holt, Reinhart and Winston, Eastbourne.

Woodward, J. (1958) *Management and Technology*, HMSO, London.

Open University courses

PT611 Structure and Design of Manufacturing Systems.
PT613 Manufacturing Management.
T241   Systems Behaviour.
T242   Systems Management.
T243   Systems Organisation: the management of complexity.
T244   Managing Organisations.
T247   Working with Systems
T301   Complexity, Management and Change: applying a systems approach.
TD342 Systems Performance: human factors and systems failures.

The Egyptian account in Figure 19.1 is from: Gardiner, A. H. (1927) *Egyptian Grammar*, Clarendon Press, Oxford.

INDEX